WORLD HEALTH ORGANIZATION

INTERNATIONAL AGENCY FOR RESEARCH ON CANCER

150, COURS ALBERT THOMAS
69008 LYON - FRANCE

INSTITUT NATIONAL DE LA SANTE ET DE LA RECHERCHE MEDICALE

101, RUE DE TOLBIAC
75645 PARIS CEDEX 13 - FRANCE

INSERM Symposium Series Vol. 52
IARC Scientific Publications N° 13

© Editions INSERM, Paris 1976
ISBN 2-85598-147-6

LES ÉDITIONS DE L'INSTITUT NATIONAL DE LA SANTÉ
ET DE LA RECHERCHE MÉDICALE

ENVIRONMENTAL POLLUTION AND CARCINOGENIC RISKS

POLLUTION DE L'ENVIRONNEMENT ET RISQUES CANCEROGENES

Lyon, 3-5 Nov. 1975

Edité sous la responsabilité de / Edited by

Claude ROSENFELD et Walter DAVIS

CONTENTS / *SOMMAIRE*

Pages

Allocution de Madame Simone VEIL
Avant-Propos des Editeurs,
Claude ROSENFELD et Walter DAVIS

Exposé introductif par John HIGGINSON :
The importance of environmental factors in cancer / *L'importance des facteurs d'environnement dans le cancer* 15

I — ATMOSPHERIC POLLUTION
POLLUTION ATMOSPHERIQUE

Lawther P.J. and Waller R.
Coal fires, industrial emissions and motor vehicles as sources of environmental carcinogens
Feux de charbon, rejets industriels et véhicules à moteurs — sources cancérogènes de l'environnement 27

Higgins I.T.T.
Epidemiological evidence on the carcinogenic risk of air pollution
Indices épidémiologiques du risque cancérogène dû à la pollution de l'air 41

Shabad L.M. and Smirnov G.A.
Aviation and environmental benzo(a)pyrene pollution
Aviation et pollution de l'environnement par le benzo(a)pyrène 53

Stupfel M. and Mordelet-Dambrine M.
Penetration of carcinogens through respiratory airways
Pénétration des cancérogènes dans les voies respiratoires 61

Leuchtenberger C. and Leuchtenberger R.   Pages

Significance of oxides of Nitrogen (NO) and SH reactive components in pulmonary carcinogenesis
Importance des oxydes d'azote (NO) et des composants SH réactifs en cancérogénèse pulmonaire 73

II — WATER POLLUTION
POLLUTION DES EAUX

Stich H.F., Acton A.B. and Dunn B.P.

Carcinogens in estuaries, their monitoring and possible hazard to man
Les cancérogènes dans les estuaires : surveillance et risque éventuel pour l'homme 83

Aubert J.

Pollutions des eaux et risques cancérogènes
Water pollution and carcinogenic risks 95

III — OCCUPATIONAL POLLUTION
POLLUTION D'ORIGINE PROFESSIONNELLE

Gilson J.C.

Asbestos cancers as an example of the problem of comparative risks
Cancers dus à l'amiante — un exemple du problème des risques comparatifs 107

Jones J.S.P., Pooley F.D. and Smith P.G.

Factory populations exposed to crocidolite asbestos — a continuing survey
Populations de travailleurs exposés à la crocidolite — enquête continue .. 117

Cross A.A.

Effect of changed working techniques on asbestos dust levels in the working environment
Effet des changements de techniques sur les concentrations de poussières d'amiante dans le milieu de travail 121

Maltoni C.

Occupational chemical carcinogenesis : new facts, priorities and perspectives
Cancérogénèse chimique professionnelle : faits nouveaux, priorités et perspectives 127

Pages

Bolt H.M., Kappus H., Kaufmann R., Appel K.E., Buchter A. and Bolt W.
Metabolism of ^{14}C-vinyl chloride in vivo and in vitro
Métabolisme du chlorure de vinyle marqué au ^{14}C in vivo et in vitro 151

Thony C., Thony J., Lafontaine M. and Limasset J.C.
Hydrocarbures polycycliques aromatiques cancérogènes dans les produits pétroliers : préventions possibles du cancer des huiles minérales
Carcinogenic polycyclic aromatic hydrocarbons in petroleum products — ways of preventing mineral oil cancer 165

Henschler D., Bonse G. and Greim H.
Carcinogenic potential of chlorinated ethylenes — tentative molecular rules
Pouvoir cancérogène des éthylènes chlorés : règles moléculaires provisoires 171

IV — RADIATION POLLUTION
POLLUTION PAR LES RADIATIONS

Latarjet R.
Les risques cancérigènes liés à la pollution par les radiations
Carcinogenic risk linked with radiation pollution 179

Mole R.H.
Radiation cancer, safety standards and current levels of exposure
Cancer dû aux rayonnements, normes de sécurité et niveaux actuels d'exposition 191

V — IDENTIFICATION OF CARCINOGENS
IDENTIFICATION DES CANCEROGENES

Chouroulinkov I. et Lasne C.
Cancérogénèse chimique en culture de tissus : Critères et tests de transformation
In vitro chemical carcinogenesis : Cellular transformation and tests 207

Bartsch H.
Mutagenicity tests in chemical carcinogenesis
Valeur des épreuves de mutagénicité pour prévoir l'éventuel risque cancérogène des substances chimiques 229

Shubik P. and Clayson D.B.

Application of the results of carcinogen bioassays to man
Application à l'homme des résultats des épreuves biologiques des cancérogènes 241

Salmon J.M., Viallet P., Kohen E. et Zajdela F.

Capacité de cellules en culture d'accumuler des hydrocarbures polycycliques à un niveau décelable microspectrofluorimétriquement — résultats préliminaires
Capacity of cells in culture to accumulate polycyclic aromatic hydrocarbons at a level sufficient to be detectable by spectrofluorimetric methods — Preliminary results 253

Tomatis L.

Extrapolation of test results to man
Extrapolation des résultats des tests à l'homme 261

Venitt S. and Searle C.E.

Mutagenicity and possible carcinogenicity of hair colourants and constituents
Mutagénicité et cancérogénicité des colorants capillaires et de leurs constituants 263

VI – ASSESSMENT OF CARCINOGENIC RISK
EVALUATION DU RISQUE CANCEROGENE

Flamant R.

Les aspects épidémiologiques dans l'évaluation des risques cancérigènes dus à la pollution
Epidemiological aspects in evaluating carcinogenic risks arising from pollution 275

Muir C.S., Mac Lennan R., Waterhouse J.A.H. and Magnus K.

Feasibility of monitoring populations to detect environmental carcinogens
Possibilité de surveillance des populations pour la détection des cancérogènes de l'environnement 279

VII – MEASURING POTENTIAL CARCINOGENS IN THE ENVIRONMENT
MESURE DES CANCEROGENES POTENTIELS DANS L'ENVIRONNEMENT

Sawicki E.

Analysis of atmospheric carcinogens and their co-factors
Analyse des cancérogènes de l'environnement et de leurs co-facteurs 297

Chovin P.

Echantillonnage et traitements préliminaires des hydrocarbures aromatiques polycycliques en vue de leur analyse ultérieure
Sampling and clean-up of aromatic polycyclic hydrocarbons 355

Fontanges R.

Méthodes de dosage proprement dites
Final methods of determination 369

Griciute L.

Composés N-nitrosés — analyse et carcinogénicité éventuelle chez l'homme
N-nitroso compounds — their analysis and possible carcinogenicity in man 375

VIII — INDUSTRIAL AND LEGAL ASPECTS
ASPECTS INDUSTRIELS ET JURIDIQUES

Epstein S.S.

Regulatory aspects of occupational carcinogens : contrasts with environmental carcinogens
Aspects réglementaires des cancérogènes professionnels : contrastes avec les cancérogènes de l'environnement 389

Westerholm P.

Administrative aspects on regulation of carcinogenic hazards in occupational environment
Aspects administratifs de la réglementation des risques cancérogènes dans l'environnement professionnel 403

Munn A.

Control of carcinogenic hazards in industry
Prévention du risque cancérogène dans l'industrie 417

Blyghton A.

The trade-unionist's view of occupational cancer
Point de vue d'un syndicaliste britannique sur le cancer professionnel ... 425

Escanez J.

Point de vue d'un syndicaliste français sur le cancer professionnel
A French trade-unionist's view of occupational cancer 431

Catton J.A.
Legislative framework of control of occupational carcinogens in the United Kingdom
Cadre législatif du contrôle des cancérogènes professionnels en Grande-Bretagne 435

List of Participants / *Table des Participants* 443
List of INSERM Publications 457
List of IARC Publications 471

Madame Simone VEIL
Minister of Health of France
Ministre de la Santé en France

Le Ministre de la Santé PARIS, le 15 octobre 1975

En 1775, Sir Percival POTT, a montré que le cancer des jeunes ramoneurs était dû à la présence dans la suie des cheminées de substances chimiques.

Cette découverte a été remarquable à bien des points de vue. Pour la première fois était démontrée l'action cancérigène de substances présentes dans notre environnement naturel. Depuis lors, bien d'autres substances ont été identifiées, poussières d'amiante, aflatoxine... et bien sûr la fumée du tabac.

Il est possible et même probable que d'autres facteurs cancérigènes existent dans notre environnement naturel ou industriel. Démontrer un pouvoir cancérigène quelconque sur l'homme est un problème extrêmement difficile. Les méthodes épidémiologiques se sont révélées les plus sûres. Malheureusement elles sont très coûteuses et il faut souvent de nombreuses années de recherches avant d'obtenir un résultat.

La nécessité d'identifier rapidement ces substances dans notre environnement quotidien, de disposer de méthodes sûres et rapides en la matière est un problème important de santé publique, problème dont la solution revêt un très grand caractère d'urgence.

Le symposium organisé conjointement par le Centre International de Recherche sur le Cancer de l'O.M.S. à Lyon et l'Institut National de la Santé et de la Recherche Médicale devrait stimuler les travaux dans ce domaine. Par la même occasion, il serait souhaitable de confronter les aspects réglementaires en usage dans les différents pays. Les responsables de la santé en France attendent beaucoup de cette réunion dont il m'est agréable de féliciter les organisateurs.

INTRODUCTION

The manifold problems arising from pollution of the environment are arousing general concern particularly in the industrialized countries. Considerable efforts are being made to reduce the level of ecological damage and to eliminate, or at least control, pollutants producing acute harmful effects.

The late toxic effects of environmental pollutants may, however, be more serious and the need to extend our knowledge of the cancer hazard inherent in environmental pollution is urgent. But the problem is complex. It is first necessary to identify potential carcinogens in the environment. With the thousands of synthetic chemicals that have been added to man's environment in the last two decades and the increased risk of exposure to ionizing radiations, this presents a formidable task. Then it is vital to provide sound criteria for deciding whether a potential carcinogen provides an actual cancer hazard for man and sound methods for control and protection.

There are many different scientific disciplines involved in tackling these problems. Epidemiology, analytical chemistry, biological chemistry, pharmacology, radiobiology must all contribute – and the list is by no means complete. The scientists must provide as complete data as possible and help in their interpretation, since interpretation is vital if social action is to follow. Who are going to be involved in that social action ? Obviously the public health authorities play the most important role, but to be effective, they must act in concert with those responsible for producing environmental pollution – and that is largely the industrial enterprises – and those exposed to the effects of the pollution – the workers in industry and the general public.

Control of pollution may often prove costly and therefore the economic implications for industry, and ultimately society, have some weight in the design of any limiting regulations. The views of those exposed to risks must also be given their weight and it is therefore essential to have a well-informed public. Since factory workers are so often subjected to much higher exposures and therefore much higher risks, their participation through representatives in the control mechanisms, particularly of the environment within and surrounding their work-place, is equally essential.

It was for all these reasons that the organizers strove to include as participants in the Symposium on Environmental Pollution and Carcinogenic Risks, the scientists who must provide the data, the representatives of public health authorities who must frame control regulations, the industrial management representatives who must apply the controls and the trade unionists who, with their workshop experience, contribute much to ensuring the effectiveness of any control or protection applied.

It was evident during the Symposium that the level of development of control mechanisms varied greatly from country to country. Discussions of the problems in an international meeting can serve not only to exchange experiences but, we hope, stimulate other countries to emulate the best examples of sound, scientifically-based efforts to improve man's environment, reducing all health hazards that may arise from pollution and, in particular, the risk of cancer.

This Symposium was the first to be organized jointly by the International Agency for Research on Cancer and the French National Institute of Health and Medical Research (INSERM), and it is hoped that this joint publication will contribute to national and international efforts in this vitally urgent field.

The EDITORS

AVANT-PROPOS

Les multiples problèmes qu'engendre la pollution de l'environnement sont un sujet général de préoccupation, dans les pays industrialisés en particulier. Des efforts considérables sont aujourd'hui déployés pour diminuer l'importance des dommages écologiques et éliminer ou, tout au moins, neutraliser les polluants qui ont des effets nocifs aigus.

Toutefois, les effets toxiques lointains des polluants environnementaux pouvant être plus graves encore, il est urgent d'améliorer notre connaissance du risque cancérogène inhérent à la pollution du milieu. Mais le problème est complexe. Il faut d'abord identifier les cancérogènes potentiels dans l'environnement. Etant donné les milliers de produits synthétiques ajoutés à l'environnement humain au cours des vingt dernières années, et le risque accru d'exposition aux rayonnements ionisants, la tâche s'annonce redoutable. Il est ensuite essentiel d'établir de solides critères pour déterminer si un cancérogène potentiel comporte un risque réel pour l'homme, et d'élaborer des méthodes sûres de lutte et de protection.

L'étude de ces problèmes fait appel à maintes disciplines scientifiques différentes. L'épidémiologie, la chimie analytique, la chimie biologique, la pharmacologie et la radiobiologie ont toutes une contribution à apporter et cette énumération n'est nullement exhaustive. Les chercheurs doivent fournir des données aussi complètes que possible et aider à les interpréter, car l'interprétation des données est capitale si elle doit être suivie d'une action sociale. Mais qui va participer à cette action sociale ? De toute évidence, c'est aux autorités de la santé publique qu'incombe le rôle le plus important, mais pour que leur action soit efficace, elle doit s'exercer en accord avec les responsables de la pollution de l'environnement, à savoir essentiellement les entreprises industrielles — et les personnes exposées aux effets de la pollution — c'est-à-dire les travailleurs de l'industrie et la population en général.

La lutte contre la pollution pouvant souvent s'avérer coûteuse, ses incidences économiques pour l'industrie, et donc la société, ne sauraient être négligées dans l'élaboration de toute réglementation limitative. On doit également prendre en considération les vues des individus exposés au risque, et il est donc indispensable qu'un public bien informé puisse exercer normalement des pressions politiques. Comme les travailleurs sont souvent soumis à des expositions bien plus fortes, et courent donc des risques bien plus grands, il est tout aussi indispensable qu'ils participent, par l'entreprise de leurs représentants, aux mécanismes de contrôle de l'environnement, sur leur lieu de travail et aux alentours en particulier.

C'est pour toutes ces raisons que les organisateurs se sont efforcés d'inviter au Symposium sur la Pollution de l'Environnement et les Risques Cancérogènes les chercheurs qui doivent fournir les données, les représentants des autorités de santé publique qui doivent élaborer les réglementations, les représentants du patronat qui doivent appliquer les mesures de contrôle et les syndicalistes qui, forts de leur expérience pratique, contribuent considérablement à l'efficacité de toute opération de lutte ou de protection.

Il est apparu au cours du Symposium que le niveau de développement des mécanismes de contrôle variait sensiblement d'un pays à l'autre. L'examen de ces problèmes dans le cadre d'une réunion internationale peut permettre l'échange de données d'expérience mais aussi, nous l'espérons, inciter d'autres pays à rivaliser avec les meilleurs exemples d'actions s'appuyant sur des connaissances scientifiques pour améliorer l'environnement de l'homme, et atténuer ainsi tous les dangers que la pollution peut comporter pour la santé et notamment le risque de cancer.

Ce symposium était le premier organisé conjointement par le Centre international de Recherche sur le Cancer et l'Institut National de la Santé et de la Recherche Médicale (INSERM), et l'on espère que cette publication commune sera une contribution aux efforts entrepris à l'échelon national et international dans ce domaine d'importance capitale.

Les EDITEURS

IMPORTANCE OF ENVIRONMENTAL FACTORS IN CANCER

John HIGGINSON
Director

International Agency for Research on Cancer
150, cours Albert Thomas, 69008 Lyon, France

INTRODUCTION

Until the mid 19th century physicians, following the writings of Hippocrates, regarded environmental factors of major importance in the causation of many diseases which were accordingly ascribed to obnoxious gases, smells, miasmas from marshlands, urban over-crowding, etc. Later, as individual bacteria were gradually identified as the cause of specific diseases, the environment tended to be considered only in terms of microbiology, since the latter readily explained the relationship between over-crowding, inadequate water supplies, and such communicable diseases as tuberculosis, scarlet fever, measles, etc. Although acute and chronic poisonings by heavy metals had been long recognized, the concept that the chemical environment could be dangerous to human health developed only slowly. Yet, the golden age of bacteriological research overlapped with the rise of the modern industrial state, with its continual growth in the use of new chemical compounds for industrial, agricultural and therapeutic practice. Further, since the beginning of the century there was increasing recognition that certain chemical compounds in the environment might pose potential hazards for human health.

Today, the important role of the chemical environment in disease is generally accepted especially in the case of cancer, although possible mutagenic and teratogenic hazards are receiving increasing attention.

THE ENVIRONMENT

The environment may be classified as:

Macro-environment

This term applies to the individual's total general environment, e.g. air and water pollution, general food supplies, etc. The individual cannot usually significantly modify the macro-environment, which is largely the responsibility of the appropriate governmental and other authorities.

Micro-environment

This term refers to the personal environment created by the individual. It includes cultural habits, e.g. cigarette smoking and individual drinking and eating habits. It may also include occupation where, however, the degree of individual control varies according to local political and socio-economic conditions.

THE ROLE OF THE EPIDEMIOLOGICAL METHOD IN THE IDENTIFICATION AND CONTROL OF ENVIRONMENTAL HAZARDS

Difficulties associated with animal experimentation

Although animal testing is the only method available at present for the prevention of the entrance of new hazardous compounds into the environment, extrapolation from animal experience is far from an exact science and remains largely problematic. Only five of those stimuli known to be carcinogenic to man were first identified in the experimental animal ; diethylstilbestrol, vinyl chloride, 4-amino-biphenyl, aflatoxin and mustard gas (Table 1).

STIMULI RECOGNIZED AS CARCINOGENIC TO MAN

SPECIES IN WHICH CARCINOGENIC NATURE FIRST IDENTIFIED

MAN		ANIMALS
COAL TAR & SOOT	TOBACCO	4 AMINO-BIPHENYL
CREOSOTE	BETEL QUID	VINYL CHLORIDE
AROMATIC AMINES	IONIZING RADIATION	MUSTARD GAS
MINERAL OILS	ULTRA VIOLET LIGHT	STILBOESTROL
PETROLEUM WAXES	BURNS	AFLATOXIN
ISOPROPYL OIL	HEAVY METALS	
BENZOL	ASBESTOS	
ALCOHOLIC DRINKS	PARASITES	
RESIDUES OF PETROLEUM		

Table 1. Stimuli recognized as carcinogenic to man. Species in which carcinogenic nature first identified.

While the majority of human carcinogens have been shown to be carcinogenic to animals, the converse is not true and to date evidence of a carcinogenic effect in man has not been demonstrated for certain recognized animal carcinogens, e.g. isoniazide and sodium penicillin. Furthermore, no rational methods yet exist for calculating a non-effect dose in man from animal data (1). If 'socially acceptable' levels of exposure for man are described, such levels are to a large extent dependent either on epidemiological data, e.g. aflatoxin (2) or ionizing radiation (3), or on 'guestimates' in which the dose carcinogenic for animals is treated according to certain mathematical formulae (4). Nonetheless, the latter, no matter how sophisticated the mathematical treatment, remains dependent for confirmation on epidemiological investigations.

Potential of the epidemiological method

Thus, in the absence of adequate human data and the relatively small number of cancers for which the aetiology is known, the public health situation for human cancer is not dissimilar to that confronting the public health official attempting to control communicable diseases in the mid 19th century without a knowledge of bacteriology.

While the major objective of the epidemiological method is to identify carcinogenic hazards within the environment, with a view to their control (5), epidemiology may also contribute to other aspects of environmental research. These include the determination of the socially 'acceptable risk', which is of great importance relative to legislation. Further, human data represent the only objective criteria against which *in vivo* and *in vitro* laboratory screening techniques may be evaluated.

THE IMPORTANCE OF CANCER AS A DISEASE

The importance of cancer in terms of human health can be calculated according to several parameters: frequency, reduced expectation of life, economic costs, etc. As a result of considering only age-specific rates per 100 000 per year the concept has arisen that cancer is relatively rare. In fact, in industrial states a male child at birth has approximately a one in four chance of developing cancer during his lifetime (6) and a one in five chance of eventually dying from the disease (7). Calculation of economic cost to the community is difficult, but all who have attempted this exercise are in accord that in industrial states a significant proportion of national resources is directed to the care and treatment of the disease (6). In terms of absolute numbers and expectation of life, degenerative arterial vascular disease may be more important than cancer. However, whereas there is evidence that cancer is largely dependent on environmental factors and is thus theoretically preventable, our present knowledge of the pathology of old age (8) would suggest that the majority of the other diseases of later life can only be retarded.

THE IMPORTANCE OF ENVIRONMENTAL FACTORS IN HUMAN CANCER

Apart from specific situations, e.g. cancers due to occupation or specific drugs, evidence of the importance of environmental factors in human cancer is based largely on geographical variations in incidence and changes in cancer incidence among migrant populations. However, there is a very considerable body of animal data available which adds strong circumstantial biological support to the human investigations.

The significance of geographical variations

As knowledge of the geographical distribution of cancer incidence, on both the national and international level, developed in the 1950s, wide variations in cancer patterns were demonstrated (9). Furthermore, changes in incidence of cancers at one site were in general unrelated to variations at other sites, suggesting that separate aetiological factors were involved (10). It became possible to calculate from such data the incidence of cancer in a hypothetical population in which the incidence of cancer was theoretically the lowest possible, by summating the minimal rates at each site from several different geographical areas.

Utilizing this approach, there is circumstantial evidence that 80 - 90% of all cancers are dependent, directly or indirectly, on environmental factors, using the term in its widest connotation. Doll (11) and Boyland (12), using other approaches, have reached similar conclusions, the latter believing that at least 90% of these exogenous factors are chemical in nature, ascribing a direct role for viruses only to certain hematopoietic and soft-tissue tumours.

The significance of migrant studies

In 1944, Kennaway (13) reported that primary cancer of the liver, a very common cancer in Africa, was rare in blacks in the United States and concluded that this cancer was dependent on exogenous and not racial factors. Similar studies were later carried out by Haenszel and co-workers (14, 15) on cancer patterns in Japanese and other migrants to the United States, in a series of classical papers. These investigations strongly supported the view that for most major cancers, racial or hereditary factors were of relatively little importance in comparison to the environment. Thus, a marked fall in incidence was demonstrated for cancer of the stomach in Japanese migrants to the United States over three generations whereas an increase in colon and breast cancer was observed occurring within the lifetime of the migrant. These facts indicated that different carcinogenic stimuli were involved operating at different intensities. Fraumeni (16) in a review of hereditary factors also concluded that cancers due to either autosomal dominance or recessive genes are rare.

The role of individual susceptibility

The above comments do not imply that individual susceptibility is unimportant. Thus, only one in eleven individuals who smokes two packs of cigarettes a day in the United Kingdom will die of cancer of the lung. Apart, however, from skin pigmentation and certain rare conditions, e.g. agamma-globulinemia, the nature of such susceptibility is not understood, nor have suitable objective methods and parameters been developed for measuring such susceptibility.

It should be noted that individual susceptibility can be modified also by exposure to environmental factors, as illustrated by the effects of cigarette smoking in asbestos workers (17). Thus, no significant increase in primary lung cancer (excluding mesothelioma) has been reported in asbestos workers who are non-smokers as compared to the general population. On the other hand, asbestos exposure increases the risk of lung cancer several fold over the usual risk observed among the general population.

CLASSIFICATION OF CANCER BY AETIOLOGICAL BACKGROUND

In addition to site, cancer may be classified according to aetiology (18).

1. ### Cancers due to the cultural environment

The most important cancers of known aetiology are those which are dependent on a cultural habit. Thus, cancer of the lung due to cigarette smoking accounts for approximately 40% of cancer deaths in males in the United Kingdom or 11% of all deaths. In India 30% of all cancers occur in the mouth and pharynx and are related to the habit of chewing betel. Excessive ingestion of alcoholic beverages has been shown to lead to cancer of the mouth, oesophagus and liver. Despite their primary importance, attempts so far to control cultural habits have largely been unsuccessful and further sociological research must therefore be directed to the nature of habits and their modifications.

2. ### Cancer due to the occupational environment

Although a relatively small proportion of all cancers (1-3%), this group is important since control and prevention may be possible, e.g. bladder cancer in the rubber industry (19). It should be noted that what may be initially an occupational hazard with a potential high risk may eventually involve large population groups at lower levels. Thus, the existence of high risk groups has been of importance in identifying the actual risk in man. However, other occupational risks identified include: nasal cancer in nickel workers or woodworkers; leukaemia in benzene workers; lung cancer in workers exposed to chromium or bis(chloromethyl ether); bone sarcomas in workers exposed to radium, where the risk has been sometimes so high as to be identifiable even without sophisticated epidemiological studies. The possibility of new carcinogens entering the environment, e.g. vinyl chloride (20) has increased in recent years with the continual increase of new compounds in industrial practice.

3. Iatrogenic factors in cancer

Until recently, cancers of this type were largely limited to certain specific agents, e.g. nitrogen mustards and radiation. Thus, 12% of individuals treated with chloronaphazine (21) developed cancer of the bladder. However, in the last two decades there has been an increase in the long-term use of certain drugs in healthy persons, e.g. contraceptives and tranquillizers. Although such drugs may have been of great benefit to many, their continual use has given rise to concern regarding potential hazards. While drugs theoretically represent the least complex situations for calculating risk/benefit situations, unfortunately adequate data are all too often unavailable (22). Slight increases in the absolute numbers of common cancers are very difficult to identify, in contrast to an increase of the same number for a rare tumour which may be easily apparent. Thus, even when the total number observed may be insufficient to calculate statistical significance, the aetiological association may sometimes be apparent beyond reasonable doubt, e.g. adenocarcinoma of the vagina, benign adenoma of the liver, mesothelioma.

4. Congenital and hereditary factors

As indicated, apart from skin pigmentation and individual susceptibility these are of little significance (16).

5. Idiopathic cancers

At present, the aetiology of approximately only 40% (excluding skin) of human cancers has been identified. Of the remainder, between half and two-thirds and possibly more are believed, for the reasons above (10, 14, 15), to be related directly or indirectly to environmental factors. It should be noted, however, that the cancer patterns of today are dependent on agents entering the environment over 20 years ago, and that those new agents at present giving rise to the greatest disquiet are unlikely to be involved. Furthermore, it should not necessarily be assumed that the modern industrial environment is more dangerous than that in the past, as indicated, for example, by the significant fall in gastric cancer that has occurred in recent years (23).

THE MEANING OF THE TERM 'RISK'

A major requirement for the modern public health officer in the control of cancer is to determine cost/benefit situations, for which accurate epidemiological data are necessary. In making comparisons it is probably desirable to present the risk of developing a cancer within a lifetime, as this permits assessment of the risk in better perspective than if only age-specific rates are examined. It is most important to distinguish between a marked increase in the relative risk of a rare tumour affecting only a small proportion of the population, and in which the potential risk to society or the group may be quite small, and a relatively moderate increased risk of a common tumour which may be very much more important in terms not only of absolute numbers in society but also for the individual himself. Data of this type are summarized in Table 2, which is calculated on a hypothetical male population, according to statistics available from recent epidemiological studies and which can be considered as reasonably representative of the actual situation in a modern industrial state. Although the figures have been rounded off and the potentiality of certain competing risks eliminated, the effect of these more sophisticated calculations would probably not significantly affect the conclusions. The above figures have been prepared for me by Drs C. S. Muir and J. A. H. Waterhouse (personal communication).

	INCIDENCE PER 10^5	NUMBER OF CASES OBSERVED	RELATIVE RISK RATIO
BRONCHOGENIC CARCINOMA IN			
NON-SMOKERS (SAY 30% POPULATION)	8	2 500	1
GENERAL POPULATION	60	60 000	7
ALL SMOKERS (SAY 70% POPULATION)	85	57 500	10
HEAVY SMOKERS (SAY 25% POPULATION)	130	32 500	15
NON-SMOKING ASBESTOS WORKERS (SAY 20 000 PERSONS)^a	8	2.5	1
SMOKING ASBESTOS WORKERS (SAY 20 000 PERSONS)	560	400	70 ±

a While the incidence of lung carcinoma in non-smoking asbestos workers is not greater than that in the general population, the data are insufficient to say with certainty that lung cancer is not increased as compared with non-smokers in the general population. Mesotheliomas have been excluded.

Table 2.   Comparison between incidence, number of cases in population and relative risks.

This table compares incidence rates per 100,000 population, the total number of cases observed and the relative risk ratio in a hypothetical population of 100 mill males. The relative ratios selected are based on the average data available for smokers and non-smokers in the UK and for asbestos workers from several series (17) and from recent reports on vinyl chloride in the US. While the figures may change with the basic data used, the general principles illustrated will essentially remai the same.

The risk of dying before the age of 65 is approximately three in ten for the average male. During this period, for a non-smoker the chance of dying from lung cancer is relatively small. However, the increase in risk may be very much greate for a heavy smoker. On the other hand, workers exposed to very high levels of vin chloride have been stated to have an increased risk of 400 times that of the genera population for developing angiosarcoma. However, since the latter is a very rare cancer, the absolute risk is still very much less than that of dying from lung canc in the general population. A number of other occupational risks, as indicated by incidence data, are given in Table 3 as calculated by Pochin (24). Such calculati may assist appropriate management and labour executives to place the risks in perspective for the individual worker as compared to the general population.

OCCUPATION	CAUSE OF FATALITY	RATE (D/M/Y)
WORKERS WITH CUTTING OILS - BIRMINGHAM	CANCER OF THE SCROTUM	60
SHOE INDUSTRY (PRESS AND FINISHING ROOMS)	NASAL CANCER	130
PRINTING TRADE WORKERS	CANCER OF THE LUNG AND BRONCHUS	ABOUT 200
WOOD MACHINISTS	NASAL CANCER	700
URANIUM MINING	CANCER OF THE LUNG	1500
COAL CARBONIZERS	BRONCHITIS AND CANCER OF THE BRONCHUS	2800
RUBBER MILL WORKERS	CANCER OF THE BLADDER	6500
MUSTARD GAS MANUFACTURING (JAPAN 1929-45)	CANCER OF THE BRONCHUS	10400
NICKEL WORKERS (BEFORE 1925)	CANCER OF THE LUNG	15500
β-NAPHTHYLAMINE	CANCER OF THE BLADDER	24000

Table 3. Estimated rates of fatality (or incidence) of disease attributed to types of chemical or physical exposure (24).

The taking of some sort of risk seems to be an inherent part of modern life, but the data permitting exact calculation of such risks are often unavailable. The recent paper by Pochin (24) on this subject is important as providing an estimate of excess risks due not only to cancer, but also to other hazards, including occupation, sports and cultural habits. The cancer risk, however, tends often to outweigh all other risks due to the mystique that the disease possesses. Pochin's calculations further emphasize the paramount importance of cigarette smoking (a personal risk), in relation to other (non-personal) risks.

It is the duty of the scientist to ensure that the risk data on which decisions can be made are developed and comprehended. It is desirable that such data should permit accurate comparison between risks, in order that the individual may place his own situation in perspective. Furthermore, attention must be drawn to the fact that not all risk-carrying chemicals are equally important. Thus, food colorants without benefit offer no excuse for taking the slightest risk. On the other hand, drugs should be considered separately and one must be exceedingly careful of delaying unnecessarily or preventing the use of an effective medicament, since the delay period or non-use may itself carry a potential risk of death or ill-health for untreated individuals (22). This is particularly true when we are considering certain of the parasitic diseases so widespread in the tropics and elsewhere.

REFERENCES

(1) IARC Monographs on the Evaluation of Carcinogenic Risk of Chemicals to Man, Volume 1. International Agency for Research on Cancer, 1972, Lyon, France.

(2) IARC Monographs on the Evaluation of Carcinogenic Risk of Chemicals to Man, Volume 10. International Agency for Research on Cancer, 1976 (in press), Lyon, France

(3) Upton A.C. : Comparative observations on radiation carcinogenesis in man and animals. Carcinogenesis. A Broad Critique: 631-675, The Williams and Wilkin Company, Baltimore, 1967

(4) Schneidermann M.A., Mantel N. and Brown C.C. : From Mouse to Man - or How to Get from the Laboratory to Park Avenue and 59th Street. Annals of the New York Academy of Sciences, 1975, 246, 237-248

(5) Higginson J. : Developments in Cancer Prevention through Environmental Control In: Advances in Tumour Prevention, Detection and Characterization. Volume 2: Cancer Detection and Prevention. C. Maltoni, ed., pp. 3-18, Excerpta Medica, Amsterdam

(6) Cancer Facts and Figures. American Cancer Society.

(7) World Health Statistics Annual. Volume I : Vital Statistics and Causes of Death, 1975.

(8) McKeown F. : De Senectute (The F. E. Williams Lecture) Journal of the Royal College of Physicians, 1975, 10, 79-99

(9) Cancer Incidence in Five Continents. Vol. II. R. Doll, C.S. Muir and J. A. Waterhouse, eds., 1970, Springer Verlag, Berlin, Heidelberg, New York

(10) Higginson J. : Present Trends in Cancer Epidemiology. In: Proc. 8th Canadian Cancer Conference, Honey Harbour, Ontario, 1969, 40-75

(11) Doll R. : Prevention of Cancer - Pointers for Epidemiology. The Rock Carling Fellowship 1967 - Nuffield Provincial Hospitals Trust 1967, 1967, Whitefriars Press Ltd., London

(12) Boyland E. : A Chemist's View of Cancer Prevention. Proc. Royal Society of Medicine, 1967, 60, 93-99

(13) Kennaway E.L. : Cancer of the Liver in the Negro in Africa and in America. Cancer Research, 1944, 4, 571-577

(14) Haenszel W. : Cancer Mortality among the Foreign Born in the United States. J. Natl. Cancer Inst., 1961, 26, 37-132

(15) Haenszel W. and Kurihara M. : Studies of Japanese Migrants. I : Mortality from Cancer and Other Diseases among Japanese in the United States. J. Natl. Canc Inst., 1968, 40, 43-68

(16) Fraumeni J.F. : Genetic Factors. In: Cancer Medicine, J.F. Holland and E. Fre eds., 1973, pp. 7-15, Lea & Febiger, Philadelphia

(17) Biological Effects of Asbestos. P. Bogovski, J.C. Gilson, V. Timbrell and J.C. Wagner, eds., IARC Scientific Publications No. 8, International Agency fo Research on Cancer, Lyon, France

(18) Higginson J. and Muir C.S. : Epidemiology. In: Cancer Medicine, J.F. Holland and E. Frei, eds., 1973, pp. 241-306, Lea & Febiger, Philadelphia

(19) Case R.A.M., Hosker M.E., McDonald D.B. and Pearson J.T. : Tumours of the urinary bladder in workmen engaged in the manufacture and use of certain dyestuff intermediates in the British Chemical Industry. Part I : The role of aniline, benzidine, alpha-naphthylamine and beta-naphthylamine. Brit. J. Industr. Med., 1954, 11, 75-104

(20) Heath C.W., Falk H. and Creech J.L. : Characteristics of cases of angiosarcoma of the liver among vinyl chloride workers in the United States. Annals of the New York Academy of Sciences, 1975, 246, 231-236

(21) Higginson J. : Cancer Etiology and Prevention. In: Persons at High Risk of Cancer - An Approach to Cancer Etiology and Control. J.F. Fraumeni Jr., ed., 1975, pp. 385-398, Academic Press Inc., New York, San Francisco, London

(22) Wardell W.M. and Lasagna L. : Regulation and Drug Development. Evaluative Studies 21, 1975, American Enterprise Institute for Public Policy Research

(23) Segi M., Murihara M and Matsuyama T. : Cancer Mortality for Selected Sites in 24 countries. No. 5 (1964-1965). 1969, Sendai, Japan, Department of Public Health, Tohoku University School of Medicine

(24) Pochin E.E. : Occupational and Other Fatality Rates. Community Health, 1974, 6, 2-13

SUMMARY

The present paper reviews the nature of epidemiological studies and their importance as a background for determining the hazard of cancer. An attempt is made to explain how risks in the cancer area may be compared by the use of cumulative data over a lifespan as compared to use of age-specific rates. In addition, the classification of the aetiological factors so far identified is summarized and it is pointed out that to date, the most important environmental risks relate to those of cultural habits. Of the 85% of cancers for which the causes have not yet been established, there is circumstantial evidence that the greater part of these are dependent directly or indirectly on environmental factors.

RESUME

Le présent article examine la nature des études épidémiologiques et leur importance fondamentale pour déterminer le risque de cancer. Il s'efforce d'expliquer comment on peut comparer les risques cancérogènes par l'utilisation de données accumulées au cours d'une durée de vie, comme par celles de taux par âge. La classification des facteurs étiologiques jusqu'ici identifiés est en outre résumée, et il est souligné qu'à ce jour les risques environnementaux les plus importants sont liés aux habitudes culturelles. Pour les 85% des cancers dont la cause n'a pas encore été établie, on a de bonnes raisons de croire, sur la base de preuves indirectes, que la plupart sont tributaires, directement ou indirectement, de facteurs d'environnement.

I

ATMOSPHERIC POLLUTION
POLLUTION ATMOSPHÉRIQUE

Chairman / *Président :* Pr. R. LATARJET

COAL FIRES, INDUSTRIAL EMISSIONS
AND MOTOR VEHICLES
AS SOURCES OF ENVIRONMENTAL CARCINOGENS

P.J. LAWTHER and R.E. WALLER

MRC Air Pollution Unit, St. Bartholomew's Hospital, Medical College
London EC1M 6 BQ, United Kingdom

The association of products from the incomplete combustion of fuels with the development of cancer can be traced back 200 years, to the original observations of Percivall Pott (1) on the occurrence of scrotal cancer among chimney sweeps. The wider implications of this finding were however recognised only slowly. In a review of the situation in 1892, Butlin (2) observed that the burning of soft coal in open grates appeared to be one of the factors responsible for the relatively frequent occurrence of chimney sweeps' cancer in Britain, and in 1922, nearly 150 years after Pott's observations, Passey (3) demonstrated experimentally the carcinogenicity of extracts of domestic soot. By then much evidence had accumulated, not only on the occurrence of scrotal cancer through contact with soot, but also on a wider range of skin cancers among workers exposed to coal tar products (4), and there followed an intensive study of the carcinogenicity of tar fractions by Kennaway and his co-workers (5), leading finally in 1933 to the identification and isolation of the potent carcinogen benzo(a)pyrene (6). Although it was realised then that the peculiarly British habit of burning coal inefficiently on open grates led to the dispersion of much tarry material into the air, the possible relevance of this to the development of lung cancer emerged only gradually as the incidence of this disease increased and as attention was drawn to the higher mortality in urban as compared with rural areas (7). This fact was the stimulus for further studies on the potentially carcinogenic properties of coal smoke; benzo(a)pyrene was detected and determined in samples of domestic soot (8) and by 1952 the distribution of this compound had been examined in a number of towns in Britain (9). The dominant source appeared to be the open coal fires that were still widely used, but contributions from industrial sources and motor vehicles were also recognised. Since that time there has been an intensification of interest in the role of carcinogens from these sources in relation to the growing problem of lung cancer throughout the world, but in view of the now well-established and clearly overwhelming effect of cigarette smoking on the development of this disease (10), it may be time to pause and reflect on the findings to date.

Smoke and benzo(a)pyrene in urban air

Benzo(a)pyrene occurs together with a wide range of other polycyclic aromatic hydrocarbons in tar from the incomplete combustion of hydrocarbon fuels, and it is usually mixed with carbon and various inorganic components of the suspended particulate matter in urban air (referred to as "smoke" in Britain). The range of polycyclic compounds that may be present, together with their structures and notes on nomenclature (benzo(a)pyrene was known earlier as 3:4-benzpyrene) and carcinogenicity has been well reviewed in a recent monograph (11). There is no direct way of determining whether the mixture of pollutants, or any specific component of it, is carcinogenic to man when inhaled. The carcinogenic properties of the various polycyclic hydrocarbons present have been determined mainly from the results of experiments in which massive doses have been applied to the skins of animals. Benzo(a)pyrene is not the only component which, on this basis, is regarded as a potent carcinogen, but it is commonly taken as an index of that class of compound.

Some guide to the concentrations of benzo(a)pyrene and related compounds can be obtained from the relatively crude measurements of smoke, as determined from the blackness of samples collected on filter papers (12) particularly where one type of source dominates, as the coal fire did in British towns prior to about 1960. This allows some idea of long-term trends to be obtained, and in Table I results obtained at a small selection of sites are shown for a 12 month period in 1935-36, compared with a survey made just 25 years later (13).

	Annual mean smoke concentrations, $\mu g/m^3$	
	1935-36	1960-61
London (County Hall)	295	153
Cardiff	160	54
Coventry	450	82
Glasgow	475	150
Stoke-on-Trent	830	250

Table I Concentration of smoke at selected sites in Great Britain, 1935-36 and 1960-61.

One notable feature is the very high concentration, in the earlier period, in Stoke-on-Trent, a city that became notorious for its pall of black smoke, coming from the coal-fired kilns of the local pottery industry as well as from houses. The other important feature to note is the substantial reduction in smoke concentrations that had been achieved in large towns in Britain, even before the Clean Air Act of 1956 had been fully implemented. During the 25 year period cited here lung cancer death rates had been rising rapidly (10), and there was no reason, from consideration of time trends, to associate the rise with air pollution. The decline in smoke concentrations continued further during the 1960's, particularly in London (Fig.1) where the values are now no higher than those in small towns in some parts of the country.

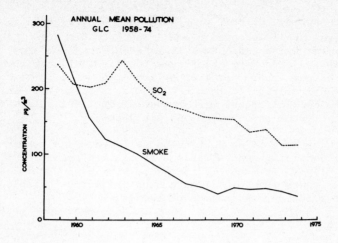

Fig. 1   Annual mean concentrations of smoke and sulphur dioxide, London, 1958-74 (mean of seven sites).

The contrast in concentrations that at one time existed between large cities, smaller towns, and rural areas, was however the key to the interest then shown in the possible relationship between lung cancer mortality and benzo(a)pyrene in the air, and when measurements were made in Northern England as part of a large survey in 1958 (14), a wide range of values was found (Table II).

	Benzo(a)pyrene $\mu g/1000m^3$
Sites in major conurbations	
Salford (Regent Rd.)	108
Newcastle (Warncliffe St.)	75
Leeds	42
Liverpool (Edge Lane)	38
Other large towns	
Warrington	34
Burnley	32
York	24
Small towns	
Ripon	15
Wetherby	11

Table II   Annual mean concentration of benzo(a)pyrene at sites in Northern England, 1958.

Large scale surveys of that type have not been repeated in recent years, but if trends in smoke concentrations are taken as a guide, as suggested above, it is clear now from the analysis of 10 years National Survey results (15) that throughout Great Britain there has been a large reduction in airborne polycyclic hydrocarbons during the 1960's, particularly in areas where concentrations were highest, resulting in a much reduced contrast between large and small towns.

Although there is no continuous series available to follow the declining concentrations of benzo(a)pyrene directly, results from several separate studies in London over the past 25 years (9, 16, 17) have been collected in Table III to provide a further indication of trends.

Period	Sampling site	Benzo(a)pyrene $\mu g/1000m^3$
1949-51	County Hall	46
1953-56	St. Bartholomew's Hospital	17
1957-64	County Hall	14
1972-73	St. Bartholomew's Medical College	4

Table III Concentrations of benzo(a)pyrene in air at sites in Central London, 1949-73, based on 24 hour samples aggregated for yearly periods.

Although there are differences in sampling site and in analytical methods within this series, the results suggest that the concentration of benzo(a)pyrene in the air is now only of the order of one tenth of what it was 25 years ago. If the earlier concentrations had been one of the factors related to the urban excess of lung cancer, then this substantial change might eventually be reflected in the mortality statistics. Since people dying now could still have been exposed to relatively high concentrations of smoke earlier in their lives, and since other factors apart from urban air pollution could be related to the urban/rural differences in lung cancer, the interpretation of current trends in mortality are difficult, but there are signs that the urban excess of lung cancer is declining (18, 19).

Industrial sources of benzo(a)pyrene

In the early studies of benzo(a)pyrene in the air (9), there were indications that in some localities the incomplete combustion of coal for heating and power generation in industry made an appreciable contribution to urban concentrations, but in general, the benzo(a)pyrene content of industrial smoke is less than that of domestic smoke (20). The open burning of refuse may contribute some benzo(a)pyrene to the air, but in Britain such sources, together with all dark smoke from industrial chimneys, are now sufficiently well-controlled to make these contributions very small. In some localities there may be emissions of smoke or hydrocarbon mixtures containing benzo(a)pyrene from oil refineries (20), but these are unlikely to make any appreciable contribution to urban concentrations.

In the coal gas and coke producing industry in which tarry material containing benzo(a)pyrene is dispersed within and around the works, the risks of skin cancer among men working with coal-tar have long been recognised, and more recent studies have shown some excess of lung cancer among men employed in gas-works retort houses (21), and at coke ovens (22). Although the emissions of benzo(a)pyrene from these sources into the general air of towns is small, measurements in and around these works are of interest in relation to the exposure of the men employed there. An example of the vast range of concentrations found within gas-works (23) is shown in Table IV.

	Benzo(a)pyrene $\mu g/1000m^3$
Maximum in fumes escaping from retorts	2,330,000
Air above horizontal retorts	216,000
Mean determined from mask samples (worn by workers)	3,000
Mean of long-period air samples within works	3,000

Table IV   Concentrations of benzo(a)pyrene within gas-works.

In this table the concentrations have been expressed in the units used for urban air values to allow direct comparison to be made with Tables II or III, and it can be seen that the concentrations to which men may be exposed in the course of their work (the mask samples in particular being related directly to their normal exposure), are several orders of magnitude higher than the average levels in urban air. It is difficult to extrapolate from the experience of these occupational groups to the general population, but the findings suggest that the present-day concentrations of benzo(a)pyrene in urban areas of Great Britain are unlikely to have any substantial effect on lung cancer death-rates.

Vehicular sources of benzo(a)pyrene

The emission of polycyclic hydrocarbons from motor vehicles varies with the condition of the engine and with the operating conditions, but benzo(a)pyrene can be present in the exhaust from both petrol and diesel engines (24). Some early work gave a misleading impression of the extent to which motor vehicles especially diesel-powered (25), might contribute to urban concentrations, and to assess the situation in towns, with an existing background of benzo(a)pyrene from domestic fires and other stationary sources, measurements have been made in streets and in other locations where there is much pollution by vehicle exhaust. An essential difference between the vehicular and stationary sources is that the latter generally emit the pollution well above breathing level, so that it is usually diluted and dispersed to provide a reasonably uniform background concentration, whereas that from vehicles is emitted at a very low level, and the concentrations to which people are exposed vary greatly with the distance from traffic. The maximum contribution to such exposures can then be assessed by making measurements very close to traffic, and comparing them with measurements made at a control site some distance away.

The effect of exhaust fumes from well-maintained diesel engines has been assessed by making measurements inside London Transport bus garages (26) and at control sites nearby (Table V).

	Benzo(a)pyrene, $\mu g/1000m^3$	
	Garage	Outside
Merton garage, April	460	470
Merton garage, June	26	14
Dalston garage, October	24	23
Dalston garage, June	9	5

Table V Concentrations of benzo(a)pyrene at London Transport bus garages (peak periods), 1956-57.

The overwhelming influence of background sources was seen in the first set of observations in this table, obtained during a short period when the weather was cold, with smoke from coal fires in adjoining houses drifting towards the garage; in warmer weather in June it could be seen that the exhaust from the buses, which contained much black smoke, made some contribution to the benzo(a)pyrene content of the garage air. These observations were followed by others in tunnels, where the mixed diesel- and petrol-engined traffic, undoubtedly including some with engines in poor condition, made an appreciable contribution to the benzo(a)pyrene content of the air in the tunnel itself (27). A series of measurements was then made in a busy London street (28). These were limited to the day-time hours only, when there was much traffic. Results averaged over a whole year are shown in Table VI.

Location of sampler	Benzo(a)pyrene $\mu g/1000 m^3$
Centre of street, very close to traffic	39
Control site, 46 metres from street	20
St.Bartholomew's Hospital (away from traffic)	26

Table VI   Concentrations of benzo(a)pyrene at sites in the City of London, daytime periods, 1962-63.

These figures are not directly comparable with those in Table III, since weekend and night-time periods, when pollution is relatively low, were omitted, but they showed again that when very close to the emissions from motor traffic there is some enhancement of benzo(a)pyrene concentrations. However it was concluded then that with coal fires still having an important influence on background concentrations, the overall contribution of traffic to the benzo(a)pyrene content of the air in London was small. Elsewhere the relative contributions of vehicular and stationary sources may be different; in Los Angeles in particular motor vehicles are the only major contributor to air pollution, and ooncentrations of benzo(a)pyrene are low compared with those in coal-burning areas of the United States (11).

Raffle (29) was unable to demonstrate any excess of lung cancer among workers in London Transport diesel bus garages when compared with other workers within the same organisation. The limited relevance of these findings, and of our analyses of the air in garages (26) to the occurrence of lung cancer in town dwellers who breathe air polluted by diesel vehicles was carefully noted; the morbidity and mortality statistics relating to London Transport workers in all categories is kept under close review.

Other carcinogens in urban air

Industry uses and produces an immense variety of compounds some of which have been shown to cause cancer; it is implicit that others exist or are yet to be produced which will ultimately be demonstrated to have carcinogenic properties. some will escape into the ambient air and may affect members of the general public who have suffered no special exposure by virtue of their occupation; some prime consumer products may be carcinogenic.

Among workers exposed over several years to any of the various types of asbestos, lung cancer is found more frequently than among the general population (30) though the excess is rapidly diminishing. The uses to which asbestos minerals are put are legion. Hendry (31) claimed that there were over 1,000 industrial uses apart from its well known incorporation in cement and in fire proofing materials. Speil and Leinweber (32) have made a more detailed technical review of its nature and industrial uses. Asbestos fibres of any sort can pollute the communal air by escaping from factories, from waste disposal and from the abrasion of brake linings, clutch facings and of other asbestos containing materials. The topic has been elegantly and concisely reviewed by Martin (33). The demonstration of asbestos bodies in the sputum and tissue of non-occupationally exposed town dwellers are regarded as being evidence merely of inhalation of asbestos fibres and not of disease; there has been much debate (34) as to whether all these bodies do indeed contain asbestos and whether they ought to be called ferruginous bodies, but there is no evidence that exposures to the concentrations of asbestos encountered in town air cause lung cancer.

A new era in work on asbestos followed the demonstration by Wagner and his colleagues (35) of the association between pleural mesothelioma and exposure to crocidolite asbestos acquired merely by living near to the mines and works where the mineral was processed. The prevalence and distribution of these tumours have been studied by many workers in Great Britain and a similar association with occupation or residence near asbestos works has been demonstrated (36). There is good evidence that exposure need be only minute and that the tumours resulting therefrom may not be clinically manifest for periods as long as 40 years.

Urban air contains many metals in trace amounts, and some of these have been associated with cancer among occupational groups. The nasal and lung cancer risks in the nickel industry appear to be associated specifically with the refining process (37), and although nickel was among the suspect trace metals determined along with polycyclic hydrocarbons in the survey of potential carcinogens referred to earlier (14), the results showed a large scatter not consistently related to size of town, and produced no evidence of hazard to the general public. Other metals included in that same survey were chromium and beryllium, because of associations with cancer in occupational groups, but again there was no evidence from the results, nor from any subsequent studies, that the risks associated with specific processes in industry could be extrapolated in any way to the exposure of the general population to these materials as trace components of urban air pollution.

Much more recently, concern about the escape of industrial carcinogens to pollute the communal air has been stimulated by the recognition of the effect of vinyl chloride monomer in causing angio-sarcoma of the liver in process workers. It is probable that the doses to which they were exposed in the past were relatively massive; it is certain that the concentrations which may be found in town air will be very small. But the precedent created by crocidolite will serve to warn us to keep the role of town air in the aetiology of cancer under continuous careful scrutiny.

Conclusion

The interest in carcinogens dispersed in the air, primed by Pott's astute observations made 200 years ago, but exploited only in relatively recent years, may to a certain extent have been misplaced or exaggerated. The whole issue as to whether urban air pollution is a relevant factor in accounting for the higher incidence of lung cancer commonly found in towns as compared with rural areas has been obscured by the effect of cigarette smoking and differences in the smoking histories of town and country dwellers, yet there is no doubt that the interest in this urban factor, aroused principally by the late Percy Stocks, was one of the stimuli that led to the investigation of the effects of smoking itself. Many years may elapse yet before it becomes clear whether urban air pollution has played or is playing any substantial role in the aetiology of lung cancer, either independently or in combination with smoking, but the massive reductions in smoke concentrations achieved in the major cities of Britain during the last two decades may be creating a natural experiment that will help to elucidate the problem.

SUMMARY

One of the most widely studied carcinogenic agents in the environment is the polycyclic hydrocarbon, benzo(a)pyrene. As a component of soot from the inefficient combustion of coal, its association with cancer can be traced back 200 years, but its possible relevance to lung cancer as a widely distributed air pollutant has been investigated only during the past 25 years. Domestic coal fires have been shown to be important sources, and smaller amounts come from industrial sources and from motor vehicles. There is evidence now that the concentration of benzo(a)pyrene in large towns in Britain has decreased by a factor of about ten during the last few decades, as a result of changing heating methods and smoke control. In view of the overwhelming effect of cigarette smoking, it is difficult to determine whether the benzo(a)pyrene content of the air has had any important effect on the development of lung cancer, but careful analysis of trends in mortality may now throw some light on this.

Among other materials with carcinogenic properties that may be dispersed into the general air, asbestos is the one that has been investigated most thoroughly. The association between exposure to asbestos and the development of lung cancer and mesothelioma of the pleura has been clearly demonstrated among people occupationally exposed to the dust, but as far as the general public is concerned, any risk may be limited to the immediate vicinity of major sources. These and other hazards demonstrated among occupational groups serve as a warning however to maintain careful scutiny of urban air pollutants in relation to the aetiology of cancer.

Résumé

Le benzo(a)pyrène, hydrocarbure polycyclique, est l'un des agents cancérogènes les plus étudiés dans l'environnement. En tant que composant de la suie résultant de la combustion incomplète du charbon, son association avec le cancer remonte à 200 ans, mais ses relations éventuelles avec le cancer du poumon, en tant que polluant de l'air largement répandu, ne sont étudiées que depuis 25 ans. On a montré que les feux de charbon domestiques sont des sources importantes de cet agent, alors que les industries et les véhicules à moteur en produisent de moindres quantités. Il apparaît maintenant que dans les grandes villes britanniques la concentration de benzo(a)pyrène est devenue environ dix fois moindre au cours des dernières décennies, en raison des nouvelles méthodes de chauffage et de la lutte contre les fumées. Etant donné les effets prédominants de l'usage de la cigarette, il est difficile de déterminer sir la concentration de benzo-(a)pyrène dans l'air a eu quelque influence importante sur le développement du poumon, mais une analyse minutieuse de l'évolution de la mortalité peut maintenant donner quelques indications à ce sujet.

Parmi les autres substances cancérogènes susceptibles d'être dispersées dans l'air en général, l'amiante est celle qui a fait l'objet des recherches les plus approfondies. L'association entre l'exposition à l'amiante et l'apparition du cancer du poumon et du mésothéliome de la plèvre a été nettement mise en évidence chez les sujets professionnellement exposés aux poussières, mais en ce qui concerne la population générale, tout danger peut être limité au voisinage immédiat des principales sources d'émission. Toutefois, ces risques, ainsi que d'autres mis en évidence dans certains groupes professionnels, montrent la nécessité de maintenir un contrôle rigoureux des polluants de l'air dans les villes, en tant que facteurs étiologiques du cancer.

References

(1) Pott, P.: Cancer scroti. In Chirurgical Observations, 1775, pp.63-68. Hawes, Clarke and Collins, London.

(2) Butlin, H.T.: Cancer of the scrotum in chimney sweeps and others. Brit.med.J., 1892, 1, 1341-1346, 2, 1-6, 66-71.

(3) Passey, R.D.: Experimental soot cancer. Brit.med.J., 1922, 2, 1112-1113.

(4) Hueper, W.C.: Occupational tumours and allied diseases. 1942, pp.79-87, Charles C. Thomas, Springfield, Illinois.

(5) Cook, J.W., Heiger, I., Kennaway, E.L. and Mayneord, W.V. The production of cancer by pure hydrocarbons. Proc. Roy.Soc.B., 1932, 111, 455-484.

(6) Cook, J.W., Hewett, C.J. and Heiger, I.: The isolation of a cancer producing hydrocarbon from coal tar. J.chem. Soc., 1933, 395-405.

(7) Stocks, P.: Regional and local differences in cancer death rates. 1947, H.M.Stationery Office, London.

(8) Goulden, F. and Tipler, M.M.: The identification of 3:4-benzpyrene in domestic soot by means of the fluorescence spectrum. Brit.J.Cancer, 1949, 3, 157.

(9) Waller, R.E.: The benzpyrene content of town air. Brit.J. Cancer, 1952, 6, 8-21.

(10) Royal College of Physicians: Smoking and Health Now. 1971, Pitman, London.

(11) National Academy of Sciences: Particulate Polycyclic Organic Matter. 1972, N.A.S., Washington D.C.

(12) British Standards Institution: Methods for the measurement of air pollution. B.S.1747 Part 2, 1969. B.S.I. London.

(13) Waller, R.E.: The interpretation and use of data on air pollution in epidemiological research. The statistician, 1966, 16, 45-58

(14) Stocks P., Commins, B.T. and Aubrey, K.V.: A study of polycyclic hydrocarbons and trace elements in smoke in Merseyside and other northern localities. Int.J.Air Water Poll., 1961, 4, 141-153.

(15) Warren Spring Laboratory: National Survey of Air Pollution, 1961-71. Vol.1, 1972, H.M.Stationery Office, London.

(16) Commins, B.T. and Waller, R.E. Observations from a 10-year study of pollution at a site in the City of London. Atmos.Environ., 1967, 1, 49-68.

(17) Commins, B.T. and Hampton, L. Changing pattern in concentrations of polycyclic aromatic hydrocarbons in the air of central London. In preparation.

(18) Waller, R.E. Tobacco and other substances as causes of respiratory cancer. In Symposium on the Prevention of Cancer, Raven, R.W., ed., 1971, pp.17-28. Heinemann, London.

(19) Higgins, I.T.T.
See page 41 this publication

(20) Hangebrauck, R.P., von Lehmden, D.J. and Meeker, J.E.: Sources of polynuclear hydrocarbons in the atmosphere. Public Health Service Publication 999-AP-33. 1967, U.S. Department of Health, Education and Welfare, Cincinnati.

(21) Doll, R., Fisher, R.E.W., Gammon, E.J., Gunn, W., Hughes, G.O., Tyrer, F.H. and Wilson, W.: Mortality of gasworkers with special reference to cancers of the lung and bladder, chronic bronchitis and pneumoconiosis. Brit.J.industr.Med., 1965, 22, 1-20.

(22) Lloyd, J.W.: Long-term mortality study of steelworkers. V. Respiratory cancer in coke plant workers. J.occup. Med., 1971, 13, 53-68.

(23) Lawther, P.J. Commins, B.T. and Waller, R.E. A study of the concentrations of polycyclic aromatic hydrocarbons in gas works retort houses. Brit.J.industr.Med., 1965, 22, 13-30.

(24) Begeman, C.R., and Colucci, J.M. Polynuclear aromatic hydrocarbon emissions from automotive engines. SAE Trans., 1970, 79, 1682-1698.

(25) Kotin, P., Falk, H.L. and Thomas M.: Aromatic hydrocarbons. III. Presence in the particulate phase of diesel engine exhaust and the carcinogenicity of exhaust extracts. Arch.industr.Hlth., 1955, 11, 113-120.

(26) Commins, B.T., Waller, R.E. and Lawther, P.J.: Air pollution in diesel bus garages. Brit.J.industr. Med., 1957, 14, 232-239.

(27) Waller, R.E., Commins, B.T. and Lawther, P.J.: Air pollution in road tunnels. Brit.J.industr.Med., 1961, 18, 250-259.

(28) Waller, R.E., Commins, B.T. and Lawther, P.J.: Air pollution in a City street. Brit.J.industr.Med., 1965, 22, 128-138.

(29) Raffle, P.A.B.: The health of the worker. Brit.J. industr.Med., 1957, 14, 73-80.

(30) Doll, R.: Mortality from lung cancer in asbestos workers. Brit.J.industr.Med., 1955, 12, 81-91.

(31) Hendry, N.W.: The geology, occurrences and major uses of asbestos. Ann.N.Y. Acad.Sciences, 1965, 132, 12.

(32) Speil, S. and Leineweber, J.: Asbestos minerals in modern technology. Env.Res., 1969, 2, 166-208

(33) Martin, A.E.: Asbestos in the environment; possible hazards in the general population. Health Trends, 1970, 2, 19-21.

(34) Lawther, P.J.: Occupational Health. The Medical Annual, 1970, pp.315-319, Bristol, Wright.

(35) Wagner, J.C., Skeggs, C.A. and Marchand, P.: Diffuse pleural mesothelioma and asbestos exposure in the North Western Cape Province. Brit.J.industr.Med., 1960, 17, 260-271.

(36) Newhouse, M.L. and Thompson, H.: Mesothelioma of pleura and peritoneum following exposure to asbestos in the London area. Brit.J.industr.Med., 1965, 22, 261-269

(37) Doll, R., Morgan, L.G. and Speizer, F.E.: Cancers of the lung and nasal sinuses in nickel workers. Brit.J.Cancer, 1970, 24, 623-632

EPIDEMIOLOGICAL EVIDENCE ON THE CARCINOGENIC RISK OF AIR POLLUTION

I.T.T. HIGGINS

Departments of Epidemiology and Environmental and Industrial Health,
School of Public Health, The University of Michigan
Ann Arbor, Michigan, 48104, U.S.A.

I am going to confine my remarks to respiratory cancer, specifically to lung cancer, or more correctly to cancer of the trachea, bronchus and lung, I.S.C. 162/163 (6th and 7th revision) and I.S.C. 162 (8th revision). There is a suggestion that other cancers -- stomach, prostate, oesophagus and skin -- may be influenced by the general air pollution, but the evidence is scanty and I shall ignore it.

There can now be little doubt that personal pollution from cigarettes is the main cause of respiratory cancer. Figure 1 shows dose/response relationships between daily cigarette consumption and lung cancer mortality ratios (death rates for non-smokers = 1.0) from the main prospective studies (1,2,3,4,5,6).

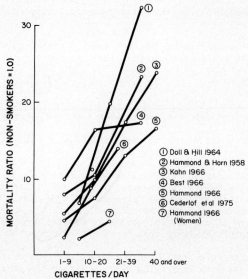

Figure 1. Dose/response relationships: Cigarettes per day and lung cancer mortality

As a rough generalization, the risk of dying of lung cancer is between 8 and 15 times as high in a cigarette smoker as it is in a non-smoker. This may be a useful yardstick with which to compare the risks of other factors.

There is excellent evidence that certain occupational exposures (Table 1)

1. RADIOACTIVE ORE MINING
2. CHROMATE PRODUCTION
3. NICKEL REFINING
4. ASBESTOS MINING AND USE
5. COKE OVEN WORK
6. GAS WORKS OPERATORS
7. MUSTARD GAS MANUFACTURE
8. IRON ORE (HAEMATITE) MINING
9. ARSENIC EXPOSURES
10. PLASTICS SYNTHESIS. BIS(CHLOROMETHYL)ETHER

Table 1: Occupational exposures causing respiratory cancer.

contribute to the development of lung cancer. This topic will be discussed by othe
My purpose is to review the evidence that general air pollution may contribute to t
disease.

Differences in lung cancer rates between town and country.

Morbidity and mortality rates from lung cancer have long been known to be consistently higher in urban than in rural areas (7,8,9,10,11,12,13,14,15,16,17,18,

The most comprehensive analyses of urban/rural differences in lung cancer mortality for the United States were carried out by Haenszel and his colleagues (20 21). Figure 2 shows age standardized mortality ratios (SMRs) for urban and rural residents according to cigarette smoking. Again you see the 8 fold and 30 fold increase in risk in the two categories of regular smokers. In contrast note the mu smaller increased risk (less than two fold) of urban compared with rural residence.

The urban/rural ratios of mortality were higher in men than in women (Table and they increase with increasing years spent in the particular areas. On the othe hand, the SMRs did not increase consistently with the duration spent in urban areas and in fact were particularly low in lifetime residents. This seems incompatible with any simple hypothesis that the longer one lives in a polluted area the higher one's lung cancer death rate. These studies showed that lung cancer was in some wa related to mobility. Thus it was found that the SMRs increased with the number of residences reported (Table 3). They were also particularly high in the foreign bor and in those who were born on farms and who moved to urban areas.

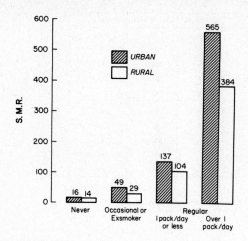

Figure 2. Standardized mortality ratios for lung cancer in U.S. white males in 1958 according to smoking and residence (20).

	Current Residence	All Durations	Under 1	1-9	10-39	40 & Over	Lifetime
Men	Urban	113	166	107	117	117	100
	Rural	79	154	88	83	75	50
	Ratio	1.43	1.08	1.22	1.41	1.56	2.00
Women	Urban	108	163	82	118	124	86
	Rural	85	205	117	79	86	49
	Ratio	1.27	0.80	0.70	1.49	1.44	1.76

Table 2: Standardized Mortality Ratios for Urban and Rural Areas by Current Residence and Duration of Residence in a Specific Community (20,21).

Further difficulties to accepting the urban/rural differences as being due to air pollution have often been noted: They are often largest where there is not much pollution, in Norway and Denmark, for example; they appear to affect men more than women, whereas air pollution would be expected to affect the two sexes equally; those most exposed through their work to polluted city air such as persons working with vehicles, policemen or others in the streets appear to have no excess risk; and finally specific occupational exposures to suggested carcinogens, such as benzpyrene, may be very much greater than those resulting from air pollution yet they result in only a small (60%) increased risk of lung cancer.

		All Deaths	Native Born	Foreign Born
Men	Total	100	93	138
	One	92	82	136
	Two	104	100	145
	Three or more	165	162	218
	Other*	104	95	136
Women	Total	100	93	141
	One	95	81	150
	Two	104	100	143
	Three or more	133	133	133
	Other	98	101	79

*Other: Unknown or incomplete histories; migrants with 6 or more residences.

Table 3: Standardized Mortality Ratios by Number of Exposure Residence and Nativit U.S. White Men and Women Aged 35 and Over Adjusted for Age and Smoking (20,21).

Various alternative explanations for the urban/rural differences have been offered. More efficient diagnosis in towns may have been important in earlier year but seems unlikely to contribute much to the difference now. Nor can specific occu pational exposures which are more likely to occur in towns account for more than a fraction of the urban excess. Urban/rural differences in smoking habits are well recognized (22,23) and have sometimes been thought to explain all the urban/rural d ferences in lung cancer (12). Doll (24) has also noted that since the habit of cig rette smoking spread from towns to countryside, comparisons of mortality based on c rent smoking habits will underestimate the differences in lifetime exposure between areas and will therefore overestimate the effect of other factors.

Experience of Migrants.

Eastcott (25) first noted that emigrants from Britain to New Zealand had death rates which were lower than those of the indigenous population of New Zealand The rates were higher in migrants who left Britain aged 30 years and over than in those who left at earlier ages. The differences could not be accounted for on the basis of differences in smoking habits. Subsequently similar observations have bee made on migrants from Britain to South Africa and Australia (26,27), the U.S. (28,2 and Canada (30). The findings in British and Norwegian migrants to the U.S. and in the populations concerned, from the study by Reid and his colleagues (31) are shown in Table 4. The rates are lowest in the Norwegians and highest in the British stay ing in their own countries with the rates for U.S.-born residents in between. Amon the men, migrants had rates between their country of birth and country of adoption. But this is less obvious for the women. These findings are usually interpreted as indicating the importance of environmental exposures possibly to air pollution befc migration. However, it is well established that migrants are not a representative sample of the country from which they migrate (even less perhaps of that they migra to). Nor has any consideration been given to the jobs which migrants adopt after migrating (32).

Birthplace and Residency	Men	Women
Norwegian-born residents of Norway	31	6
Norwegian-born residents of the U.S.	48	11
U.S.-born residents of the U.S.	72	10
British-born residents of the U.S.	94	12
British-born residents of Britain	151	19

Table 4: Age Standardized Lung Cancer Death Rates per 100,000 per Annum for Migrants from Norway and Britain to the United States (31).

Surveys of Lung Cancer and Smoking.

Most surveys of lung cancer and smoking which have looked into the urban/rural differences have found, like Haenszel and his colleagues, that these are small compared with the differences attributable to cigarette smoking. Thus in the ten year follow-up of British doctors (1) Doll and Hill reported that the death rates in those under 65 years were 40/100,000 in rural areas compared with 64/100,000 in the conurbations. Since this study was confined to a single occupational group and the comparison was standardized for smoking, occupational exposures, smoking and socio-economic factors, which might contribute to the urban/rural differences were eliminated. Table 5 shows Hammond and Horn's findings (2). In this study, the differences between

Area	Never Smoked	Cigarette Smokers
Rural	0	65
Suburbs or small towns	5	72
Cities of 10,000-50,000 inhabitants	9	71
Cities of 50,000 or more inhabitants	15	85

Table 5: Death Rates per 100,000 from Lung Cancer by Place of Residence and Smoking Habits in Men Aged 50 to 69 Years in the U.S. (2).

urban and rural lung cancer mortality were 27% higher in urban areas for all cases of lung cancer and 33% higher for well established cases exclusive of adeno-carcinoma.

Earlier retrospective surveys of smoking and area of residence, notably tho[se] of Stocks and Campbell (33) in north east England including Liverpool and Dean (34) and Wicken (35) in and around Belfast, northern Ireland, also produced evidence for the importance of urban residence in the genesis of lung cancer after allowing for smoking. On the basis of the findings in Belfast, Buck and Wicken (36), calculated that if the death rate from lung cancer for a symptomless non-smoker in a rural are[a] is taken as a minimum, the risk of death for a man (non-smoker) living in an urban area would be about double while it would be increased about 20 fold for a smoker o[f] 20 cigarettes a day.

A number of studies have failed to show a relationship between general air pollution and lung cancer. For example in the extensive studies conducted by the U.S. Public Health Service in Nashville, Tenn. (37), no relation between several indices of air pollution and respiratory cancer mortality could be demonstrated. Again in the Buffalo-Erie County study (38,39) no relationship between suspended particulates and lung cancer was found after social factors have been allowed for.

There is uncertainty if urban living enhances lung cancer death rates more [in] non-smokers than in cigarette smokers. Stocks and Campbell (33) found the differenc[es] were greatest in non-smokers and declined with increasing tobacco consumption. Haenszel and his colleagues (20) on the other hand found only minimal differences among non-smokers. In a study of smoking and air pollution in two Japanese cities [in] the vicinity of Osaka, Hitosugi (40) confirmed that mortality from lung cancer increased with increasing smoking. He also found that death rates increased slight[ly] with increasing pollution among smokers but not among non-smokers. He suggested th[at] this might indicate synergism between smoking and air pollution.

Correlations Between Lung Cancer Death Rates and Measures of Pollution.

Attempts have often been made to relate lung cancer mortality to various indices of air pollution such as smoke, SO_2 hydrocarbons and trace elements. Or, i[n] the absence of exact measurements, to more indirect estimates such as fog frequency or coal consumption. The evidence has often been contradictory due mainly to the difficulty of allowing adequately for population density and socio-economic status. On the basis of his studies in north east England Stocks (33) suggested that benzpyrene might be one of the urban factors involved. Subsequently Stocks (41) related SMRs for lung cancer to smoke and various hydrocarbons (including benzpyren[e]) at various locations in North Wales. He found high positive correlations. He also noted a positive correlation between lung cancer mortality and solid fuel consumpti[on] in 20 different countries (42).

Carnow and Meier (43) extended Stocks' analysis to include age specific dea[th] rates from lung cancer and cigarette consumption for 19 of the 20 countries. They also analyzed lung cancer death rates in relation to smoking (based on cigarette sales) and benzpyrene (based on a weighting of urban and rural values) in 48 contig- uous states of the United States. They concluded from their analysis that a 5% increase in lung cancer mortality will be produced by an increase of 1 $\mu g/1000m^3$ benzpyrene. The validity of benzpyrene levels for any area as large as a state is, however, questionable. Furthermore, even if these are accepted as valid, the corre[-] lations are unimpressive and appear to be due mainly to a few rural states having particularly low values.

On the assumption that benzpyrene measurements for smaller areas might be more valid than statewide estimates I have correlated benzpyrene concentrations giv[en] for the period January to March, 1959, by Sawicki (44) for certain Standardized Metropolitan Statistical Areas (SMSAs) with the age adjusted lung cancer death rate[s] for those areas in the three year period 1959 to 1961 (45). I also correlated measurements of total suspended particulates and sulphates as given by the Environmental Protection Agency for the years 1957 to 1961 (46) and 1958 (47) respectively with the lung cancer death rates in some 50 SMSAs. The correlations (Table 6) with

Measure of Pollution	Death Rates from Lung Cancer in 1959-1961			
	Males		Females	
	White	Non-White	White	Non-White
Total Suspended Particulates[2]	-0.03	0.25	-0.09	-0.02
Sulphates[3]	0.42	0.39	0.19	-0.05
Benzo(a)pyrene[4]	0.17	0.00	0.10	-0.11

Table 6: Correlation Coefficients Between Measures of Air Pollution and Age Standardized Death Rates from Lung Cancer (ISC 162/3) in Approximately 50 Standard Metropolitan Statistical Areas of U.S.

total suspended particulates and with benzpyrene were unimpressive. But surprisingly there was a moderately high correlation with sulphates in white men and women and non-white men. I present this as information which requires further study. I wonder if it is in some way related to the cocarcinogenic action of SO_2 which has been suggested in animal experiments (48).

Trends in Lung Cancer Related to Trends in Factors.

It has been pointed out (49) that coal utilization in Britain, which was at a peak during the first quarter of this century, has been declining during the time when lung cancer mortality has been increasing. The report also noted that the lung cancer rise preceded rather than followed the increased usage of diesel fuel. Diesel smoke, therefore, seems unimportant. Observations that those occupationally exposed to diesel smoke have no increased lung cancer risk support this conclusion. Finally, although a rise in petrol fuel usage preceded the increase in lung cancer mortality, the later rise in women and the absence of any increase risk in those occupationally exposed to gasoline again suggest this can have little influence on the disease.

Effect of a Reduction in Air Pollution on Lung Cancer Mortality.

One way in which the impact of any factor on disease can be assessed is to see what happens when the factor is reduced. Particulate pollution in London has been reduced remarkably since the late 1950s. It occurred to me to see if lung cancer death rates in London had declined more than in other parts of England and Wales where pollution had declined less (50). Figure 3 shows the rates for men aged 25 to 74 for London and elsewhere in England and Wales. At ages under 65 years the lung cancer death rates in London do seem to have declined somewhat more than elsewhere. Among women there was some approximation of the rates in the 45 to 54 year age group, but no convincing trends in any of the other age groups. I do not think that these

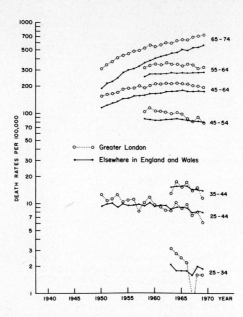

Figure 3. Age specific death rates from lung cancer in men aged 25-74 in greater London and elsewhere in England and Wales from 1950 to 1969.

trends can be explained by greater changes in cigarette smoking in London compared with elsewhere.

I learned sometime after I had looked into these trends that Waller (51) ha also studied this question. Waller observed that from 1931 to 1965 the lung cancer death rates of men in Greater London decreased, whereas in other urban areas of England and Wales they remained relatively constant and in rural areas they actuall increased. During this period the ratios between the London and the rural rates declined from 3.2 to 1.7 indicating that the disease was becoming more uniform over the whole country. Waller suggested a number of possible explanations for these trends:

1. A decrease in pollution in large towns;
2. Smoking habits may have become more uniform;
3. Diagnostic standards may have become more uniform;
4. The more mobile town dwellers may have moved to the country and vice versa.

By cohort analysis he was able to show that some of the pattern could be accounted for by changes in smoking habits. But he considered that there was no wa of determining from analyses of mortality if pollution had played a part in the urban/rural differences. While not wishing to underestimate the difficulties I believe on the contrary that further observations on the trends in lung cancer mortality in England and Wales in relation to changes in air pollution and cigarett smoking may still be worth while.

In summary, higher lung cancer morbidity and mortality in urban than in rural areas and the presence of carcinogens in polluted air suggest that air pollution may play a role in this disease. Surveys of lung cancer have shown that differences in smoking habits and specific occupational exposures cannot account for all the urban excess. Positive correlations between lung cancer death rates and indices of pollution in different places, the experience of migrants and a possible decline in lung cancer mortality with decline in pollution provide support for the view that air pollution is a factor in this disease. But the effect of pollution cannot be large. It is likely to be a small fraction (possibly a tenth) of the effect of cigarette smoking.

References

(1) Doll R. and Hill A.B.: Mortality in relation to smoking: 10 years observations of British doctors. Part I. Brit. Med. J., 1964, 1, 1399-1410.

(2) Hammond E.C. and Horn D.: Smoking and death rates--report on 44 months of follow-up on 187,783 men, part 2: Death rates by cause. J. Amer. Med. Assoc., 1958, 166, 1294-1308.

(3) Kahn H.A.: The Dorn study of smoking and mortality among U.S. veterans: Report on 8 1/2 years of observation. In: Epidemiological approaches to the study of cancer and other diseases, Haenszel, W. ed., 1966, Monog. 19, pp. 1-25, U.S. Public Health Service, Nat. Cancer Institute, Bethesda.

(4) Best E.W.R.: A Canadian study of smoking and health, 1966, 137 pp., Department of National Health and Welfare, Ottawa.

(5) Hammond E.C.: Smoking in relation to the death rates of 1 million men and women. In: Epidemiological approaches to the study of cancer and other diseases, Haenszel W. ed., 1966, Vol. 19, pp. 127-204, U.S. Public Health Service, Nat. Cancer Institute, Bethesda.

(6) Cederlof R., Friberg L., Hrubec Z. and Lorich U.: The relationship of smoking and some social coveriables to mortality and morbidity. A ten year follow-up in a probability sample of 55,000 Swedish subjects age 18-69. 1975, S-104-01, Dept. Environm. Hyg., Karolinska Institute, Stockholm, Sweden.

(7) Stocks P.: Studies on medical and population subjects. Regional and local differences in cancer death rates, 1947, No. 1, H.M.S.O., London.

(8) Clemmesen J., and Nielsen A.: Geographical and racial distribution of cancer of the lung. Schweiz. Ztschr. Allg. Path., 1955, 18, 803-819.

(9) Clemmesen J., Nielsen A. and Jensen E.: Mortality and incidence of cancer of the lung in Denmark and some other countries. Acta Un. Int. Canc., 1953, 9, 603-636.

(10) Curwen M.P., Kennaway E.L., and Kennaway N.M.: The incidence of cancer of the lung and larynx in urban and rural districts. Brit. J. Cancer, 1954, 8, 181-198.

(11) Hoffman E.F. and Gilliam A.G.: Lung cancer mortality. Geographical distribution in the United States for 1948-1949, Publ. Hlth. Rep., Washington, 1954, 69, 1033-1042.

(12) Kreyberg L.: Lung Cancer and Tobacco Smoking in Norway. Brit. J. Cancer, 1955, 9, 495-509.

(13) Saxen E.: Report from the Finnish Cancer Registry. Schweiz. Z. Allg. Path., 1955, 18, 556.

(14) Mancuso T.F., MacFarlane E.M., and Porterfield J.D.: Distribution of cancer mortality on Ohio. Am. J. Pub. Health, 1955, 45, 58-70.

(15) Griswold M.H., Wilder C.S., Cutler S.J. and Pollack E.S.: Cancer in Connecticut. 1935-1951. Hartford, Conn. Connecticut State Department of Health, 1955.

(16) Haenszel W., Marcus S.C., and Zimmerer E.G.: Cancer morbidity in urban and rural Iowa. Pub. Hlth. Monograph No. 37, U.S. Government Printing Office, 19

(17) Mills C.A. and Porter M.M.: Tobacco smoking, motor exhaust fumes, and general air pollution in relation to lung cancer incidence. Cancer Res, 1957, 17, 981-990.

(18) Mills C.A.: Motor exhaust gases and lung cancer in Cincinnati. Am. J. Med. Sco., 1960, 239, 316-319.

(19) Levin M.L., Haenszel W., Carroll R.E., Gerhardt P.R., Handy V.H., and Ingraham S.C. II. Cancer incidence in urban and rural areas of New York State J. Nat. Cancer Institute, 1960, 24, 1243-1257.

(20) Haenszel W., Loveland D.B., and Sirken M.G.: Lung cancer mortality as related to residence and smoking histories. White Males. J. Nat. Cancer Institute, 1962, 28, 947-1001.

(21) Haenszel W. and Taeuber K.E.: Lung cancer mortality as related to residence and smoking histories. White females. J. Nat. Cancer Institute, 1964, 32, 803-838.

(22) Haenszel W., Shimkin M.B., and Miller H.P.: Tobacco smoking patterns in the United States, 1956, Public Health Service Publication 463, Washington.

(23) Todd G.F.: Statistics of smoking in the United Kingdom, 5th edition, 1969, pp. 58-61, Tobacco Research Council, London.

(24) Doll R.: Atmospheric pollution. In: Carcinoma of the lung. Bignall J.R. ed 1958, pp. 81-94, E.S. Livingstone Ltd., Edinburgh & London.

(25) Eastcott D.F.: The epidemiology of lung cancer in New Zealand. Lancet, 1956 1, 37-39.

(26) Dean G.: Lung cancer among white South Africans. Brit. Med. J., 1959, 2, 852-857.

(27) Dean G.: Lung cancer in South Africans and British immigrants. Proc. Roy. Soc. Med., 1964, 57, 984-987.

(28) Mancuso T.F., and Coulter E.J.: Cancer mortality among native white, foreign-born white, and non-white male residents of Ohio: Cancer of the lung, larynx bladder and central nervous system. J. Nat. Cancer Institute, 1958, 20, 79-1

(29) Haenszel W.: Cancer mortality among the foreign-born in the United States. J. Nat. Cancer Institute, 1961, 26, 37-132.

(30) Coy P., Grzybowski S., and Rowe J.F.: Lung cancer mortality according to bir place. Can. Med. Assoc. J., 1968, 99, 476-483.

(31) Reid D.D., Cornfield J., Markush R.E., and Siegel D.: Studies of disease among emigrants and native populations in Great Britain, Norway, and the United States. III. Prevalence of cardiorespiratory symptoms among migrants and native born in the United States. 1966, Nat. Cancer Institute Monograph No. 19, 321-346.

(32) Wynder E.L., and Hammond E.C.: A study of air pollution carcinogenesis. Cancer, 1962, 15, 79-92.

(33) Stocks P., and Campbell J.M.: Lung cancer death rates among non-smokers and pipe and cigarette smokers. Brit. Med. J., 1955, 2, 923-929.

(34) Dean G.: Lung cancer and bronchitis in Northern Ireland. Brit. Med. J., 1966, 1, 1506-1514.

(35) Wicken A.J.: Environmental and personal factors in lung cancer and bronchitis mortality in Northern Ireland; 1960-1962. 1966, Tobacco Research Paper No. 9.

(36) Buck S.F., and Wicken A.J.: Models for use in investigating the risk of mortality from lung cancer and bronchitis. Appl. Statist., 1967, 16, 185-210.

(37) Zeidberg L.D., Horton R.J.M., Landau E., Hagstrom R.M. and Sprague H.A.: The Nashville air pollution study. V. Mortality from diseases of the respiratory system in relation to air pollution. Arch. Environm. Hlth., 1967, 15, 214-224.

(38) Winkelstein W., Jr.: The Erie County air pollution-respiratory function study. J. Air. Pollut. Control Assoc., 1962, 12, 221-222.

(39) Winkelstein W., Jr.: The relationship of air pollution and economic status to total mortality and selected respiratory system mortality in man. Arch. Environm. Hlth., 1967, 14, 162-171.

(40) Hitosugi M.: Epidemiological study of lung cancer with special reference to the effect of air pollution and smoking habits. Inst. Publ. Health Bull., 1968, 17, 237-256.

(41) Stocks P.: On the relations between atmospheric pollution in urban and rural localities and mortality from cancer, bronchitis and pneumonia, with special reference to 3:4 Benzopyrene, beryllium, molybdenum, vanadium, and arsenic. Brit. J. Cancer, 1960, 14, 397-418.

(42) Stocks P.: Lung cancer and bronchitis in relation to cigarette smoking and fuel consumption in twenty countries. Brit. J. Prev. Soc. Med., 1967, 21, 181-185.

(43) Carnow B.W., and Meier P.: Air pollution and pulmonary cancer. Arch. Environm. Hlth., 1973, 27, 207-218.

(44) Sawicki E., Elbert W.C., Hauser T.R., Fox F.T., and Stanley T.W.: Benzo(a)pyrene content of the air of American cities. Amer. Industr. Hyg. Assoc. J. 1960, 21, 443-451.

(45) Duffy E.A., and Carroll R.E.: United States Metropolitan Mortality 1959-1961. P.H.S. Publication No. 999-AP-39, 1967, National Center for Air Pollution Control, Cincinnati.

(46) Spirtas R., Levin H.J.: Characteristics of particulate patterns 1957-1966. 1970, pp. 10, 11, National Air Pollution Control Administration, Raleigh, North Carolina.

(47) Air pollution measurements of the National Air Sampling Network: Analysis of suspended particulates 1957-1961. Public Health Service Publication No. 978, 1962.

(48) Kuschner M.: The causes of lung cancer. Amer. Rev. Resp. Dis., 1968, 98, 573-590.

(49) Royal College of Physicians. Air Pollution and health. 1970, 48-57, Pitman Medical, London.

(50) Higgins I.T.T.: Trends in respiratory cancer mortality in the U.S. and in England and Wales. Arch. Environm. Hlth., 1974, 28, 121-129.

(51) Waller R.: In: The prevention of cancer. Raven R.W., and Roe F.J.C., eds., 1967, pp. 181-186, Butterworths, London.

RESUME

En résumé, le fait que la morbidité et la mortalité par cancer du poumon soient plus fortes dans les villes que dans les régions rurales et la présence de cancérogènes dans l'air pollué conduisent à penser que la pollution de l'air joue peut-être un rôle dans cette maladie. Les enquêtes sur le cancer du poumon ont montré que les différences dans l'usage du tabac et les expositions professionnelles particulières ne peuvent expliquer la surfréquence urbaine dans sa totalité. Les corrélations positives entre la mortalité par cancer du poumon et les indices de pollution en différents lieux, l'évolution observée chez les migrants et une éventuelle diminution de la mortalité par ce cancer lorsque la pollution diminue, renforcent l'opinion selon laquelle la pollution de l'air serait un facteur dans cette maladie. Mais les effets de la pollution ne peuvent être considérables. Ils ne représentent sans doute qu'une petite fraction (un dixième peut-être) des effets de la cigarette.

Environmental Pollution and Carcinogenic Risks
Pollution de l'environnement et risques cancérogènes

AVIATION AND ENVIRONMENTAL BENZO(A)PYRENE POLLUTION

L.M. SHABAD and G.A. SMIRNOV

Institute of Experimental and Clinical Oncology
Academy of Medical Sciences of USSR
Kashirskoye shosse, 6 115478 Moscow, USSR

INTRODUCTION

In 1959 benzo[a]pyrene (BP), a carcinogenic polycyclic aromatic hydrocarbon (PAH), was found in samples of soil from a big city (Leningrad) (1). Later, Blumer (2) in the USA, Borneff & Fischer (3) in the Federal Republic of Germany, Mallet & Héros (4) in France and Zdrazil & Picha (5) in Czechoslovakia also detected this substance in soil samples taken from different regions. A number of systematic investigations of soil from different geographic areas - mountains and plains, woods and steppes, etc. - were carried out in our laboratory (6). Of more than 1000 samples investigated, all contained some BP; this finding led to the suggestion that such pollution had a common source.

The presence of PAH, and especially of BP, in motor car exhausts naturally suggested the possibility of their occurrence in the exhausts of aircraft engines, which had not yet been studied from this point of view. The identification and quantitative evaluation of PAH can be accomplished by spectral-fluorescent methods based on the Shpolsky (7) effect i.e., on the appearance of quasi-linear spectra in frozen crystalline solutions of normal paraffin at low temperatures ($77^\circ K$). This method, with modifications and similar to that of Muel & Lacroix (8), was elaborated by Khesina (9) and has been used by many authors in the USSR.

BP IN SOOT AND IN EXHAUST GASES OF AIRCRAFT ENGINES

Soot was obtained by scraping the nozzles of turbine engines and the exhaust pipes of piston engines, and from soot-collectors of exhaust systems of piston and turbo-jet engines at the test stands. The amounts of BP found in all soot samples were from 50 µg to 30 µg.

Secondly, the BP content of exhaust gases from turbo-jet and turbo-prop engines was measured on the test stand (Fig. 1). Knowing the volume of gas passing through the absorbents and its BP content we can calculate the amount of BP in the total volume of gas discharged from the nozzle. It was found that each turbo-jet or turbo-prop engine could discharge from 2-4 mg BP/min. into the atmosphere, the amount was dependent on the working regime of the engine and was greater on starting than in flight (14, 15).

EXPERIMENTS IN ANIMALS

Since this was the first study of aircraft engines as a source of environmental carcinogenic pollution, and taking into consideration that soot from different aircraft engines may contain various substances other than BP - carcinogenic,

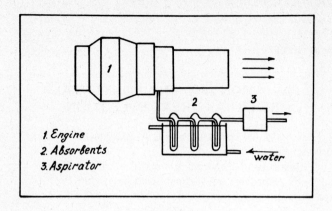

FIG. 1   MEASUREMENT OF BP CONTENT OF EXHAUST GASES

cocarcinogenic and/or carcinogen-inhibiting agents - we decided to test samples of aviation soot for their carcinogenic activity in animals (10). The experiments were carried out in hybrid $F_1(C_{57bl}XCBA)$ mice, with the test substance applied from a pipette onto the depilated interscapular skin. Benzene extracts of soot samples containing 0.1% BP from jet and piston engines were used in Groups 1 and 2; the "technical control" group (Group 3) was treated with 0.1% pure BP in benzene; and the control animals (Group 4) were given pure benzene. In all the groups each mouse received 50 applications. The results are summarized in Tables I and II.

In the first three groups all the animals developed tumours; almost all the tumours proved to be malignant (96.5% in Group 1, 100% in Group 2 and 93% in Group 3), and metastases were observed in 6 out of 52 mice in groups 1 and 2. The prevailing form of malignant tumour obtained was epidermoid carcinoma. One mouse developed a sarcoma, and four developed carcinosarcomas. Similar results were obtained in the "technical control" group. No tumours were observed in the pure control group.

The numbers of animals with tumours, their character and time of appearance were virtually the same in all the three experimental groups of mice. This observation, together with many other facts, led us to consider BP as an indicator of carcinogenic PAH. Thus, evidence of the occurrence of the carcinogenic hydrocarbon BP in aviation soots was provided not only by physico-chemical but also by morphological investigations.

BP IN SOIL, VEGETATION AND SNOW SAMPLES TAKEN IN THE REGION OF AIRPORT

Our investigations on BP content in soil and vegetation in the airport area were undertaken in order to study the possibility that exhausts from aircraft engines pollute the environment with this carcinogenic hydrocarbon. The ground chosen for investigation was a runway at one of the Moscow airports (11). The samples were collected in the directions shown in Fig. 2: I-I and II-II, perpendicular to the runway and crossing it, at intervals of 50 m; III-III, parallel

TABLE I Effects of skin applications to mice of BP from various sources: total experimental results after 9 months

Animal group	Extract applied	Original number of animals	Time of appearance of 1st tumours (weeks)	Number of animals surviving at time the 1st tumour appeared	Number of animals with tumours at the end of the experiment	
					Total	Number with malignant tumours
I	Soot from turbo-jet engine (with 0.1%BP)	33	13	29	29 (100%)	28 (96.5%)
II	Soot from piston engine (with 0.1% BP)	33	11	29	29 (100%)	29 (100%)
III	0.1% BP in benzene	34	13	29	28 (96.5%)	26 (93%)
IV	Control	20	-	20	0	0

Table II Types of tumours induced

Group	Papilloma	Malignant tumours			Metastases
		Epidermoid carcinoma	Sarcoma	Carcino-sarcoma	
I	1	26	0	2	4
II	0	26	1	2	2
Total for Groups I & II	1	52	1	4	6
III Technical control	2	23	0	3	1

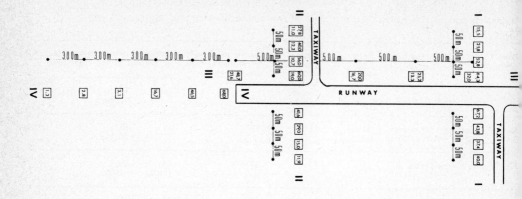

FIG. 2   PLOT OF BP CONCENTRATIONS AT A MOSCOW AIRPORT

to the runway 5 m from its edge, at intervals of 500 m; IV-IV, from the end of the runway, in the directions of take-off and landing, at intervals of 300 m.

Two numbers given for the same point indicate samples taken from a depth of 5-7 cm (upper number) and samples taken from 10-12 cm (lower number). The BP content in the lower layer is usually 2-3 times less than in the upper one, which can be attributed to the thick grass and turf through which the BP has to filter.

It can be seen from Fig. 2 that there are certain regularities in the distribution of BP in the soil: in samples taken in directions I-I and II-II the BP content decreased from 64.3 to 15.5 µg/kg of dried soil as the collection point moved away from the runway, the centre of carcinogenic discharge. It should be noted that samples taken in these directions from sections which were parallel to the taxiways yielded a higher concentration of BP (67.2 µg/kg) than those situated far from the taxiways (e.g. 41.6 µg/kg).

In samples taken in direction III-III the BP content is higher in those collected at the ends of the runway (64.3 and 48.7 µg/kg) than in the middle. This is due to the fact that the end parts of the runway are those at which the plane takes off and at which, therefore, its engine is operating at the maximum working regime. In samples taken in direction IV-IV the BP content decreased with the distance from the end of the runway: at 1.5 km the BP concentration was 1.3 µg/kg, which is almost equal to the "general background" level. Control soil samples collected in three rural areas between Moscow and the airport had BP concentrations of 0.3, 0.5 and 0.8 µg/kg.

At the same time we studied the BP content of samples of vegetation collected from the polluted area. The collection was made in two directions: II-II and III-III. The results show that BP pollution of vegetation varies from 21.3 to 5.4 µg/kg for plants and from 7.0 to 3.1 µg/kg for the roots of these plants. The regularities in BP distribution seen in the soil samples are not so clearly manifested here and this is probably due to the variety of the plants growing at the airport and to differences in the shape of their leaves, stems, etc. Nevertheless, a greater concentration of BP was found near the taxiways than in other airport areas. Control samples of vegetation showed considerably lower BP contents: 0.5-0.7 µg/kg in plants and 0.2 and 0.3 µg/kg in the roots of these plants.

We assumed that the samples of soil and vegetation from the area around the airport runways were polluted with carcinogenic substances sedimented from the air. To further validate this assumption we investigated samples of snow, which represent another kind of sedimentation (12). Snow samples were collected in the same directions as were those from the soil, and regularities of BP distribution in snow cover at the area of runway were found to be similar to those in the soil, although the BP content of the snow cover proved to be greater. It decreased with the distance between the collection point and the runway. Three control samples, taken at 10-15 km from the airport, showed BP contents of 0.03, 0.04, 0.1 µg/m^2 in 24 h.

ON THE SPREADING OF BP WITH AVIATION EXHAUSTS

Do aviation exhausts induce only local BP pollution? To answer this question, soil samples were taken from under two "traffic corridors" in the area surrounding an airport near Moscow. The distance between the points of collection was within 1.5-2.0 km, and 10 samples were taken from each "corridor", thus covering about 20 km. At the same time, a total of 32 control soil samples were taken within a 20 km radius around the Moscow airport. The concentration of BP in the soil samples taken from under the corridors varied from 1.17 to 8.6 µg/kg of dried soil; the control samples also showed a BP pollution of 1.17 to 8.6 µg/kg.

The irregular and sometimes high levels of BP in both control soil samples and in the samples taken under the "corridors" can be attributed to local sources; and it can be concluded that these local sources of carcinogenic discharge have a greater influence on environmental pollution with BP than do aviation exhausts. However, there is still no doubt that aviation contributes to the environmental pollution of the region investigated.

The role of aviation in the carcinogenic pollution of the environment is thus difficult to establish in highly industrialized districts with heavy traffic, such as the Moscow region. For this reason, the second part of our study was an investigation of 10 soil samples taken within a 20 km radius of the Pavlodar airport in Northern Kazakhstan, where industry and traffic are much less prevalent than in the suburbs of Moscow. BP content of soil samples taken under the "corridors" near Pavlodar airport (up to 5.5 µg/kg) was several times greater than the background pollution observed in the control samples. On the basis of this study, we concluded that aviation exhausts do affect the BP pollution of the environment.

And what about the case where an airport is situated inside the city? Studies performed in Riga (12) showed that the BP content of soil and snow samples collected in the vicinity of that airport was 8-9 times higher than that in samples from other parts of the same city.

THE INFLUENCE OF THE WORKING REGIME OF ENGINES AND VARIOUS KINDS OF FUEL ON BP DISCHARGE

Estimations were made of the BP content of the exhaust gases of aircraft engines under various operating conditions. The investigations were performed in turbo-jet and turbo-prop engines on the test stand. The amount of BP discharged by the engine was found to be fully dependent on its working regime, increasing as a turbo-jet changes to a higher working regime involving more revs/min., higher temperature and greater air and fuel consumption with a concomitant increase in the volume of gas released out from the nozzle. At the maximum regime the amount of BP discharged is more than 10 mg/min for turbo-jet engines and more than 3 mg/min for turbo-prop engines, corresponding to an increase of about 5.5 times that for low speed for a turbo-jet and 3 times that for a turbo-prop.

We have also studied the influence on BP release of de-aromatized oil fuel and of fuel with magnesium additives. The experiment was performed in a turbo-jet engine using three types of fuel: the two mentioned above and a normal T-1 oil fuel of the kerosene type.

It was shown (Fig. 3) that the use of fuel with magnesium additives leads to an average of 31.5% reduction in the amount of BP discharged by the engine for all the regimes in comparison with the amount discharged with the usual fuel. The use of de-aromatized oil fuel resulted in an even greater decrease, to 59.5%.

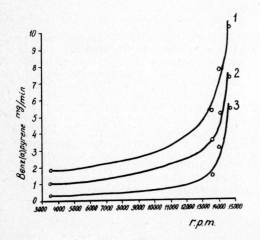

1 Usual fuel oil
2 With magnesium additives
3 De-aromatized fuel oil

FIG. 3   EFFECT OF ADDITIVES ON BP CONTENT OF EXHAUST GASES

CONCLUSION

The data presented show that aviation engines are a newly-discovered source of environmental pollution with BP. Aviation engines release considerable amounts of BP into the atmosphere. The carcinogenicity of aviation exhausts has been established in experiments with animals, the results of which confirmed those of the physico-chemical analysis of aviation soot.

An approximate calculation can be made of the amount of BP discharged by aircraft into the atmosphere. An aircraft engine discharges several mg of this substance per minute. A big modern airport receives or sends an average of one plane every 3-4 minutes. Thus, about 4 mg BP per minute, or nearly 2 kg per year, would be discharged. This amount is less than that discharged in an automobile park in a big city, since one automobile engine discharges about 10 μg/min BP (16), i.e. 600 μg/hr or 3.5 mg per 6 working hours. One automobile can discharge over 1 kg BP in a year. Additionally, it should be taken into consideration that release of automobile exhausts takes place within the zone of human inhalation, whereas BP discharge from airplanes takes place at a certain height where it undergoes degradation by UV rays and ozone, although the latter degradation is likely to decrease for ultrasonic aviation.

The significance of aviation exhausts for BP environmental pollution is less considerable than that of automobiles; however, its spread and distribution over a wide area is much more important. It is no exaggeration to say that our whole planet is polluted by the carcinogenic hydrocarbon, BP, and that a considerable share of this pollution falls on aviation.

Thus, the possibility of further spreading of BP in the environment has been stated, and its approximate scale has been determined. The connection between the amount of BP in aviation exhausts and the working regime of the engines as well as the kind of fuel used points to one of the possible ways of preventing BP environmental pollution from this source.

SUMMARY

Spectrofluorescent methods of analysis have shown that soot and exhaust products of aviation engines, both piston and turbine, contain benzo[a]pyrene (BP). A modern aircraft engine releases into the atmosphere from 2-10 mg BP per min. Extracts of aviation engine soot applied to the skin of mice induced malignant tumours in almost all treated animals. The ground within airports is polluted with BP, its level diminishing with distance from the runway. The concentration of carcinogenic hydrocarbons in aircraft exhausts is dependent on the working regime of the engine and on the character of fuel combustion.

RESUME

Les méthodes d'analyse par spectrofluorescence ont montré que la suie et les produits d'échappement des moteurs d'avion, à piston ou à turbine, contiennent du benzo[a]pyrène (BP). Un moteur d'avion moderne libère dans l'atmosphère 2 à 10 mg de BP par minute. Des extraits de suie de moteur d'avion, appliqués sur la peau de souris, ont induit des tumeurs malignes chez presque tous les animaux traités. Dans les aéroports, le sol est pollué par le BP, dont le taux diminue à mesure qu'on s'éloigne de la piste. La concentration d'hydrocarbures cancérogènes dans les gaz d'échappement d'avion dépend du régime du moteur et de la combustion du carburant.

REFERENCES

(1) Shabad L.M. and Dikun P.P.: Air pollution by carcinogenic hydrocarbon - Benz(a)pyrene. 1959, Medgiz, Moscow (Russ.).

(2) Blumer M.: Benzpyrenes in soil. Science, 1961, 134, 3477, 474.

(3) Borneff I. and Fischer R.: Kanzerogene Substanzen in Wasser und Boden. VIII. Untersuchungen an Filter - Aktivkohle nach Verwendung im Wasserwerk. Arch. Hyg. Bakt., 1962,

(4) Mallet L. and Héros M.: Pollution de terres végétales par les hydrocarbures polybenzéniques du type benzo-3,4-pyrène. Compt. Rend. Acad. Sci. (Paris), 1962, 254, 5, 958-960.

(5) Zdrazil I. and Picha F.: The occurrence of the carcinogenic compounds 3,4-benzpyrene and arsenic in the soil. Neoplasma, 1966, 13, 11, 49.

(6) Shabad L.M., Cohan Y. L., Ilnitsky A.P. et al.: The carcinogenic hydrocarbon benz(a)pyrene in the soil. J. Natl Cancer Inst., 1971, 47, 1179-1191.

(7) Shpolsky E.V.: Line fluorescence spectra of organic compounds and their applications. Usp. Fiz. Nauk, 1960, 215-242 (Russ.).

(8) Muel B. and Lacroix G.: Characterisation et dosage du 3,4-benzopyrène par spectrophotométrie de luminescense à 190°. Bull. Soc. Chim. France, 1960, 11-12, 2139-2147.

(9) Khesina A.Y.: On the quantitative evaluation of polycyclic hydrocarbons in complex mixtures by linear spectra of luminescence. Vsesoyuznoe Soveschchanie po Luminestzenzii. Materialy, 1964, 13, 113-114 (Russ.).

(10) Linnik A.B., Smirnov G.A. and Shabad L.M.: Carcinogenic effect of aviation soot engines, studied in animals. Bull. Exp. Biol. Med., 1971, 2, (Russ.).

(11) Smirnov G.A.: The study of benz(a)pyrene content in soil and vegetation in the airfield region. Vop. Onkol., 1970, 16, 5, 83-86 (Russ.).

(12) Smirnov G.A.: Pollution of airdrome territory with benz(a)pyrene. Gig. i Sanit., 1970, 8, 126-127 (Russ.).

(13) Audere A.K., Lindberg Z.Y., Smirnov G.A. and Shabad L.M.: An experience of studying influence of airdrome situated inside the city on the level of BP environmental pollution. Gig. i Sanit., 1973, 9, 90 (Russ.).

(14) Shabad L.M. and Smirnov G.A.: 3,4-benzpyrene containing in the soot and exhaust gases of turbo-jet and turbo-prop engines. Gig. i Sanit., 1969, 2, 98-99 (Russ.).

(15) Shabad L.M. and Smirnov G.A.: Aircraft engines as a source of carcinogenic pollution of the environment (Benzo(a)pyrene studies). Atmph. Env., 1972, 6, 153-164.

(16) Shabad L.M., Khesina A.Y. and Khitrovo S.S.: On carcinogenic hydrocarbons in exhaust gases of automobile engines and the possibility of their decrease. Vestnik AMN SSSR, 1968, 6, 12 (Russ.).

Environmental Pollution and Carcinogenic Risks
Pollution de l'environnement et risques cancérogènes

PENETRATION OF CARCINOGENS THROUGH RESPIRATORY AIRWAYS

Maurice STUPFEL and Madeleine MORDELET-DAMBRINE

Groupe de recherches INSERM U.123
44, Chemin de Ronde, 78110 Le Vésinet, France
and
Département de Physiologie, Faculté Necker-Enfants-Malades,
156, rue de Vaugirard, 75015 Paris, France

If it is admitted that 60 % of carcinogenic etiology is related to carcinogenic environmental substances (1) it becomes of interest to examine the penetration of the carcinogenic airborne substances into the respiratory airways. One reason is that it is generally admitted that the respiratory mucosa offers not such a good protection against external aggressors as the gastrointestinal mucosa. One example is provided by experiments demonstrating that asbestos fibers do not penetrate tissues when ingested (2) instead that millions of these fibers may invade the lungs of city dwellers.

As a previous review : "penetration of pollutants in the airways" has been recently published (3), this actual brief survey will try to analyze, in the scope of this Symposium, attempts made to explain lung carcinogenesis by physical processes, enzymatic activation or inhibition, physiological and immunological reactions taking place during the travel of the potential carcinogenic substance through the respiratory airways.

PHYSICOCHEMICAL AND BIOCHEMICAL PROCESSES

Volatility and solubility determine the penetration of gases and aerodynamic size limits the depth of deposition of liquid and solid aerosols in the respiratory tract.

Vinylchloride monomer and bis (chloromethyl) ether are gases which have tumorogenic properties. However their sites of action are quite different. In a group of 30 Sprague Dawley rats discontinuously exposed to bis (chloromethyl) ether during 659 days, LASKIN et al. (4) reported a high incidence of esthesio-neuroepithelioma of the olfactory epithelium and squamous cell carcinomas of the lungs. Twenty six out of 50 mice repeatedly exposed during 82 days to vapor of this same gas by LEONG, MACFARLAND and REESE (5) developed lung tumors. Vinylchloride monomer when inhaled gives oncogenic responses mostly liver angiosarcoma while lung tumors have been rarely obtained and these in rats according to VIOLA, BIGOTTI and CAPUTO (6).

It is generally considered that particles of a diameter superior to 10 microns are stopped in the nose and pharynx by inertia and impaction, and that deposition by sedimentation of particles of a diameter between 0.5 and 5 microns occurs in tracheobronchial tree and alveoli. Particles below 0.5 micron penetrate into alveoli by diffusion and Brownian movements. Nevertheless fibers of asbestos of 3 microns of diameter and of more than 5 microns of lenght are discovered in distal parts of the lung parenchyma. However the different varieties of asbestos fibers : chrysotile, amosite, crocidolite, anthophyllite having different lenghts and diameters and diverse resistances to fracture penetrate into various parts of the respiratory air tract with as a consequence distinct cancer sites : bronchial, mesothelial and pleural (7).

Cigarette smoke and motor automotive exhaust which are the most common environmental air aggressors are composed of gases, liquid and solid aerosols. Their most carcinogenic constituents are apparently polycyclic aromatic hydrocarbons which are found in air, either as dissolved on tar, either as absorbed on solid particles. These last ones can carry metals possibly carcinogenic. Most inhaled particles are constituted by 99 % of minerals and 1 % of organic matters of which only traces are polycyclic aromatic hydrocarbons.

The metabolism of these polycyclic hydrocarbons, the carcinogenic activity of which has been amply demonstrated have been extensively investigated since 1933 when COOK, HEWETT and HIEGER (8) isolated the tumorigenic 3,4 benzpyrene from coal tar. Among the variety of these carcinogenic polycyclic hydrocarbons (9, 10) the metabolic transformations of the benzo(a)pyren is well known. It is hydroxylated by an enzyme the benzpyren hydroxylase (an aryl hydrocarbon hydroxylase) that can be induced in various tissues, among which the lungs (11), by different substances. The hydroxylation resulting from its action is supposed to decrease the carcinogenic potential of the polyaromatic hydrocarbons. Recently (1974) GROVER et al. (12) discovered a new enzymatic metabolic transformation of the benzo(a)pyren in contact with microsomes of rats liver and lung preparations. After induction of a microsomal "epoxide hydrase" activity by 3-methylcholanthrene, epoxide intermediates to hydrodiols, to phenols and to glutathione conjugated were obtained from the initial benzo(a)pyren. According to these experimenters this metabolic activation to effective epoxides may be involved in the reported induction of tumors of the respiratory airways in the rats by polycyclic hydrocarbons.

Concerning the finding that long-term exposure of humans to the dust of fiber glass does not cause demonstrable macroscopic or microscopic pulmonary damage (13) whereas asbestos fibers air pollution results in cancer (14, 15), it has been considered that trace elements possibly carcinogenic polycyclic hydrocarbons (16) and/or trace metals (17) absorbed on these fibers of asbestos should act as the effective carcinogens.

Of interest is the respiratory penetration in the air tract of the particles of the metals which are now usually admitted as carcinogenic or cocarcinogenic : nickel, beryllium, chromium and arsenic (18, 20). Their active ionic form is an electrophilic one which can react with various nucleophiles and may also be formed *in vivo* (42).

Gaseous nickel carbonyl is the most dangerous of nickel compounds. As it is highly volatile and in addition soluble in lipids this form is presumably able to penetrate cellular membranes (43). Metastasizing pulmonary tumors have been experimentally induced in rats by inhalation of nickel carbonyl by SUNDERMAN and DONNELLY (44). Besides inhibition of the induction of pulmonary benzpyrene hydroxylase in rats by inhalation of nickel carbonyl has been reported by SUNDERMAN (45). This should cause a retention of 3,4-benzpyrene in the lungs and prolong its carcinogenic activity. Nickel dust, nickel sulfide, nickel carbonate, nickel oxide, nickel carbonyl and nickelocene are sparingly soluble in water at 37°C. In men these nickel compounds provoke cancers of the nasal cavities (46, 47) and do not apparently accumulate in the lungs (19). However BERRY, GALLE and STUPFEL (48) found nickel in lungs of rats exposed during all their life to a dilution of automotive exhaust with a very high concentration (23 ppm) of nitrous oxides (NOx). It has been reported that the chrysotile variety of asbestos contains significant amounts of nickel salts (49) and that nickel carbonyl is present in tobacco smoke (50).

Inhalation of beryllium compounds mostly of oxides which are insoluble deposit into the lungs of exposed workers and cause a bronchoalveolitis or a chronic granulomatosis resembling sarcoidosis. Studies in dogs have revealed that beryllium compounds can produce pulmonary lesions (21) though these results have not been confirmed by CONRADI et al. (22). Recent studies (1973) of JACQUES and WITSCHI (23) on rats intratracheally injected with beryllium have shown but a small lowering of the aryl hydrocarbon hydroxylase detoxifying 3-OH-benzo(a)pyrene and but a small prevention of the induction of this hydroxylase in rats pulmonary tissue by methylcholanthrene.

LASKIN, KUSCHNER and DREW (24) have produced lung cancers in rats with implanted intrabronchial pellets of several different chromium salts. Moreover respiratory cancer have been reported in male workers of the U.S. chromate industry (25, 26). The responsibility of the role of arsenicals in the development of lung cancer is still debated though observations have been reported in men (20, 27). Arsenic does not seem to accumulate in human lungs but its concentration is higher in lymph pulmonary glands than in pulmonary parenchyma (28). Though the use of arsenical pesticides could lead to the contamination of cigarettes (29) it seems that in animals no respiratory tumors have been experimentally induced by inhalation of arsenic compounds.

PHYSIOLOGICAL AND IMMUNOLOGICAL REACTIONS

When entering into the respiratory airways, potential carcinogens as other air inhalants can induce a great variety of physiological reactions. Some of them are mostly mechanic as increase of nasal resistance by congestion of nasal mucosa, modifications of the calibers of the tracheobronchial tree, or elimination by the mucociliary escalator. Others appear more specific such as pulmonary alveolar macrophages phagocytosis, lymphatic transports and immunological reactions. The details of these processes have been exposed by different authors (3, 30-38).

At present time no specific studies have been made on eventual physiological responses of the airways to air carcinogens. However they could be considered as conditioned essentially by their physical form : gas or aerosols. At the end of this review, when the role of the associated air pollutants will be considered examples of these different physiological mechanisms regulating the inhalants penetration in the airways will be given.

There is also an apparent lack of immunological investigations in regard to the possible immunological effects of carcinogenic air pollutants. However experiments have been performed in order to see if in cigarette smokers carcinogenicity could not be explained by the escape of airway tumors from recognition and control by the immune system. In fact the experiments published till nowadays in the literature and concerning mice discontinuously exposed to cigarette smoke do not give gases concentrations measurements and the immunological results appear inconclusive or contradictory (39 - 41).

ASSOCIATED RESPIRATORY PATHOLOGY

A paralysis of the ciliary movements and a squamous metaplasia of a chronically irritated epithelium lining the respiratory tract and resulting from chronic respiratory infection makes this epithelium highly vulnerable to the action of inhaled carcinogens (51). A decrease of local pulmonary cellular defences and of immunity could also be of importance to the penetration of air pollutants. RYLANDER (52, 53) has experimentally demonstrated in guinea pigs increases of the lungs macrophages and leucocytes numbers by the particulate phase of the cigarette smoke. Moreover alterations of alveolar macrophages function have been reported in smokers (54) and as an effect of tobacco smoke in vitro (55, 56).

Augmentation of sensibility towards experimental infection results from inhalation of nitrogen-oxides (57-59), or ozone (59, 60). This could contribute to explain the enhancement of chronic bronchial infection in smokers which could constitute an important factor of carcinogenicity.

Similarly in asbestos exposed workers cigarette smoking has proved to increase considerably (92 times) the risk of dying of bronchogenic carcinoma (61).

Different arguments are stressing out the importance of genetic factors in respiratory pathology. Mice experiments have pointed out strains variations to a similar ozone exposure (62), and the importance of heredity in asthma is known since a long time (63, 64). The influence of inherited respiratory pathological factors have been scientifically proved by the discovery that a genetic deficiency of antitrypsin in the blood is associated to pulmonary chronic obstructive lung disease (65 - 71). If it is not demonstrated that these genetic variations have a direct role in the penetration of carcinogenic substances it can be supposed that they have an indirect action in conditioning pulmonary diseases which could influence airways sensitivity to these substances.

ASSOCIATION OF AIR POLLUTANTS

Epidemiological surveys as well as animal experiments point out that individuals react very differently to similar pollutants exposures. If we have previously examined some biological causes of individual discrepancies, it must be also considered that the toxic aerial and alimentary environment varies widely from one person to another. The great number of active components though usually at minute concentrations in polluted air is susceptible to influence most of the physiological and immunological mechanisms regulating penetration and retention of airborne carcinogens. Nevertheless the number and the amount of air pollutants emitted either by industrial urban or automotive sources are so much diluted in the ambient atmosphere or so associated to other oncogenic environmental factors that it looks very difficult to estimate their *per se* (at their real concentrations) pulmonary carcinogenic effects (72, 73). On the contrary tabacco smoke which is directly inhaled in the respiratory tract has been proved to be the most frequent and greatest risk for lung cancers. So it could appear realistic to take cigarette smoke as an example of an association of air pollutants. Therefore a brief survey will be made of the way by which some of the well definite principal constituents of cigarette smoke can influence penetration of their most important carcinogens which nowadays are still supposed to be the polycyclic aromatic hydrocarbons and principally benzpyrene (74). The principal components of tobacco smoke will be considered by order of decreasing concentrations. RONDIA (75) has measured a decrease of 10-25 % of rats'liver benzpyrene hydroxydase activity by concentrations of 60-100 ppm of carbon monoxide. MORDELET-DAMBRINE et al. (76) observed bronchospasm in urethanized and artificially ventilated guinea pigs for a threshold of 1000 ppm of carbon monoxide. With the same experimental protocol they noted thresholds of tracheal pressure modifications for 0.5 % carbon dioxide, 0,5 ppm of nitrous oxides (NOx) and 2 ppm of sulfur dioxide (77, 78). An augmentation of tracheal pressure can be related to the increase in pulmonary resistance by inhalation of total tabacco smoke that SAINDELLE (79) has observed in guinea pigs by this same technique and that many authors have reported in smoking men (80-85). Nitrous dioxide another noxious constituent of tobacco smoke

has been studied in man by VON NIEDING et al. (86) who have
obtained an increase in airway resistance for concentrations of
1.5 ppm NO_2. It looks however difficult to teleologically accept
the prospective value of these "bronchoconstriction" and "increase in pulmonary resistance" provoked by tobacco smoke intruding the tracheobronchial tree. The modification of the mucociliary transport by tobacco smoke has been proved in animals (87-91) and studied in humans (92-94). Part of this action on mucociliary clearance can be attributed to the nitrogen dioxide (95)
or to the aldehydes (87, 96) content of tobacco smoke.

CONCLUSION

Trying to gather some specific mechanisms without a
sufficient knowledge should be looked as hazardous though it
has the merit to try to put some signification in a preventive
and eventually a therapeutic attitude by a better comprehension
of toxicological effects and mostly before asserting a chemical
air environmental carcinogenic substance one must have the assurance of its penetration into the body either of itself or of
its active forms. The field of investigations is widely open.
Nowadays cellular researchs could help inquiring about the local
reaction to the penetration of a specific pollutant and much
remain to be done particularly in immunology.

We acknowledge with gratitude the assistance of Pr.
R. FONTANGES, Centre de Recherches du Service de Santé des Armées,
Lyon, for the informations he provided us.

RESUME
PENETRATIONS DES SUBSTANCES CANCEROGENES DANS LES VOIES AERIENNES

Les auteurs qui ont précédemment traité (Bull. Physio.-
Pathol. Resp., 1974, 10, 481-509) la pénétration des polluants
atmosphériques dans les voies aériennes envisagent le devenir
des substances cancérogènes qui pénètrent dans le tractus respiratoire. Des réactions physiques, physicochimiques, biochimiques,
physiologiques et immunologiques se produisent aux divers étages
de l'arbre respiratoire suivant l'identité de l'agresseur :
gaz [bis (chlorométhyl) éther, chlorure de vinyl], aérosol liquide ou solide [hydrocarbures polycycliques aromatiques, métaux
(nickel, beryllium, chrome, arsenic), fibres d'amiante]. En
fait, bien souvent, l'agent cancérogène est transporté sur un
vecteur et agit conjointement avec d'autres substances chimiques
(fumée de tabac, rejets industriels, gaz d'échappement d'automobiles) atmosphériques. Les associations physicochimiques et physiopathologiques susceptibles de moduler la pénétration ont un
rôle important et rendent compte de l'effet activateur cocancérogène. De plus, pour élucider la cancérogénésité d'une substance chimique de l'environnement aérien, il convient de mettre
en évidence sa pénétration dans le tractus respiratoire et sa
ou ses formes actives. Beaucoup est à faire.

REFERENCES

(1) Higginson J. : The importance of environmental factors in cancer. This Symposium, 1975, Lyon.
(2) Gross P., Harley R.A., Swinburne L.M., Davis J.M.G. and Greene W.B. : Ingested mineral fibers do they penetrate tissue or cause cancer? Arch. Environ. Health, 1974, 29, 341-347.
(3) Stupfel M. and Mordelet-Dambrine M. : Penetration of pollutants in the airways. Bull. Physio-Pathol. resp., 1974, 10, 481-509.
(4) Laskin S., Kuschner M., Drew R.T., Cappiello V.P. and Nelson N. : Tumors of the respiratory tract induced by inhalation of bis (chloromethyl) ether. Arch. Environ. Health, 1971, 23, 135-136.
(5) Leong B.K.J., Macfarland H.N. and Reese W.H. : Induction of lung adenomas by chronic inhalation of bis (chloromethyl) ether. Arch. Environ. Health, 1971, 22, 663-666.
(6) Viola P.L., Bigotti A. and Caputo A. : Oncogenic response of rat skin lungs and bones to vinyl chloride. Cancer Res., 1971, 31, 516-519.
(7) Gilson J.C. : Asbestos cancers as an example of the problem of comparative risks. This Symposium, 1975, Lyon.
(8) Cook J.W., Hewett C.L. and Hieger I. : The isolation of a cancer-producing hydrocarbon from coal tar. Parts I, II and III. J. Chem. Soc., 1933, 395-405.
(9) Pullman A. and Pullman B. : Cancérisation par les substances chimiques et structure moléculaire, 1955, Masson, Paris.
(10) Herndon W.C. : Theory of carcinogenic activity of aromatic hydrocarbons. Trans. N.Y. Acad. Sci., 1974, 36, 200-217.
(11) Wattenberg L.W., Leong J.L. and Galbraith A.R. : Induction of increased benzpyrene hydroxylase activity in pulmonary tissue in vitro. Proc. Soc. Exp. Biol. Med., 1968, 127, 467-469.
(12) Grover P.L., Hewer A. and Sims P. : Metabolism of polycyclic hydrocarbons by rat lung preparations. Biochem. Pharmacol., 1974, 23, 323-332.
(13) Gross P., Tuma J. and De Treville R.T.P. : Lungs of workers exposed to fiber glass. A study of their pathologic changes and their dust content. Arch. Environ. Health, 1971, 23, 67-76.
(14) Selikoff I.J., Churg J. and Hammond C.E. : Asbestos exposure and neoplasia. J.A.M.A., 1964, 188, 22-26.
(15) Selikoff I.J., Nicholson W.J. and Langer A.M. : Asbestos air pollution. Arch. Environ. Health, 1972, 25, 1-13.
(16) Commins B.T. and Gibbs G.W. : Contaminating organic material in asbestos. Brit. J. Cancer, 1969, 23, 358-362.
(17) Roy-Chowdhury A.K., Mooney T.F. and Reeves A.L. : Trace metals in asbestos carcinogenesis. Arch. Environ. Health, 1973, 26, 253-255.
(18) Stocks P. : On the relations between atmospheric pollution in urban and rural localities and mortality from cancer, bronchitis and pneumonia, with particular reference to 3:4 benzopyrene, beryllium, molybdenum, vanadium and arsenic. Brit. J. Cancer, 1960, 14, 397-418.

(19) Schroeder H.A. : A sensible look at air pollution by metals. Arch. Environ. Health, 1970, 21, 798-806.
(20) Louria D.B., Joselow M.M. and Browder A.A. : The human toxicity of certain trace elements. Ann. Intern. Med., 1972, 76, 307-319.
(21) Robinson F.R., Schaffner F. and Trachtenberg E. : Ultrastructure of the lungs of dogs exposed to beryllium containing dusts. Arch. Environ. Health, 1968, 17, 193-203.
(22) Conradi C., Burri P.H., Kapanci Y., Robinson F.R. and Weibel E.R. : Lung changes after beryllium inhalation. Ultrastructural and morphometric study. Arch. Environ. Health, 1971, 23, 348-358.
(23) Jacques A. and Witschi H.R. : Beryllium effects on aryl hydrocarbon hydroxylase in rat lung. Arch. Environ. Health, 1973, 27, 243-247.
(24) Laskin S., Kuschner M. and Drew R.T. : Studies in pulmonary carcinogenesis. In : Hanna M.G. Jr., Nettesheim P. and Gilbert J.R., Inhalation carcinogenesis, 1970, 321-351, U.S. Atomic Energy Commission.
(25) Gafafer W.M. et al. : Health of workers in chromate producing industry. Health Service Publication n° 192, 1953, U.S. Public Health Service.
(26) Clayson D.B. : Chemical carcinogenesis, 1962, Little Brown, Boston.
(27) Ott M.G., Holder B.B. and Gordon H.L. : Respiratory cancer and occupational exposure to arsenicals. Arch. Environ. Health, 1974, 29, 250-255.
(28) Molokhia M.M. and Smith H. : Trace elements in the lung. Arch. Environ. Health, 1967, 15, 745-750.
(29) Lee B.K. and Murphy G. : Determination of arsenic content of American cigarettes by neutron activation analysis. Cancer, 1969, 63, 1315-1317.
(30) Hatch T. and Gross P. : Pulmonary deposition and retention of inhaled aerosols, 1964, Academic Press, New York.
(31) Proctor D.F., Swift D.L., Quinlan M., Salman S., Takagi Y. and Evering S. : The nose and man's atmospheric environment. Arch. Environ. Health, 1969, 18, 671-680.
(32) Du Bois A. : Mechanism of bronchial response to inhalants. In : "Physiology, environment and man", Lee D.K. and Minard D. eds., 1970, 80-85, New York.
(33) Bouhuys A. : Airway dynamics, 1970, Thomas, Springfield (Ill.).
(34) Chrétien J. and Masse R. : L'épuration broncho-pulmonaire des particules non-organiques. Mécanismes et implications pratiques. Acquisitions médicales récentes, 1973, 117-146, Sandoz, Paris.
(35) Molina C. : Immunopathologie bronchopulmonaire, 1973, Masson, Paris.
(36) Brain J.D. and Valberg P.A. : Models of lung retention based on ICRP task group report. Arch. Environ. Health, 1974, 28, 1-11.
(37) Kilburn K.H. : Pulmonary reactions to organic materials. Ann. N.Y. Acad. Sci., 1974, 221, 1-390.

(38) Lauweryns J.M. and Baert J.H. : The role of the pulmonary lymphatics in the defenses of the distal lung : morphological and experimental studies of the transport mechanisms of intracheally instillated particles. Ann. N.Y. Acad. Sci., 1974, 221, 244-275.
(39) Esber H.J., Menninger F.F. Jr., Bogden A.E. and Mason M.M. : Immunological deficiency associated with cigarette smoke inhalation by mice. Primary and secondary hemagglutinin response. Arch. Environ. Health, 1973, 27, 99-104.
(40) Thomas W.R., Holt P.G. and Keast D. : Cellular immunity in mice chronically exposed to fresh cigarette smoke. Arch. Environ. Health, 1973, 27, 372-375.
(41) Thomas W.R., Holt P.G. and Keast D. : Humoral immune response of mice with long-term exposure to cigarette smoke. Arch. Environ. Health, 1975, 30, 78-80.
(42) Miller J.A. : Carcinogenesis by chemicals : an overview. Cancer Res., 1970, 30, 559-576.
(43) Hackett R.L. and Sunderman F.W. Jr. : Pulmonary alveolar reaction to nickel carbonyl. Ultrastructural and histochemical studies. Arch. Environ. Health, 1968, 16, 349.
(44) Sunderman F.W. and Donnelly A.J. : Studies of nickel carcinogenesis : metastasizing pulmonary tumors in rats induced by the inhalation of nickel carbonyl. Am. J. Path., 1965, 46, 1027-1041.
(45) Sunderman F.W. Jr. : Inhibition of induction of benzpyrene hydroxylase by nickel carbonyl. Cancer Res., 1967, 27, 1595.
(46) Doll R. : Cancer of the lung and nose in nickel workers. Brit. J. Ind. Med., 1958, 15, 217-233.
(47) Sunderman F.W. Jr. : Nickel carcinogenesis. Dis. Chest., 1968, 54, 527-534.
(48) Berry J.P., Galle P. and Stupfel M. : Analyse des dépôts intra-pulmonaires chez des rats exposés au gaz d'échappement de moteur automobile. Nouv. Presse Méd., 1973, 2, 1856-1858.
(49) Dixon J.R., Lowe D.B., Richards D.E., Cralley L.J. and Stokinger H.E. : The role of trace metals in chemical carcinogenesis : asbestos cancer. Cancer Res., 1970, 30, 1068-1074.
(50) Sunderman F.W. and Sunderman F.W. Jr. : Nickel poisoning, XI. Implication of nickel as a pulmonary carcinogen in tobacco smoke. Am. J. Clin. Path., 1963, 39, 549.
(51) Frost J.K., Gupta P.K., Eroza Y.S., Carter D., Hollander D.H., Levin M.L. and Ball W.C. Jr. : Pulmonary cytologic alterations in toxic environmental inhalation. Human pathol., 1973, 4, 521-536.
(52) Rylander R. : Influence of infection on pulmonary defense mechanisms. Ann. N.Y. Acad. Sci., 1974, 221, 281-289.
(53) Rylander R. : Pulmonary cell responses to inhaled cigarette smoke. Arch. Environ. Health, 1974, 29, 329-333.
(54) Mann P.E.G. et al. : Alveolar macrophages : structural and functional differences between nonsmokers and smokers of marijuana and tobacco. Lab. Invest., 1971, 25, 111-120.
(55) Lentz P. and Di Luzio N. : Influence of tobacco smoke on the phagocytic, bactericidal and metabolic activities of rat alveolar macrophages. J. Reticuloendothel. Soc., 1972, 2, 425-426.

(56) York G.K., Arth C., Stumbo J.A., Cross C.E. and Mustafa M.G. : Pulmonary macrophage respiration as affected by cigarette smoke and tobacco extract. Arch. Environ. Health, 1973, 27, 96-98.
(57) Ehrlich R. : Effect of nitrogen dioxide on resistance to respiratory infection. Bact. Rev., 1966, 30, 604-614.
(58) Goldstein E., Eagle C.M. and Hoeprich P.D. : Effect of nitrogen dioxide on pulmonary bacterial defense mechanisms. Arch. Environ. Health., 1973, 26, 202-204.
(59) Goldstein E., Warshauer D., Lippert W. and Tarkington B. : Ozone and nitrogen dioxide exposure. Murine pulmonary defense mechanisms. Arch. Environ. Health, 1974, 28, 85-90.
(60) Coffin D.L. and Blommer E.J. : The influence of cold on mortality from streptococci following ozone exposure. J. Air. Poll. Contr. Assoc., 1965, 19, 523-524.
(61) Selikoff I.J., Hammond C.E. and Churg J. : Asbestos exposure, smoking and neoplasia. J.A.M.A., 1968, 204, 106-112.
(62) Goldstein B.D., Lai L.Y., Ross S.R. and Cuzzi-Spadar R. : Susceptibility of inbred mouse strains to ozone. Arch. Environ. Health, 1973, 27, 412-413.
(63) Salter H.H. : On asthma : its pathology and treatment, 1868, London. In : Schwartz M. "Heredity in bronchial asthma" Med. Thesis, 1952, Copenhagen.
(64) Parrot J.L. and Saindelle A. : L'hérédité du terrain allergique et les données de la clinique. Rev. Fr. étud. Clin. Biol., 1963, 8, 570-586.
(65) Eriksson S. : Studies in alpha-1-antitrypsin deficiency. Acta med. scand., 1965, 177, suppl., 1-85.
(66) Fagerhol M.K. : Quantitative studies on the inherited variants of alpha-1-antitrypsin. Scand. J. clin. lab. invest., 1969, 23, 97-103.
(67) Kueppers F., Fallat R. and Larson R.K. : Obstructive lung disease and alpha-1-antitrypsin deficiency gene heterozygosity. Science, 1969, 165, 899-901.
(68) Ostrow D.N. and Cherniack R.M. : The mechanical properties of the lungs in intermediate deficiency of alpha-1-antitrypsin. Am. Rev. Resp. Dis., 1972, 106, 377-383.
(69) Lieberman J. : Involvement of leukocytes proteases in emphysema and antitrypsin deficiency. Arch. Environ. Health, 1973, 27, 196-200.
(70) Mittman C., Barbela T. and Lieberman J. : Antitrypsin deficiency and abnormal protease inhibitor phenotypes. Arch. Environ. Health, 1973, 27, 201-206.
(71) Martin J.P. : L'alpha-1-antitrypsine et le système P_i, INSERM Symp., 1975, INSERM, Paris.
(72) Stern A.C. : Air pollution, 1962, Academic Press, New York.
(73) Stupfel M. : Recent advances in the investigations of autoexhaust toxicity. Environ. Health Perspect., 1975, in press.
(74) Chrétien J., Hirsch A. and Thieblemont M. : Pathologie respiratoire du tabac. L'expérimentation animale dans le monde. Objectifs et méthodologie, 1973, Masson, Paris.
(75) Rondia D. : Abaissement de l'activité de la benzopyrène hydroxylase hépatique in vivo après inhalation d'oxyde de carbone. C. R. Acad. Sci., 1970, 271, 617-619.

(76) Mordelet-Dambrine M., Stupfel M., Romary F. and Parrot J.L.:
Etude comparative des effets de l'oxyde de carbone et de
l'hypoxie sur la bronchomotricité du Cobaye anesthésié.
Arch. intern. physiol. biochim., 1973, 81, 673-688.
(77) Mordelet-Dambrine M., Stupfel M. and Parrot J.L. : Réactions
bronchopulmonaires du Cobaye exposé de façon aiguë à divers
polluants atmosphériques. In:INSERM Symp. "Réactions bronchopulmonaires aux polluants atmosphériques", Sadoul P.
éd., 1974, 29, pp. 245-258, INSERM, Paris.
(78) Stupfel M., Mordelet-Dambrine M. and Parrot J.L. : The
acute actions of automotive exhaust gas and its components
on guinea pig tracheal pressure. Toxicol. appl. Pharmacol.,
1975, 33, 401-413.
(79) Saindelle A. : Contribution à l'étude de la libération
d'histamine. Thèse Sci. Nat., 1967, Paris.
(80) Loomis T.A. : A broncho-constrictor factor in cigarette
smoke. Proc. Soc. exp. Biol.,   1956, 92, 337-340.
(81) Nadel J.A. and Comroe J.H. Jr. : Acute effects of inhalation
of cigarette smoke on airway conductance. J. Appl. Physiol.,
1961, 16, 715-716.
(82) Damoiseau J., Petit J.M., Troquet J., Pirnay F.E. and
Marcelle R. : Action de la fumée de tabac sur les résistan-
ces du courant aérien. J. Physiol. (Paris), 1962, 54, 318-
319.
(83) Aviado D.M. and Samanek M. : Bronchopulmonary effects of
tobacco and related substances. Arch. Environ. Health, 1965,
11, 141-176.
(84) Clarke B.G., Guyatt A.R., Alpers J.H., Fletcher C.M. and
Hill I.D. : Changes in airways conductance on smoking a
cigarette. A study of repeatability and of the effect of
particulate and vapour phase filters. Thorax, 1970, 25,
418-422.
(85) Gayrard P., Orehek J., Grimaud C. and Charpin J. : Bronchoconstriction due à l'inhalation de fumée de tabac : effets
comparés chez le sujet normal et l'asthmatique. Bull. Physio.
Pathol. resp., 1974, 10, 451-461.
(86) Von Nieding G., Wagner M., Krekeler H., Smidt U. and Muysers
K. : Grenzwertbestimmung der akuten NO_2-Wirkung auf den
respiratorischen Gasaustausch und die Atemwegswiderstände
des chronisch lungenkranken Menschen. Int. Arch. Arbeitsmed., 1971, 27, 338-348.
(87) Dalhamn T. : Mucous flow and ciliary activity of healthy
rats exposed to respiratory irritant gases (SO_2, H_3N, HCHO).
Acta physiol. scand., 1956, 36, suppl. 123.
(88) Guillerm R., Badré R. and Vignon B. : Effets inhibiteurs de
la fumée de tabac sur l'activité ciliaire de l'épithélium
respiratoire et nature des composants responsables. Bull.
Acad. Méd., 1961, 145, 416-423.
(89) Albert R.E., Spiegelman J.R., Shatsky S. and Lippmann M. :
The effects of acute exposure to cigarette smoke on bronchial clearance in the miniature donkey. Arch. Environ.
Health, 1969, 18, 30-41.
(90) Dalhamn T. and Rylander R. : Ciliotoxicity of cigar and
cigarette smoke. Arch. Environ. Health, 1970, 20, 252-253.

(91) Wanner A., Hirsch J.A., Greeneltch D.E. and Swenson E.W. : Tracheal mucous velocity in beagles after chronic exposure to cigarette smoke. Arch. Environ. Health, 1973, 27, 370-371.
(92) Camner P., Philipson K. and Arvidsson T. : Cigarette smoking in man. Short-term effect on mucociliary transport. Arch. Environ. Health, 1971, 23, 421-426.
(93) Camner P. and Philipson K. : Tracheobronchial clearance in smoking discordant twins. Arch. Environ. Health, 1972, 25, 60-63.
(94) Albert R.E., Peterson H.T. Jr., Bohning D.E. and Lippmann M.: Short-term effects of cigarette smoking on bronchial clearance in humans. Arch. Environ. Health, 1975, 30, 361-367.
(95) Giordano A.M. and Morrow P.E. : Chronic low-level nitrogen dioxide exposure and mucociliary clearance. Arch. Environ. Health, 1972, 25, 443-449.
(96) Guillerm R., Badré R. and Hée J. : Détermination du seuil d'action des polluants irritants de l'atmosphère. Comparaison de deux méthodes. Ann. Occup. Hyg., 1967, 10, 127-133.

Environmental Pollution and Carcinogenic Risks
Pollution de l'environnement et risques cancérogènes

INSERM, 1976, Vol. 52, pp. 73-80

SIGNIFICANCE OF OXIDES OF NITROGEN (NO) AND SH REACTIVE COMPONENTS IN PULMONARY CARCINOGENESIS
An experimental study related to tobacco smoke

Cecilie LEUCHTENBERGER and Rudolf LEUCHTENBERGER
With the technical assistance of
Irene ZBINDEN and Elisabeth SCHLEH

Department of Cytochemistry
Swiss Institute for Experimental Cancer Research
Lausanne, Rue du Bugnon 21, Switzerland

I. Introduction

Epidemiological studies have demonstrated that chronic inhalation of cigarette smoke is one of the major causes of human lung cancer (1). Our extensive experimental studies using 2 modelsystems, such as inhalation in mice of and exposure of human and animal lung cultures to fresh cigarette smoke support the results of human epidemiological data that cigarette smoke is an environmental source that carries carcinogenic risks (2). However, the important question, which components in the cigarette smoke are responsible for enhancement of human pulmonary carcinogenesis, cannot be answered conclusively at present. There is the widely accepted concept that it is mainly "tar" in cigarette smoke which is responsible for the development of lung cancer in heavy human smokers. However, this opinion is open to criticism, because it is based on extrapolation from animal skin painting experiments with condensates or extracts of tobacco smoke (3), and not on experimental inhalation studies with fresh cigarette smoke itself, such as it is inhaled by the human smoker. Cigarette smoke has a very complex chemical composition, already over 1200 components have been identified which are present partly in form of particles (particulate phase), partly in form of gases and vapours (gas vapour phase).

We have been interested for a considerable time in the experimental exploration of the role of cigarette smoke in pulmonary carcinogenesis (2), paying special attention to the identification of carcinogenic or cocarcinogenic components in the particulate or in the gas vapour phase of cigarette smoke. Such an approach appeared to us necessary, since our own experimental studies disclosed that enhancement of pulmonary carcinogenesis in mice was not only observed after chronic inhalation of whole cigarette smoke, that is smoke containing particles, such as tar and nicotine, but also after chronic inhalation of the gas vapour phase alone (2, 4, 5). The same holds true for enhancement of malignant transformation of lung cultures.

The present correlated biological and chemical study deals with the question, whether varying quantities of certain important constituents in the particulate

Fig. 1 TIME SEQUENTIAL EFFECTS OF FRESH CIGARETTE SMOKE WITH HIGH AND LOW SH REACTIVITY ON HAMSTER LUNG CELLS.

BEFORE EXPOSURE TO SMOKE:

| WITH HIGH SH REACTIVITY | AFTER EXPOSURE TO SMOKE: | WITH LOW SH REACTIVITY |

MARKED — **STAGE I** CYTOTOXICITY — INHIBITION OF DNA, RNA, PROTEIN SYNTHESIS, MITOSIS — ABSENT

 MARKED — **STAGE II** ATYPICAL PROLIFERATION — STIMULATION OF DNA, RNA PROTEIN SYNTHESIS, ABNORMAL MITOSIS, DNA COMPLEMENT, NUMBER OF CHROMOSOMES, NUCLEI, NUCLEOLI, CYTOPLASM. — VERY MILD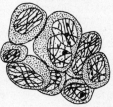

POSITIV BEFORE 6 MONTHS — **STAGE III** MALIGNANT CELL TRANSFORMATION — INVASIVE TUMORS IN NUDE MICE (FIBROSARCOMA) — POSITIV AT LATER PERIODS 12-24 MONTHS (SPONTANEOUS TRANSFORMATION OF CONTROLS)

phase or in the gas vapour phase of cigarette smoke influence atypical growth and/or malignant transformation in hamster lung cultures. In the chemical analysis attention was focussed on content of tar, nicotine, carbon monoxide (CO), oxides of nitrogen (NO), and SH reactivity of the cigarette smoke. An analysis of SH reactivity of the smoke appeared especially pertinent, since it has been reported that cigarette smoke inhibits SH dependent enzyme systems and that free radicals are responsible for this phenomenon (6-9). Furthermore, as seen in the scheme of Fig. 1, we had previously observed that only cigarette smoke with high SH reactivity evoked enhancement of malignant cell transformation of hamster lung cultures, while cigarette smoke with low SH reactivity did not have this effect (10,11).

II. Methods

For the present correlated biological and chemical study we exposed again hamster lung cultures to puffs of fresh cigarette smoke in a CSM_{12} smoking machine under standardized conditions as previously described (10). This modelsystem developed by us has proved to be a suitable bioassay to assess simultaneously the sequential alterations in morphology, growth and DNA metabolism of cultured lung cells after short and long term exposure to smoke from tobacco cigarettes (12-14).

For this study 12 different experiments comprising over 7000 hamster lung cultures were utilized. The sets comprised non exposed control cultures, and cultures exposed to 4 puffs daily (25 ml at intervals of 58 s.) of fresh smoke from 8 different types of cigarettes on 3 consecutive days per week for a period of 1 week up to 6 months. The smoke from the 8 types of cigarettes was analyzed by gas-chromatography and/or chemically, and the SH reactivity (SH index) was determined as previously described (10), (Table I).

The assessment of the biological effects of the smoke on the cultures was done in living cultures and in fixed and stained preparations from a morphological and cytochemical point of view (10, 12-14), without knowledge of the analytical results of the smoke. Cultures exposed to smoke from C_1-C_7 cigarettes were compared among each other, with those of non exposed control cultures and cultures exposed to smoke from Kentucky Standard cigarettes. The latter exposed cultures were used as a standard of reference because they displayed 3 main well defined stages of sequential morphological and cytochemical alterations, namely cytotoxicity, atypical growth, and malignant transformation (Fig. 1) within a period of 6 months (15).

III. Results

Results were reproducible in all experiments. As can be seen from the data in Table I, with exception of the cultures exposed to smoke from C_1 cigarettes which resembled control cultures, smoke from all other cigarettes produced alterations in the cultures. However, smoke from C_2-C_4 cigarettes evoked less abnormalities in growth, namely only atypical growth, than smoke from Kentucky Standard and C_5-C_7 cigarettes, which evoked malignant cell transformation. Relating these different biological effects to the analyzed chemical constituents of the smoke, it can be seen that amounts of TPM, tar and nicotine did not influence significantly the occurrence of atypical growth and/or malignant transformation in the hamster lung cultures. Smoke from C_1 cigarettes which did not evoke any alterations in the cultures had only a slightly lower tar content than smoke from C_2 and C_7 cigarettes which produced atypical growth and very marked malignant transformation respectively. Furthermore, smoke from Kentucky Standard cigarettes with a relatively high tar content led to less marked malignant transformation than smoke of C_6 and C_7 cigarettes having a significantly lower tar content. Therefore, these results ob-

TABLE I

Effects of Quantitative Differences in Constituents and SH Reactivity of Cigarette Smoke on Growth of Hamster Lung Cultures

Type of ciga-rette	PARTICULATE MATTER			GAS		VAPOUR		PHASE		SH index of smoke	BIOLOGICAL EFFECTS (sequence of eve		
	TPM mg	Tar mg	Nicotine mg	CO mg	HCN µg	NO µg	TGVP r.d.	Acetaldehyde r.d.	Acrolein r.d.		Stage I Cytoto-xicity	Stage II Atypical growth	Stage III Malignant transforma
C 1	4,8	4,4	0,4	5,0	35	33	<28	traces	traces	5,1	(+)	(+)	-
C 2	5,9	5,5	0,4	4,9	44	31	52	2,0	0,12	9,5	(+)	++	-
C 3	6,2	5,8	0,4	5,1	53	36	42	1,9	0,18	9,8	(+)	++	-
C 4	8,5	8,2	0,3	8,5	77	33	73	4,1	0,18	12,7	++ - +++	++ - +++	-
C 5	8,9	8,4	0,5	7,3	73	117	59	3,0	0,18	19,6	++++	++++	++++
Kentucky Standard	10,5	9,8	0,7	7,3	153	104	70	3,6	0,20	23,5	++++	++++	++++
C 6	6,5	6,1	0,4	7,5	51	354	59	1,9	0,08	29,9	++++++	++++++	++++++
C 7	5,9	5,6	0,3	6,8	40	378	68	2,2	0,10	35,7	++++++	++++++	++++++

- = negative   ++ = mild   ++++ = marked   TPM = total particulate matter
(+) = doubtful   ++ - +++ = moderate   ++++++ = very marked   TGVP = total gas vapour phase

r.d. = relative delivery

All analytical values are calculated for 4 puffs

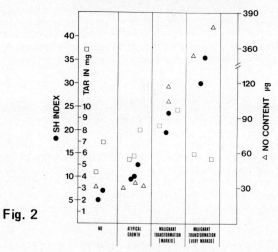

Fig. 2

ABNORMALITIES IN GROWTH OF HAMSTER LUNG CULTURES AFTER EXPOSURE TO CIGARETTE SMOKE WITH DIFFERENT SH INDEX NO, AND TAR CONTENT. (N = 12)

ALTERATIONS IN CULTURES

tained with fresh cigarette smoke do not support the widely accepted concept, based on skinpainting experiments with "tar" of cigarette smoke condensates, that it is mainly "tar" of cigarette smoke which is responsible for lung cancer of the human cigarette smoker (3). On the other hand the amounts of gas vapour phase, and especially the SH reactivity of the smoke, seemed to play a role for the occurrence of atypical growth and of malignant transformation of the hamster lung cultures. Smoke from C_1 cigarettes which had the lowest amounts of TGVP, of acrolein and acetaldehyde, and the lowest SH index did not produce any alterations, while smoke from all cigarettes with higher amounts of TGVP and higher SH index evoked atypical growth and/or malignant transformation. There exists especially a striking positive correlation between SH index and biological effects of the smoke on the cultures. The higher the SH index the greater the number and severity of the alterations, particularly of those of growth abnormalities. Smoke with relatively lower SH index produced atypical growth, while smoke with significantly higher SH index evoked malignant transformation. Furthermore, the malignant transformation was more marked and was observed at an earlier period (3 months) after exposure to smoke from C_6 and C_7 cigarettes which had a higher SH index than after smoke from C_5 and Kentucky Standard cigarettes (6 months) which had a lower SH index. It can also be seen, that there is a positive correlation between amounts of the SH reactive gas vapour phase constituent NO, and occurrence of malignant transformation. Only smoke with high amounts of NO evoked malignant transformation, while smoke with relatively low content of NO produced only atypical growth. Furthermore, the most marked transformation was observed after smoke with NO content above 350 μg.

IV. Discussion

The positive correlation between high SH reactivity and high NO content of the gas vapour phase of fresh cigarette smoke and malignant transformation (Fig. 2) is in good accordance with observations that atypical growth and enhancement of pulmonary carcinogenesis occurred not only after exposure to whole smoke but also after exposure to the gas vapour phase only (4, 5, 10, 11, 13, 14). That high NO content of the gas vapour phase appeared to play a role in the malignant transformation is of special interest. NO is considered to be a potential precursor in the formation of N-nitrosamine, a carcinogenic substance found in cigarette smoke (16). The present data indicating a positive correlation between SH reactivity and carcinogenic effects of cigarette smoke are also in accordance with recent observations that 3-4 benzpyrene carcinogenesis in mice can be reduced or completely inhibited by thiols (17), and that there is tumor rejection in animals treated with radioprotective thiols (18). The results obtained in the present study appear also to add evidence for the hypothesis that one of the causes of human lung cancer may be the inhalation of components in cigarette smoke which react with thiols, thereby removing free cysteine from the bronchial epithelium (19).

Further exploration of the carcinogenic or cocarcinogenic role of gas vapour phase constituents with high SH reactivity, such as NO, are urgently needed. The search for and identification of such health damaging factors, also in gases and vapours from other environmental sources (i.e. exhaust gases from motor vehicles, planes) may also help in attempts to eliminate carcinogenic or cocarcinogenic risks.

Acknowledgment

We thank Dr. H. Gaisch and U. Nyffeler of Fabriques de Tabac Réunies S.A., Neuchâtel, for carrying out the analyses, and the measurements of the SH index of the smoke.

This work was supported in part by a grant from the ASFC, Switzerland.

Résumé

L'exposition des cultures cellulaires de poumons d'hamsters à la fumée fraîche de 8 types de cigarettes avec quantités variables des facteurs chimiques de la phase particulaire et de la phase vapeur-gaz a resulté après une période de 3-6 mois dans une croissance atypique et/ou dans une transformation cellulaire maligne. Une correlation positive a été démontrée entre une haute réactivité SH ainsi qu'une haute teneur en NO de la phase gaz-vapeur et la transformation maligne Il n'y avait aucune correlation positive avec les autres facteurs chimiques analysés, dont la tenue en goudron.

References

(1) US Department of Health, Education and Welfare, Public Health Service, Health Services and Mental Health Administration: The Health Consequences of Smoking, a report to the Surgeon General, 1970, 1971, 1972, 1973.

(2) C. Leuchtenberger and R. Leuchtenberger: The experimental exploration of health damaging factors in cigarette smoke. Sozial- und Präventivmedizin, 1974, 19, 41.

(3) E.L. Wynder and D. Hoffmann: Tobacco and Tobacco Smoke, studies in experimental carcinogenesis, 1967, pp. 1-730, Academic Press, New York - London.

(4) C. Leuchtenberger and R. Leuchtenberger: Effects of chronic inhalation of whole fresh cigarette smoke and of its gas phase on pulmonary tumorigenesis in Snell's mice. In: US Atomic Energy Commission, Division of Technical Information, 21st AEC Symposium Series: Morphology of experimental respiratory carcinogenesis, 1970, 329-346.

(5) C. Leuchtenberger and R. Leuchtenberger: Differential Response of Snell's and C57 Black Mice to Chronic Inhalation of Cigarette Smoke. Pulmonary Carcinogenesis and Vascular Alterations in Lung and Heart. Oncology, 1974, 29, 122.

(6) P.D. Jocelyn: Biochemistry of the SH group, 1972, pp. 1-404, Academic Press, London, New York.

(7) R. Lange: Inhibiting Effect of Tobacco Smoke on Some Crystalline Enzymes. Science, 1961, 52, 134.

(8) G.M. Powell and G.M. Green: Cigarette smoke - A proposed metabolic lesion in alveolar macrophages. Bioch. Pharmac., 1972, 21, 1785.

(9) T. Sato, T. Suruki and T. Fukijama: Cigarette smoke: Mode of adhesion and haemolyzing and SH-inhibiting factors. Br. J. Cancer, 1962, 16, 7.

(10) C. Leuchtenberger, R. Leuchtenberger and I. Zbinden: Gas Vapour Phase Constituents and SH Reactivity of Cigarette Smoke Influence Lung Cultures. Nature, 1974, 247, 5442, 565.

(11) P. Davies, G.S. Kistler, C. Leuchtenberger and R. Leuchtenberger: Ultrastructural studies on cells of hamster lung cultures after chronic exposure to whole smoke or the gas vapour phase of cigarettes. Beitr. Path., 1975, 155, 168-180.

(12) C. Leuchtenberger and R. Leuchtenberger: Differential cytological and cytochemical responses of various cultures from mouse tissues to repeated exposures to puffs from the gas phase of charcoal-filtered fresh cigarette smoke. Exptl. Cell Res., 1970, 62, 161.

(13) C. Leuchtenberger, R. Leuchtenberger and A. Schneider: Effects of marijuana and tobacco smoke on human lung physiology. Nature, 1973, 241, 137.

(14) C. Leuchtenberger, R. Leuchtenberger, N. Jnui and U. Ritter: Effects of Marijuana and Tobacco Smoke on DNA and Chromosomal Complement in Human Lung Explants. Nature, 1973, 242, 403.

(15) C. Leuchtenberger and R. Leuchtenberger: Correlated cytological and cytochemical studies of the effects of fresh smoke from marijuana cigarettes on growth and DNA metabolism of animal and human lung cultures. Proceedings of the International Conference on the Pharmacology of Cannabis, Savannah, Georgia, in press (1975).

(16) D. Hoffmann and J. Vais: Analysis of volatile N-nitrosamines in unaged mainstream smoke of cigarettes. Paper presented at the 25th Tobacco Chemists' Research Conference, Louisville, Ky, USA, October 6-8, 1971.

(17) G. Kallistros: Verhinderung der 3,4-benzopyrene-Karzinogenese durch natürliche und synthetische Verbindungen. Münch. med. Wschr., 1975, 117, 391.

(18) C. Apffel, J. Walker, and S. Issarescu: Tumor Rejection in Experimental Animals Treated with Radioprotective Thiols. Cancer Research, 1975, 35, 429.

(19) M.L. Fenner and J. Braven: The mechanism of carcinogenesis by tobacco smoke. Further experimental evidence and a prediction from the thiol defence hypothesis. Brit. Journ. Cancer, 1968, 22, 474.

II

WATER POLLUTION
POLLUTION DES EAUX

Chairman / *Président* : Pr. R. MONIER

Environmental Pollution and Carcinogenic Risks
Pollution de l'environnement et risques cancérogènes

CARCINOGENS IN ESTUARIES, THEIR MONITORING AND POSSIBLE HAZARD TO MAN

H.F. STICH, A.B. ACTON and B.P. DUNN

Department of Medical Genetics, Department of Zoology
and Cancer Research Centre, University of British Columbia, Vancouver, Canada

There is compelling evidence that points to cancer as an ecological disease. A comparison of high with low incidences of tumors in different geographic regions, along with changes in tumor frequencies among immigrant groups and the clustering of many human tumors in industrialized regions in the U.S.A., suggest a strong environmental influence in cancer induction, promotion, or both (1,2,3). Thus it seems reasonable to apply ecological methodology to uncover the environmental component in cancer formation. The ecological approach may have the further advantage of being highly relevant to man, since it is economical and manageable with the presently available manpower. It has become obvious that the use of rodents for testing the carcinogenic capacity of chemicals can now hardly cope with the large number of newly synthesized compounds, with the vast array of complex mixtures, and with the staggering number of additive, cumulative, enhancing, or suppressing interactions between carcinogenic and co-carcinogenic compounds. The routine in vivo screening procedures must be supplemented by rapid and economic short-term tests on non-human organisms and cultured mammalian cells which could permit an early detection of carcinogens in man's environment.

In this paper we emphasize the methodological aspects of using biological and chemical indicators for carcinogens in estuaries that are areas of intensive aquaculture and human recreation and are the likeliest regions to suffer man-made contamination.

BIOLOGICAL INDICATOR SYSTEMS

A relatively large number of microbial organisms, plants and animals have been proposed as "built in" indicators of water quality. As a rule these organisms respond unspecifically to a wide array of chemical and physical factors that have sublethal toxic actions. However, the ideal indicator organism would detect the appearance of specific groups of compounds. In search for such an indicator we have explored the feasibility of using bottom-dwelling flatfish species as test organisms with skin tumors as the criterion. Since flatfish inhabit virtually all subtropical and tropical oceans and are common in the intertidal and subtidal zones which are prone to contamination, and since skin tumors can be readily diagnosed and counted, many of the prerequisites of an economical indicator system are met. Nevertheless, several factors in the proposed assay remain unknown and must be examined before fish tumors can be introduced and accepted as a reliable monitoring system for carcinogens in marine habitats. The unsolved issues concern mainly sampling procedures and can be summarized by the following questions:

1. What to sample?
2. When to sample?
3. Where to sample?
4. How many to sample?

1. A good indicator organism should be fairly uniformly distributed over a wide area, should survive in contaminated waters, should stay within a restricted

territory, and should be easy to collect in large quantities. Our studies have shown that the lemon (English) sole (Parophrys vetulus) appears to be the most suitable species in the northern Pacific ocean. The young fish post-metamorphosis occupy shallow mud flats and sand bars extending from Alaska to southern California. Since they usually appear in relatively large numbers, 2 to 4 bottom trawls, each of 30 min. duration, suffice to yield 100 to 200 specimens of the desired "0" age group (roughly 35 to 130 mm in length). Other suitable flatfish include the Japanese cresthead flounder (Limanda schrenki), which has been examined for skin tumors by Oishi (4), the Dover sole (Microstomus sordidus), which was used by Mearns and Sherwood (5) in their pollution studies along the coast of southern California, and starry flounders (Platyichthys stellatus), which accumulate in and around the brackish waters of estuaries that, near cities, may carry significant industrial discharges.

Apart from selecting the right species for a locality, the second crucial factor is the proper choice of age groups. In all flatfish species so far examined, skin tumors appear after metamorphosis and reach the highest prevalence within one year (6-8). To permit meaningful comparisons, it is of paramount importance to identify the peak of skin tumor prevalence and to restrict comparisons to fish populations of comparable postmetamorphosis age or to age-adjust the data.

Early in the first year, the tumors on lemon sole are small nodules (fig. 1). Later on the tumors have the histological characteristics of papillomas. It is very likely, but not yet proved, that the nodules progress to the papillomas. From a practical view there is no need to distinguish these two tumors and all presented results include both types.

2. The age dependency of tumor formation requires that samples be taken within the first year following metamorphosis. At a later stage, the prevalence of skin

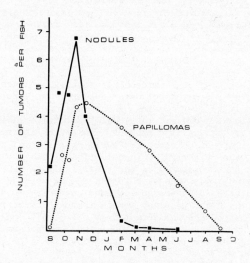

fig. 1 The average number of skin nodules and papillomas per lemon sole (Parophrys vetulus) of age group 0; sampled at Vancouver during 1973.

papillomas could be affected by a preferential loss of tumor-bearing fish to predators or by their premature death. Furthermore, it is possible that not all tumor bearers will participate in the normal migration to deeper waters which is carried out by 1 to 2-year-old lemon soles. To avoid any misleading conclusions it is advisable to sample at the peak of tumor prevalence. At this early stage in tumor formation, most afflicted fish have only small nodules which do not seem to affect their behaviour or survival as judged by holding experiments in fish tanks. To be able to sample at the peak of tumor prevalence, one must know the time when the young fish settle to the bottom following metamorphosis. Spawning periods vary within a single species and between different species. Information on these points can be obtained only by sampling monthly throughout an entire year cycle. However, once the beginning of tumor formation and the peak of tumor prevalence are established for a region, the sampling can be restricted to this period. In the case of lemon sole in the southern part of British Columbia (Canada) and the State of Washington (U.S.A.), the skin tumor prevalence reaches a peak between the late part of August and beginning of October. Samples taken during this period and screened for skin tumors should be comparable. The critical nature of the sampling time is best illustrated by fig. 2, which shows the great differences in tumor prevalence in lemon sole populations collected in various months from the same geographic location.

Sampling stations should be established at a global level to monitor contamination carried across international boundaries, and at a local level to detect and trace the source of "hot spots" of man-made or naturally-occurring carcinogens. The design of a global network of sampling stations will mainly depend on the willingness of fisheries institutes to screen, store and record samples from flatfish populations which they collect in their routine bottom trawls.

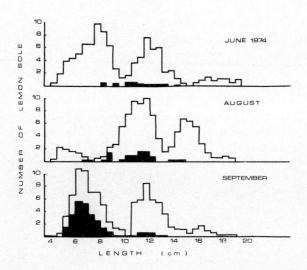

fig. 2   Prevalence of skin tumors among lemon sole (Parophrys vetulus) from the Vancouver area, solid: tumor-bearing fish; open: tumor-free fish.

No precise proposal for the establishment of local sampling stations can be made because of the great variability in shore lines, rivers, estuaries, currents, bottom constitution, etc. However, a "must" is the selection of a base-line value which should be obtained from a location devoid of human activity and exposed to steady flushing by currents.

Two examples may exemplify the distribution pattern of flatfish with skin tumors. In one study, the prevalences of skin papillomas among young lemon soles which were collected at the lower edge of the intertidal zone of relatively uncontaminated areas (Vancouver Island and the Sunshine Coast of British Columbia) were compared to those found in sole populations inhabiting the shores around Vancouver, White Rock, Bellingham, and Everett. As can be seen from fig. 3, the tumor prevalences vary from 0.0% to 58.6%, the higher values being found in areas adjacent to the cities. The prevalence of skin papillomas decreases with distance from a city, as can be seen from the frequencies in the vicinity of Everett.

In the second study, an area free of industrial discharge was selected (4). Populations of cresthead flounders (Limanda schrenki) within a poorly flushed lagoon (Furen "Lake" in Hokkaido, Japan), which contains a heavy deposit of peat

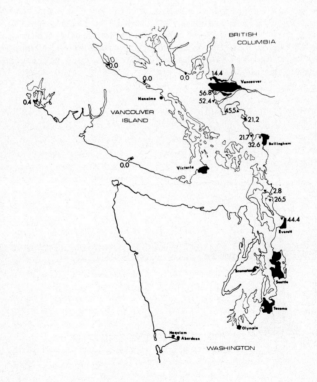

fig. 3 Prevalence of skin tumors (nodules and papillomas) of lemon sole (Parophrys vetulus) collected from different geographic locations along the coast of British Columbia and the State of Washington.

moss, were compared with those inhabiting a strait (Notsuke Strait) free of any obvious industrial contamination. The prevalence of skin papillomas among Limanda schrenki caught in commercial traps within the lagoon is about 24 times larger than that among fish from the strait.

Since we are proposing a monitoring programme, we must now face the question: how large a sample of flatfish is sufficient? The first issue to be settled is the frequency of tumors in populations that are believed not to be exposed to any carcinogens or co-carcinogens. If this frequency is in reality strictly zero, statistical estimates would be difficult to evaluate, since any population having just a single tumor-bearing fish would be significantly different from "background" (9). This, and the fact that there is no a priori reason why the background frequency should be zero, means that an estimate must be made. Our samples divide clearly into two groups: those with tumor prevalences below 0.01% and those above 1.0%. If all samples in the first group are lumped, the background frequency is 1/1383 or .000723. Statistical procedures for detecting a difference between two percentages are readily available. For ease of use, the statistical function is plotted as a graph rather than as a table. A decision has to be made about the confidence limits to be used, and it is felt that $\alpha = 0.05$ is appropriate. Fig. 4 is a graph of this function for a "true" background tumor level of 0.000723.

In practice, the operator applying the monitoring procedure would take a sample of fish from a suspected locality. Let us say that he finds 4 fish with tumors in a total catch of 40 fish. Is this sample sufficient for him to be reasonably certain that the observed tumor frequency (0.1) is greater than the observed background (0.0007)? The graph shows that for an observed frequency of

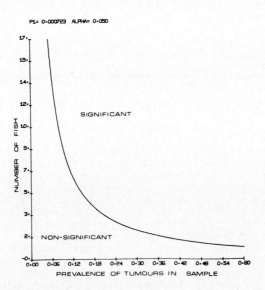

fig. 4   Graph showing size of sample required to establish that the observed tumor prevalence is significantly greater than the "true" background (0.000723).

0.1, he must look at a minimum of 9 fish. Since his sample is larger than this he need take no further samples.

CHEMICAL INDICATOR SYSTEMS

Analysis of the frequency of skin tumors in naturally-occurring fish populations gives a measure of the total carcinogenic load of the environment. A second approach to uncovering the environmental component of cancer formation is to use chemical techniques to directly estimate the amounts of known carcinogens in environmental samples. The two indicator systems are complementary - one measures the carcinogens indirectly by their effect on an indicator species, the other measures them directly in the laboratory. In developing methodology for chemical indicator systems, a number of questions need to be considered.

What to Measure

It is technically impossible to measure all carcinogens or mutagens in the environment, and logistically difficult to measure more than a few. However the estimation of several key compounds may suffice to give a simple "carcinogen index" representative of the level of contamination by many compounds. Such an index would be analogous to the use of E. coli in water quality studies as an indicator for the possible presence of dangerous coliform bacteria. The choice of which compounds to measure for a carcinogen index is influenced by several requirements: (a) the chemicals chosen for the index should be representative of a larger group of carcinogenic compounds, (b) the chemicals should be detectable in small amounts, at a reasonable cost, and in samples of a size suitable for collection, (c) the chemicals should be reasonably stable in the environment so that transient discharges can be detected.

Promising contaminents for the establishment of a "carcinogen index" are the polycyclic aromatic hydrocarbons (PAH) (10). They fulfill all of the requirements and constitute a major portion of the carcinogenic load in the marine environment. Rather than devising lengthy analysis for the numerous PAH species we have concentrated on the development of procedures for the measurement of benzo(a)pyrene.

How to Measure

Reported procedures for the isolation and measurement of PAH often leave much to be desired in terms of speed, simplicity, reliability, and cost (11,12). We have recently developed standardized procedures for the extraction and purification of PAH from environmental samples. These involve sample extraction by digestion in alcoholic KOH, and the removal of interfering materials by column chromatography and solvent extraction techniques. B(a)P is separated from other PAH by thin-layer chromatography and measured by fluorimetry. Alternatively, the purified PAH fraction may be subjected to gas chromatography to provide a finger print of the entire PAH fraction. Radioactive benzo(a)pyrene is used as a tracer in each and every sample to allow the calculation of the percentage recovery of compound and thus eliminate a major source of variability found in other techniques. The procedures as developed are straightforward, reliable, and use a minimum of expensive equipment. A variety of sample types can be processed using minor modifications of the procedure - a typical sample is 20 - 40 grams of tissue, or 10 - 20 g bottom sediments (wet weight). Reagent costs amount to approximately $10 per sample, while total labour per sample including time for preparation of reagents and laboratory maintainence amounts to 3 to 4 hours. Analysis of subsamples of bulked tissue samples indicates that the procedure has a standard deviation in the range of 5 to 10%.

What to Sample

In the marine environment there are three classes of samples, each with its own particular advantages and disadvantages.

Water samples are universally obtainable. Because of currents and tides, water samples usually have very low spatial and temporal resolution. It is very difficult to extrapolate from a limited number of samples to the time-average of pollution in a given area. Integrating the level of pollutant in a given area by analyzing a bulked sample composed of a number of subsamples taken at regular time intervals is possible, but demands either considerable labour or considerable investment in automated sampling equipment. Bottom sediments show very good spatial resolution, and are particularly relevant to the question of tumors of bottom-feeding fish. Figure 5 illustrates the level of benzo(a)pyrene in bottom sediments in the vicinity of a sewage treatment plant - note the gradient of concentration towards the sewage outfall. Bottom sediments do not, however, give good time resolution as in most cases the necessary information on sediment deposition rates and half-lives of compound is not readily available. In certain special cases, however, layered sediments may have the potential for providing historical information as to the levels of compounds in the past. Marine organisms as accumulators for carcinogens are particularly attractive. Ideally one would estimate the levels of carcinogens in the biological indicator species (e.g., lemon sole). Unfortunately, it does not appear possible to use bottom-dwelling fish as accumulator organisms because they possess enzymes capable of metabolizing PAH (13). In contrast to vertebrates, shellfish appear to lack PAH-metabolizing enzymes and therefore act as bioaccumulators of these compounds. The common mussel (Mytilus edulis or Mytilus californianus) appear to be uniquely suited to the role of a monitoring organism, being widely distributed, readily sampled, and non-migratory. Furthermore, these organisms are very hardy and

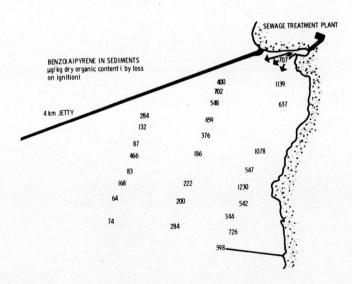

fig. 5   Variations in the benzo(a)pyrene content in bottom samples collected at various distances from the sewage outlet.

lend themselves to transplantation and laboratory experiments designed to determine uptake and discharge rates of carcinogens. Such experiments in our laboratory have indicated that the half-life of B(a)P in mussels is approximately 2 weeks. A furth point of significance is that the common mussel species, M. edulis, is cultivated a consumed as food in many areas of the world.

It has been suggested that much of the PAH in the marine environment is biosyn thetic, and that consequently background levels of PAH may be relatively high (14). This would pose severe problems for the use of the level of B(a)P as an indicator f man-made carcinogen pollution. Recent data from our studies, however, has indicate that the level of B(a)P in mussels taken from uninhabited areas of the Pacific coas line of North America is uniformly low, and that appreciable concentrations of carcinogens are only encountered as the result of man's activities (fig. 6). Most notable is the elevation in B(a)P contamination in the vicinity of docks and wharfs which appears to be the result of contamination by creosote (15).

TIER SYSTEM

The recently proposed three-tier approach to assessing a mutagenic hazard star with economical short-term in vitro tests, proceeds to the more expensive and time-consuming mammalian assays, and finally includes an estimation of the actual risk to human populations (16). The adaptation of a several-tier protocol to evalu ate the carcinogenic hazard of contaminated marine estuaries is worth careful consideration.

The first tier uses a battery of "built-in" organisms that can be readily screened for an increase in tumor prevalence. The organisms which fulfill many of

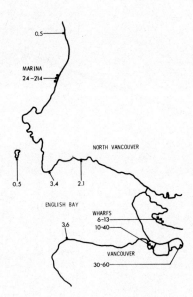

fig. 6 Variations in the benzo(a)pyrene content in mussels (Mytilus edulis) collected from the vicinity of Vancouver and Vancouver harbour.

the prerequisites demanded of an indicator are bottom-dwelling flatfish (Pleuronectidae), and gobies (Godildae). The fish-tumor system would respond to single carcinogens or complex mixtures of carcinogens and co-carcinogens within the sampled areas. Other promising organisms include molluscs which can develop neoplasms (17) and the seaweed Porphyra tenera which is cultured on rafts and has proved to be a highly successful indicator for effluents contaminated with polycyclic aromatic hydrocarbons (18,19). However, though an increased tumor frequency in fish, molluscs, or algae will not permit any conclusion to be drawn as to the molecular structure of the causative agent, it could lead to the tracing of the source of contamination.

The second tier of tests would use accumulator organisms in the detection of a selected group of key compounds that represent larger groups of carcinogens. The above-described analysis of B(a)P in mussels exemplifies this stage.

CONCLUSION AND OUTLOOK

It has been our intention to examine the feasibility of using bottom-dwelling flatfish populations and their skin tumors as an early-warning system for chemical carcinogens in subtidal or intertidal waters that are economically important and prone to urban and industrial contamination. The ease with which it is possible to collect a statistically acceptable number of flatfish, the short time required to screen fish samples for skin tumors, and the availability of well-defined methods for the computation of the significance of any increased tumor prevalence, indicate the suitability of this procedure for large-scale monitoring programmes. A trial of this procedure in British Columbia and the State of Washington has revealed a link between tumor prevalence in lemon sole (Parophrys vetuius) and urban activity. However, we would be remiss if we did not point to at least two of the issues that need to be resolved prior to the introduction of this bioassay. Firstly, skin papillomas afflict many, if not all flatfish species in the northern Pacific Ocean, but seem to appear only occasionally in flatfish populations of the Atlantic. It is likely, although hypothetical at present, that an epidemic of potentially oncogenic viruses affects Pacific flatfish species but that these "latent" viruses only become activated in the presence of chemical (or physical) carcinogens. On the other hand, the Atlantic flatfish may either lack this hypothetical virus or contain a strong viral suppressor that makes them resistant to skin tumor induction. The fish-tumor system may resemble the behaviour of oncogenic viruses in feral rodent and bird populations (20,21). The second issue concerns the lack of information about the sensitivity of fish skin to chemical carcinogens. There is convincing evidence that the response of fish to a series of carcinogens is comparable to that of rodents (22-24). However, since most of these studies are restricted to hepatomas of trout and tropical fish, the response of flatfish epidermis to chemical carcinogens remains unknown. Because they can be submitted to experimental analysis, the solution of these two issues is not insurmountable.

The possible use of marine accumulator organisms in the detection of chemical carcinogens is exemplified by the estimation of B(a)P in mussels or bottom samples of mud. The observed correlations between B(a)P levels in mussels and the levels in pilings, marinas and harbours, underlines the high sensitivity, reproducibility and economy of this assay, considering that one technician can analyse 10 samples per week.

Chemical indicator systems are also of great value in preparing detailed surveys of small geographical areas, such as would be encountered in trying to track down the source of carcinogenic pollutants detected initially by biological indicator systems. The fine geographical resolution provided by bottom sediments or the non-migratory molluscs coupled with the possibility of fingerprinting sources and contaminated samples can rapidly lead to unambiguous identification of sources of contamination. Chemical indicator systems may uncover directly a hazard to man.

Edible mussels, as well as clams and oysters are all capable of accumulating substantial amounts of PAH. Levels of B(a)P in excess of 50 µg/kg are not uncommon in mussels from areas close to urban activity. When compared with standards to be applied to smoked foods (i.e., 1 µg/kg in West Germany), it is obvious that a potential human health hazard exists.

In summary, we would like to suggest that a battery of biological and chemical assays, rather than one test, should be introduced in the detection of carcinogens in the marine environment and that the evaluation of a carcinogenic hazard should proceed through a tier system. Because the proposed assays are so economical, they can be adapted to a large-scale routine surveillance system (25).

ACKNOWLEDGMENTS

This study was supported by the National Cancer Institute of Canada. Professor H.F. Stich is a Research Associate of the N.C.I. of Canada. The authors wish to acknowledge the technical assistance of R. Scheuing and C. Lawler.

REFERENCES

(1) Higginson, J.: Present plans in cancer epidemiology. Proc. Canad. Cancer Res. Conf. 1969, 8, 40-75.
(2) Hoover, R. and Fraumeni, J.F.: Cancer mortality in U.S. counties with chemical industries. Environm. Res., 1975, 9, 196-207.
(3) Oettle, A.G.: Cancer in Africa: especially in regions south of the Sahara. J. Natl. Cancer Inst., 1964, 33, 383-439.
(4) Oishi, K., Yamazaki, F. and Harada, T.: Epidermal papilloma of flatfish in the coastal water of Hokkaido Island, Japan. J. Fish. Res. Board Canada, 1975 (in press).
(5) Mearns, A.J. and Sherwood, M.: Environmental aspects of fin erosion and tumors in Southern California Dover sole. Trans. Am. Fish. Soc., 1974, 103, 799-810.
(6) Wellings, S.R., Chuinard, R.G. and Bens, M.: A comparative study of skin neoplasms in four species of pleuronectid fishes. Ann. N.Y. Acad. Sci., 1965, 126, 479-501.
(7) Cooper, R.C. and Keller, C.A.: Epizootiology of papillomas in English sole, Parophrys vetulus. Natl. Cancer Inst. Monograph, 1969, 31, 173-185.
(8) Stich, H.F. and A.B. Acton.: The possible use of fish tumors in monitoring for carcinogens in the marine environment. Exper. Tumor Res., 1975 (in press).
(9) Sokal, R.R. and Rohlf, F.J.: Biometry, 1969. Freedman, W.H. and Co., San Francisco.
(10) Shabad, L.M.: "Carcinogen Circulation in the Environment." Medicina, Moscow, 1973.
(11) Cahnmann, H.J. and Kuratsune, M.: Determination of polycyclic aromatic hydrocarbons in oysters collected in polluted water. Anal. Chem., 1957, 29, 1312-1317.
(12) Howard, J.W., Teague, R.T., White, R.H. and Fry, B.E.: Extraction and estimation of polycyclic aromatic hydrocarbons in smoked foods. In: General Method, J. Assoc. Off. Anal. Chem., 1966, 49, 595-611.
(13) Lee, R.F., Sauerheber, R. and Dobbs, G.H.: Uptake, metabolism and discharge of polycyclic aromatic hydrocarbons by marine fish. Mar. Biol., 1972, 17, 201-208.
(14) Zobell, C.E.: Sources and biodegradation of carcinogenic hydrocarbons. Proc. Joint Conf. on Prevention and Control of Oil Spills, American Petroleum Institute, Washington, 1971, 441-451.
(15) Dunn, B.P. and Stich, H.F.: The use of mussels in estimating benzo(a)pyrene contamination of the marine environment. Proc. Soc. Exptl. Biol. Med., 1975, 150, 49-51.

16) Bridges, B.A.: The three-tier approach to mutagenicity screening and the concept of radiation-equivalent dose. Mutation Res., 1974, 26, 335-340.
17) Christensen, D.H., Farley, C.A. and Kern, F.G.: Epizootic neoplasms in the clam Macoma balthica (L.) from Chesapeake Bay. J. Natl. Cancer Inst., 1974, 52, 1739-1749.
18) Ishio, S., Kawabe, K. and Tomiyama, T.: Algal cancer and its causes. I. Carcinogenic potencies of water and suspended solids discharged to the river Ohmuta. Bull. Jap. Soc. Sci. Fish., 1972, 38, 17-24.
19) Ishio, S., Nakagawa, H. and Tomiyama, T.: Algal cancer and its causes. II. Separation of carcinogenic compounds from sea bottom mud polluted by wastes of the coal chemical industry. Bull. Jap. Soc. Sci. Fish., 1972, 38, 571-576.
20) Gardner, M.B., Officer, J.E., Rongey, R.W., Estes, J.D., Turner, H.C. and Huebner, R.J.: C-type RNA tumor virus genome expression in wild house mice. Nature, 1971, 232, 617-620.
21) Weiss, R.A.: Ecological genetics of RNA tumor viruses and their hosts in "Analytic and experimental epidemiology of cancer." Nakahara, W., Hirayama, T., Nishioka, K. and Sugano, H. eds. Univ. Park Press, Baltimore, 1973.
22) Halver, J.E.: Hepatomas in fish. In Primary Hepatoma. Ed. Burdette, W.J. Univ. of Utah Press, Salt Lake City, 1965, pp. 103-112.
23) Khudolei, V.V.: The induction of hepatic tumors by nitrosamines in aquarium fish (Lebistes reticulatus). Voprosi Onkologii, 1971, 17, 67-72.
24) Ayres, J.L., Lee, D.J., Wales, J.H. and Sinnhuber, R.O.: Aflatoxin structure and hepatocarcinogenicity in rainbow trout (Salmo gairdneri). J. Natl. Cancer Inst., 1971, 46, 561-564.

SUMMARY

Bottom-dwelling flatfish and their skin tumors can be used as an early-warning system for the pollution by chemical carcinogens in sub- and inter-tidal waters. A trial in British Columbia and the State of Washington showed a link between tumor prevalence and urban activity.

The possible use of marine accumulator organisms (mussels) in the estimation of benzo(a)pyrene levels in harbours is considered.

The combination of biological and chemical assays could provide the basis for a large-scale routine surveillance of the marine environment.

RESUME

Les poissons plats de fond et leurs tumeurs cutanées peuvent servir de système d'alerte pour la détection précoce de la pollution due aux cancérogènes chimiques dans les eaux subcotidales et intercotidales. Un essai effectué en Colombie britannique et dans l'Etat de Washington a fait apparaître une liaison entre la prévalence tumorale et l'activité urbaine.

L'emploi d'organismes accumulateurs marins (moules) pour l'évaluation des concentrations de benzo(a)pyrène dans les ports est envisagé.

L'utilisation conjointe d'épreuves biologiques et chimiques pourrait permettre une surveillance étendue et systématique de l'environnement marin.

POLLUTIONS DES EAUX ET RISQUES CANCEROGENES

Jacqueline AUBERT (*)

Groupe de Recherches INSERM U.40 de Biologie et d'Océanographie Médicale, CERBOM
Parc de la Côte, 1, avenue Jean Lorrain 06300 Nice, France

I - INTRODUCTION :

A l'heure actuelle, où il est généralement admis que 80 % des cancers sont la conséquence de "l'Environnement", par suite de la modification des qualités originelles de ses trois éléments : Terre - Air - Eau, il est du plus grand intérêt de faire le bilan des risques cancérogènes apportés par le milieu Eau (dulçaquicole ou marin), tel qu'il se présente aujourd'hui.

Notre but sera :
1 - Rechercher dans les milieux aqueux, la présence des substances polluantes -d'origine biologique ou chimique- responsables de cancérogénèse, soit directement, soit par l'intermédiaire des chaînes alimentaires, à l'exclusion des radiations ionisantes, celles-ci du fait de leur importance ne seront pas traitées au cours de cet exposé, elles font l'objet d'un autre rapport.
2 - Tenter d'évaluer le potentiel cancérogène actuel des milieux aquatiques pollués vis-à-vis de la Santé Publique.

Mais auparavant, nous voudrions insister sur de nombreux cas d'épizooties à caractère tumoral rencontrés dans la faune aquatique.

Parmi celles-ci, nous évoquerons les observations :
- de Eric R. Brown et coll. (1970) sur la fréquence des tumeurs trouvées sur les poissons pêchés dans des eaux polluées comparativement aux poissons prélevés dans des zones non polluées (augmentation des tumeurs du foie, de l'estomac, de la peau) ;
- de Mawdesley-Thomas (1970) qui a décrit l'apparition des tumeurs envahissantes de la peau chez les poissons vivants dans les sédiments pollués ;
- de D.J. Christensen et C.A. Farley qui rapportent également une épizootie à caractère néoplasique, apparue chez les clams dans la Chesapeake Bay ;
- En France, on a signalé la présence de tumeurs chez les poissons plats pêchés dans l'estuaire de la Seine ;
- Ainsi, s'il n'est pas toujours possible de démontrer de manière certaine la présence de la substance oncogène causale, il est cependant important de signaler que ces épizooties surviennent le plus souvent dans les zones où la contamination est la plus évidente.

(*) Maître de Recherche à l'INSERM

II - SUBSTANCES CANCEROGENES DE NATURE CHIMIQUE :

Avant de faire un bilan (forcément incomplet) des substances chimiques reco nues actuellement comme susceptibles de provoquer l'apparition et le développement d' cancer, nous résumerons rapidement les différents modes d'actions possibles de la gér des cancers par voie chimique tel que l'a mis en évidence l'étude expérimentale en ut lisant l'application des agents supposés cancérogènes sur des systèmes biologiques sensibles : animaux de laboratoire et cultures de cellules vivantes.

A - MODE D'ACTION :

Diverses théories ont été émises pour expliquer la conversion d'une cellule normale en cellule cancéreuse.

1 - *Théorie de la cancérisation en deux stades* :

Deux groupes de substances seraient nécessaires pour induire une tumeur cutanée :
- les substances initiatrices ou cancérogènes vraies (elles provoquent une transformation définitive des cellules, sans déterminer l'apparition de tumeurs ;
- les substances promotrices : elles révèlent l'action de l'initiatrice et achèvent le processus d'induction néoplasique.

2 - *Théorie de la délétion chimique* (Muller et Muller, 1953)(Sorof et coll. 1975)(Heidelberger, 1962).

Les cellules tumorales ne possèdent plus certaines protéines qui, dans la cellule normale, se lient aux cancérogènes : la liaison chimique de la substance cancérogène et des protéines tissulaires conduit, après quelques mitoses cellulaires, à l'apparition de cellules dépourvues des protéines, si ces protéines sont des enzymes qui interviennent dans le contrôle de la croissance, leur disparition donnera lieu à une croissance cellulaire non contrôlée, c'est-à-dire un cancer.

3 - *Enfin, la théorie de la mutation somatique* fait intervenir comme facter déterminant, l'absence ou la disparition d'un élément cytoplasmique. Elle est actuellement controversée.

N'étant pas spécialisés dans ces questions, nous nous garderons bien d'accorder une préférence à l'une ou l'autre de ces théories, qui d'ailleurs peuvent tou s'accorder avec les phénomènes cancérogénétiques que l'on peut trouver dans les milie aquatiques.

B - BILAN DES SUBSTANCES CHIMIQUES CONNUES ACTUELLEMENT OU SUSPECTES DE CANCEROGENESE PRESENTES DANS LES EAUX DOUCES ET MARINES :

Les unes ont des effets carcinogènes forts qui induisent chez les animaux de laboratoire des cancers à des doses aussi basses que celles qui peuvent être trou spontanément dans certains milieux aqueux, telles que des nitrosamines ou certaines mycotoxines, les autres ont des effets carcinogènes faibles, tels que des polluants atmosphériques, certains pesticides ou additifs alimentaires, qui échappent à la dét tion par les tests biologiques conventionnels, mais qui, par leur conjonction, peuve créer des conditions carcinogéniques.

Suivant la classification de Epstein (1), on peut définir les polluants chimiques en quatre catégories :
- Nitrosamines
- Mycotoxines
- Complexes organiques et inorganiques
- Les synthétiques chimiques.

LES NITROSAMINES :

Elles semblent actuellement être parmi les substances cancérogènes les plus répandues et les plus redoutables pour deux raisons :
- d'une part, à cause de leur large distribution dans l'environnement naturel, aggravée par leur possibilité de formation à partir de précurseurs présents dans tous les systèmes aquatiques (lacs, rivières, eaux résiduaires) ;
- et d'autre part, du fait de leur pouvoir cancérogène particulièrement puissant.

En effet, leur synthèse éventuelle se fait volontiers dans les eaux douces ou marines, plus facilement encore dans les eaux polluées, à partir soit des nitrates ou des nitrites où leur taux est élevé (régions agricoles, eaux résiduaires) et des amines secondaires présentes dans le phytoplancton et les plantes aquatiques ou les pesticides. De plus, leur persistance et leur stabilité dans les eaux facilitent leur entrée dans le cycle vital et favorisent leur extension (2).

Selon certains auteurs (Lijinsky et Epstein, 1970)(3), les Diméthylnitrosamines peuvent être formés dans l'estomac à partir d'agents nitrosants. Alan et coll. ont d'ailleurs démontré, au cours d'infections du tractus digestif et urinaire, la formation de nitrosamines, à partir d'amines secondaires et de nitrates en présence de certaines bactéries.

Des taux significatifs de Diméthylnitrosamines ont été trouvés dans des poissons salés et fumés (0,6 à 9 ppm dans des poissons provenant de Canton ; 4 à 26 ppb dans des poissons de sable et des saumons traités aux nitrates).

En pathologie expérimentale :

Leur pouvoir de cancérogénécité est particulièrement vaste et unanimement reconnu. Il a fait l'objet de recherches approfondies chez divers animaux de laboratoire. Ainsi, de nombreuses nitrosamines ont été testées, dont la plupart se sont révélées cancérogènes (4). Ainsi, à titre d'exemple, nous citerons la nitrosopipéradine qui provoque des tumeurs de l'oesophage ou d'autres organes chez le rat quelle que soit la voie d'administration (orale, intra-veineuse ou sous-cutanée), alors que la nitrosométhylamine et la nitrosooctanéthylamine induisent à la fois des carcinomes squameux du poumon et de l'oesophage quand elles sont administrées dans l'eau de boisson (6).

Parmi les nombreux tests effectués sur différents animaux de laboratoire -souris, rats, hamsters, cobayes, lapins, truites, chiens- où les Diméthylnitrosamines se sont montrés carcinogènes, nous citerons :
- les études à long terme, effectuées par Clapp et coll. (1970), où une dose de 0,4 mg/kg et par jour dans l'eau de boisson provoquait, au bout de 225 jours, l'apparition de tumeurs ;
- d'autre part, selon Teraccini et coll., 5 ppm de Diméthylnitrosamines, donnés dans un régime journalier pendant un an à des rats, sont considérés comme carcinogènes ;
- chez le hamster, Tomatis a observé des carcinomes hépatiques après administration de 0,0025 % de D.M.N. dans l'eau de boisson pendant onze semaines ;
- chez la truite, des doses variant de 200 ppm à 12 000 ppm donnés pendant six mois, ont induit des adénocarcinomes hépatiques (Ashley et Halva, 1968) ;

- enfin, nous signalerons également une expérience réalisée par Petit et
Rudaly, où des doses journalières de Diméthylnitrosamines de l'ordre de 100 à 200 γ
administrées par voie intra-péritonéale à des souris ont induit en six mois simulta
nément un néoplasme rénal et un néoplasme pulmonaire (5).

En épidémiologie humaine, elles ont été impliquées dans certaines données
statistiques.

Ainsi, l'incidence élevée de cancers gastriques au Japon, en Islande, au
a été associée avec un régime à forte consommation de poissons (les amines secondai
du poisson se combinant avec les nitrites utilisés pour leur conservation)(1).

Egalement, l'incidence élevée des cancers de l'oesophage chez les Zambies
qui boivent de l'alcool de Kachasu peut être en rapport avec une contamination élev
en nitrosamines (1). Dans une étude réalisée en Ille-et-Vilaine, en France, par le
C.I.R.C. et l'I.N.S.E.R.M., il a été mis en évidence une certaine corrélation entre
la fréquence des cancers de l'oesophage et la consommation de boissons alcoolisées
associées au tabac.

Néanmoins, des études plus poussées dans l'ordre statistique sont nécessa
pour préciser les effets à long terme, chez l'homme, des nitrosamines.

LES MYCOTOXINES :

Dans le groupe des mycotoxines, les aflatoxines se rencontrent dans les
milieux naturels et se développent facilement sur de nombreux produits alimentaires

Ainsi, l'aflatoxine B_1, mycotoxine sécrétée par *Aspergillus flavus* est le
plus puissant des hépatocarcinogènes connus à ce jour. Actif par voie orale, son r
cancérogène a été mis en évidence la première fois dans les truites d'élevage en 1
en Californie (9). Par la suite, les expériences sur les animaux de laboratoire (r
ont permis de préciser qu'une quantité égale à un pour dix millions d'aflatoxine B
donnée pendant six semaines, était suffisante pour induire un cancer du foie (6).

Par ailleurs, l'injection de 5 µg par jour d'aflatoxine peut provoquer,
bout d'un an des tumeurs chez les rats soumis à un tel régime.

En épidémiologie, on a suggéré que les incidences élevées de cancers du
observées chez les Bantous pourraient être la conséquence d'une nourriture contami
par de l'aflatoxine (1).

HYDROCARBURES ET DERIVES :

Leur répartition dans l'hydrosphère est extrêmement vaste puisqu'on éval
de 5 à 10^6 tonnes par an la quantité de pétrole brut et d'hydrocarbures déversés d
les Océans (Bluner, 1970)(7). D'autre part, leur possibilité d'accumulation dans l
organismes marins, par l'intermédiaire de la chaîne nutritionnelle marine, peut at
teindre les organismes récoltés pour l'alimentation humaine, et par cette voie rep
sente un danger potentiel pour l'homme.

Rappelons que certaines de ses fractions et de ses dérivés, tels que les
hydrocarbures aromatiques et les produits de raffinage obtenus à haute température
(entre 300 et 500°) se présentent comme des substances cancérogènes éventuelles. O
sait, en effet, que les hydrocarbures aromatiques (3-4 benzopyrène et 1-2-5-6 dibe
zanthracene, ou encore 7-12 méthylbenzanthracène, 20 méthylcalanthène) sont parmi
substances expérimentales cancérogènes les plus actives (8). En milieu marin, ceux
ont été trouvés à des concentrations importantes dans les sédiments (420 à 5 000 µ

de poids sec) dans les zones très polluées (Lalou et coll., 1962)(9)(10).

Or, selon les travaux de J.W. Anderson et J.M. Neff, les mono, di et triméthylnaphtene ainsi que les benzopyrenes sont les plus facilement concentrés dans les organismes marins (11).

A titre indicatif, Callaghan a trouvé des taux supérieurs à 1 mg/kg d'hyrocarbures aromatiques dans les huîtres récoltées en zones polluées ; Mallet et coll. trouvent des teneurs supérieurs à 9 µg/100 g de matières sèches dans les clams, ou même 100 µg/100 g dans les zones exceptionnellement polluées (12).

Bien que jusqu'à présent nous n'ayions encore que des informations épidémiologiques fragmentaires sur les effets cancérogènes directs vis-à-vis de la Santé humaine, les données expérimentales sont suffisamment nombreuses et éloquentes pour que des études plus approfondies soient réalisées dans ce domaine.

LES PESTICIDES :

Créés par l'homme, ces poisons sont répandus dans l'environnement aquatique où on les retrouve dans tous les plans d'eaux, les rivières, les estuaires et les étendues océaniques.

D'autre part, du fait de leur persistance et de leur bioaccumulation, au niveau du tissu adipeux du foie et des muscles, on les retrouve dans tous les organismes aquatiques, qu'ils soient dulçaquicoles ou marins, à des taux extrêmement importants de 10 à 3 600 ppb)(14).

Or, leur effet cancérogène a été bien démontré (en particulier pour le DDT) chez la souris et le rat, et il semble limité au parenchyme hépatique (15).

Bien que, jusqu'à présent, aucun effet évident de cancérogénécité n'ait été apporté chez l'homme, ils restent suspects.

LES POLYCHLOROBIPHENYLS (PCB) :

Derniers arrivés des produits chimiques synthétiques, ils entrent dans la fabrication de nombre de produits usuels : fluides diélectriques, échangeurs de chaleur, fluides hydrauliques, peintures et vernis, encres, adhésifs et plastiques ; leur production atteint maintenant près de 50 tonnes par an pour les seuls pays de l'O.C.D.E. (16)

Leur concentration dans les eaux est assez variable, mais cependant dans l'eau de mer, en Atlantique Nord, on en trouve en moyenne 35 ng/kg. Ils semblent s'accumuler volontiers dans les organismes aquatiques (10 à 20 ng/kg chez la carpe de certains viviers au Japon (16).

Si aucune donnée sur leurs effets chez l'homme, du point de vue de leur cancérogénécité n'a été relatée à notre connaissance, il n'en demeure pas moins que les expériences sur les animaux de laboratoire (souris et rats femelles) concordent toutes pour montrer la formation de tumeurs hépatiques (hépatomes), après injection orale de doses importantes de Kanaclor 500 (Nagasachi et coll., Towini, 1973, ITO). Une expérience sur le singe Rhésus Mâle recevant par voie orale pendant trois mois des doses égales à 300 mg/kg d'Arochlor 1248 a confirmé ce pouvoir cancérogène (17).

SUBSTANCES MINERALES ET METAUX LOURDS :

Présents spontanément dans l'hydrosphère, leur accroissement dans les milieux naturels et les chaînes nutritionnelles ne cesse de se développer du fait de leur utilisation tant dans l'industrie que dans l'agriculture.

Un grand nombre d'entre eux a été reconnu doué de cancérogénécité soit chez l'homme, soit chez les animaux de laboratoire.

Parmi ceux-ci, deux méritent d'être étudiés ici : l'Arsenic et le Cadmium.

a) L'Arsenic :

L'arsenic se rencontre normalement dans les eaux douces à très faibles doses (0,0004 mg/kg), ainsi que dans les eaux océaniques à des taux variant de 0,006 mg/kg à 0,03 mg/kg. Ces chiffres sont largement dépassés dans les zones réceptrices de rejets industriels ou dans les estuaires (atteignant fréquemment 1 mg/kg)(18).

De plus, il importe de signaler que les plantes et les animaux accumulent volontiers l'Arsenic, en particulier dans les milieux salés, où crustacés et coquillages peuvent en contenir de 1 mg à 10 mg/kg.

Son pouvoir cancérogène repose sur diverses observations consécutives soit à l'exposition de préparation à base d'Arsenic, soit à l'injection d'eau riche en Arsenic.

Ainsi, Neubauer (en 1947) signale 143 cas de cancers cutanés dus à l'injection de substances arsénicales de pharmacie.

D'autre part, des études épidémiologiques rapportées par Braun (1958) mettent en évidence une augmentation de l'incidence des cancers chez les utilisateurs d'insecticides à base d'Arsenic (cancers de la peau et des bronches en particulier).

Plus significatifs encore sont les faits pathologiques en rapport avec l'eau de boisson riche en Arsenic.

. En Argentine (dans la Province de Cordoba), où l'eau contenait des quantités d'Arsenic (largement supérieures à 0,5 mg/l), il existait une "Arsénisation Chronique Régionale Endémique" avec hyperkératose palmaire et plantaire en 2-3 ans. Or, le pourcentage des cancers cutanés d'origine arsénicale était nettement plus élevé dans les zones où l'on avait pu constater des cas "d'Arsénisation Chronique Régionale Endémique". De plus, le pourcentage de décès (de 1949 à 1959) par cancer, s'est trouvé être plus élevé dans la zone "arséniée" que dans les autres zones (24).

. A Taïwan, en Chine, on a remarqué que le taux de prévalence du cancer de la peau augmentait selon la teneur en Arsenic de l'eau du puits, en d'autres termes plus l'eau contenait de l'Arsenic, et plus le nombre de patients atteints de cancers cutanés était élevé.

Par contre, au cours des expériences sur les animaux de laboratoire (souris ou rats) soumis à un régime à base d'Arsenic, il n'a pu être mis en évidence d'apparition de cancer.

b) Le Cadmium :

Son existence dans la biosphère et en particulier dans les eaux est très variable.

Cependant, issu des activités industrielles métallurgiques (recyclage des ferrailles, récupération du cuivre), ou pétrochimiques (incinération des matières plastiques), sa principale origine dans l'eau provient de la galvanoplastie, si bien que l'eau contribue pour 5 à 10 % à l'apport quotidien à l'organisme. Mais, du fait des facteurs d'accumulation de ce métal dans certaines chaînes alimentaires (en particulier les mollusques et les crustacés, ou encore le riz), son incidence sur la

Santé Publique doit être soigneusement surveillée ; selon Schroeder (1967), les huîtres parviendraient à concentrer de 10 000 à 100 000 fois le métal à partir de l'eau de mer (18).

A l'embouchure du Rhône, nous avons trouvé au C.E.R.B.O.M., des teneurs en Cadmium très élevées dans les sédiments (92 ppm) ainsi que dans la biomasse (4,5).

Les effets cancérogènes sont considérés par certains auteurs extrêmement importants. Ainsi, Gunn et coll. (1967) estiment que le Cadmium est l'un des plus puissants métaux cancérogènes. S'appuyant sur des expériences de laboratoire sur l'animal, ils ont démontré l'apparition de fibrosarcomes dans tous les tissus du mésoderme des rats, après administration parentérale de faibles doses de Cadmium, ainsi que le développement de tumeurs des cellules interstitielles du testicule.

Malgré l'absence d'enquête épidémiologique, chez l'homme, il a été suggéré que la prévalence du cancer chez les travailleurs chroniquement exposés au Cadmium semblait plus élevée : en particulier, fréquence des carcinomes de la prostate, des tumeurs malignes du tractus gastro-intestinal, des voies respiratoires et des voies urinaires (Potts, 1965 ; Kipling, 1967). Par ailleurs, Morgan trouve des concentrations anormales de Cadmium dans le sang, le foie et les reins chez des patients porteurs de cylindromes du poumon.

Quoique certains de ces faits soient contestés, il n'en demeure pas moins que la confrontation des expériences sur l'animal et des constatations pathologiques chez l'homme doivent retenir toute notre attention.

D'autres métaux, peut-être moins répandus dans notre environnement -le Sélénium, le béryllium, le cobalt, le plomb- ont également été suspectés de possibilité cancérogène chez l'homme. Le nickel et le chrome ont fait également l'objet d'expériences positives sur les animaux ; en particulier, certains composés de chrome (acide chromique, sel de chromate, etc...) semblent être à l'origine de formations cancérogènes et d'ulcères de la peau sur des poissons marins prélevés près des aires de rejets industriels aux U.S.A. (19).

CONCLUSION :

Si les données apportées, soit par l'épidémiologie, soit par les épizooties de néoplasmes dans les organismes aquatiques, sont encore trop ténues pour affirmer la cancérogénécité de certains agents chimiques ou biologiques contenus dans les eaux et reconnus comme cancérogènes, elles sont suffisantes pour que le problème soit posé.

Il est donc urgent d'entreprendre des recherches plus approfondies sur ce sujet, étant donné l'ubiquité du phénomène hydrique dans la biosphère.

Ces recherches devront continuer de s'appuyer évidemment sur les méthodes classiques tant expérimentales qu'épidémiologiques, mais aussi s'ouvrir sur des techniques nouvelles qui tiennent compte de la multiplicité des facteurs écologiques. En particulier, nous citerons :
- les phénomènes de concentration susceptibles d'apparaître à travers les chaînes biologiques ;

- les phénomènes de biodégradation qui peuvent être à l'origine de modification de la structure chimique de certaines substances, au départ non reconnues comme cancérogènes, mais susceptibles de le devenir ;
- l'association plus ou moins réactionnelle de diverses substances polluantes susceptibles de modifier la composition initiale du milieu ;
- et enfin, des phénomènes d'ordre enzymatique liés à la vie aquatique elle-même susceptible d'induire, sous l'effet de déviations métaboliques liées à certains polluants, des troubles des constituants cellulaires et être, ainsi, à l'origine d'une évolution anarchique des cellules.

REFERENCES

(1) Epstein S. : Environmental Pathology. American Journal of Pathology, 1972, 66, 2, 352-370.

(2) Tate III R.L. and Alexander M. : Stability of Nitrosamins in Sample of Lake water Soil and Sewage. Journal of the National Cancer Institute, 1973, 54, 2, 327-330.

(3) Lvinsky W. and Epstein S. : Nitrosamins as Environmental Carcinogenes. Nature, 197 225, 3, 21-23.

(4) I.A.R.C. : Monographs on the Evaluation of Carcinogenic Risk of Chemicals to man. I.A.R.C., 1972, Lyon, 1, 95-100.

(5) Petit L. : Communication personnelle.

(6) Moreau C. : Moisissures toxiques dans l'alimentation. Messon Edit., 128-139.

(7) Blumer M. : Oil Contamination and the Living Resources of the Sea. Marine Pollution and Sea Life, 1972, Publish by F.A.O., 476-481.

(8) Suess M.J. : Polynuclear Aromatic Hydrocarbon Pollution of the Marine Environment. Marine Pollution and Sea Life, 1972, Publish by F.A.O., 568-570.

(9) Mallet L. : Investigation for 3-4 BP type polycyclic hydrocarbons in the fauna of marine environments (the Channel, Atlantic and Mediterranean). Compte-rendu Acad. Sc. Paris, 1961, 2531, 168-170.

(10) Lalou C.; Mallet L. and Heros M. : Distribution en profondeur de 3-4 BP dans un échantillon de la Baie de Villefranche-sur-Mer. Compte-rendu Acad. Sc. Paris, 1961 255, 1, 145-147.

(11) Anderson J.W. and Neff J.M. : Accumulation and release of petroleum hydrocarbons by edible marine animals. International Symposium Environment and Health CEE-EPA-W 1974, Paris.

(12) Callaghan J. : International Aspects of Oil Pollution. Trans. N. Am. Wild Nat. Res 1961, Conf., 26, 328-342.

(13) Halstead B.W. : Toxicity of Marine Organisms caused by Polluants. Marine Pollution and Sea Life, 1972, Publish by F.A.O., 584-594.

(14) Modin J.C. : Residues in fish, wild life and estuaries. Pestici. Monitg. J.3 (I), 1969, 1-7.

(15) Tomatis L. : The Carcinogenic risk for Man of Environment Chemicals. Internationa Symposium Environment and Health CEE-EPA-WHO, 1974, Paris, 157.

6) The Hazards to health and Ecological effects of persistent substances in the environment. Polychlorinated biphenyls report on a Working group. Brussels 3-7 Dec. 1973, EURO, 3109, 2.

7) Allen J.R. and Norback K.N. : Polychlorinated Biphenyls and triphenyls induced gastric Mucosal Hyperplasic in Primates. Science, 1973, 498-499.

8) O.M.S. : Risques pour la santé liés à la présence d'Arsenic, de Cadmium, de Manganèse, de Mercure et de Plomb dans l'eau. EURO, 1975, Copenhague, 3109 W, 1, 17-34.

9) Dvizhkov P.P. : Glastomogenic effects of industrial metals and their compounds. Ark. Pathl., 1967, 29, 3-11.

SUMMARY

The main subjects emphasized in this contribution are the following ones :

1 - To investigate the presence of cancerogenic compounds derived directly r by transformations within trophic chains from domestic and industrial effluents.

2 - To evaluate potential cancerogenic health hazards due to polluted quatic environment.

The following substances might be regarded as to be cancerogenic : nitroamins, mycotoxins, complex organic and inorganic compounds and synthetic organicals.

Based on epidemiological and epizoonic evidences, as well as human and xperimental pathology, the relevant importance of particular group of compounds s evaluated.

Presented evaluation is indicating that the water-borne cancerogenesis ight be of an important significance. Therefore, the needs for essential increase n relevant research, implemented by advanced methods, which can cope with complexity f ecological processed, are stressed and further efforts recommended.

III

OCCUPATIONAL POLLUTION

POLLUTION D'ORIGINE PROFESSIONNELLE

Chairman / *Président :* Dr. E. MASTROMATTEO

ASBESTOS CANCERS AS AN EXAMPLE
OF THE PROBLEM OF COMPARATIVE RISKS

J.C. GILSON

MRC Pneumoconiosis Unit, Llandough Hospital
Penarth, Glamorgan, CF6 1XW, Wales, U.K.

Occupational exposures to several dusts have caused an excess of respiratory tract cancers - for example, arsenic, chromates, fluorspar, haematite, nickel, uranium. Here the agent is thought to be the element or its salt, or ionising radiation, or the two operating together. Professor Maltoni and Dr. Mole will be discussing examples of these.

I will limit my remarks to asbestos and some other mineral fibres for two reasons. First, present evidence suggests that the physical properties of the fibres rather than their chemical composition are especially important in the cancers they induce; and second, because "asbestos" is a commercial term of a small sub-group of fibrous silicates, but there are many other fibrous silicates and other fibrous minerals* widely distributed over the surface of the earth, so that the dust from these - even though in very small amounts - has been in the general air since time immemorial and may be seen in airborne dust and in lung residues. (1).

These simple observations indicate that even though there may not be a threshold level below which no effect is produced in a strict biological sense, there is a practical level below which no serious disease is produced.

Our knowledge of the carcinogenic effects of asbestos is almost entirely derived from occupational and para-occupational exposures in the past. The question is, therefore, do we yet know enough about the effect of varying intensity and type of exposure in industry to specify what may be acceptable conditions within industry in the future, as well as to the general public?

Types of Cancers

Table I shows the types of asbestos and the cancers with which they are associated. Nearly all of it is chrysotile (about 4 million tons/year), but it should not be assumed, as used to be done, that the biological effects of all types are the same. The cancers widely accepted as due to asbestos are bronchial and mesotheliomas of the pleura and peritoneum. But an excess of gastro-

---

* These include calcium silicates (woolastomite, cement dust); sepiolite (hydrous magnesium silicate); hornblende (amphibole mineral variety); diatomaceous earth (amorphous silica); fibrous clay minerals (kaolinites, bentonites); and naturally occurring fibrous minerals, such as pyroxene minerals; amphibole minerals (other than commercial varieties); serpentine minerals (antigorite); and oxide minerals (brucite, magnesium hydroxide).

intestinal tumours has been reported in several large cohort mortality studies of asbestos workers (2,3,4,), though the excess has been much smaller than for the lung cancers. Larynx, pancreas, and lymphomas have also been suggested. There is still uncertainty about the causal relationship of these other cancers with asbestos.

Cancer Sites	Types of Asbestos
Bronchial	Chrysotile (95%)
Mesothelial	Amphiboles
Pleural	Amosite
Peritoneal	Crocidolite
	Anthophyllite
G.I. Larynx Others } ?	

Table I: Asbestos and Cancers

A pattern of recent research

The last 15 years has seen the results of many epidemiological studies into the incidence of asbestos-related cancers (5), and also parallel animal experimental studies to explore mechanisms as to how such remarkably stable and inert minerals are carcinogenic. The pattern of this work illustrates one of the themes Dr. Higginson has been developing at the IARC; a combined epidemiological and experimental programme of research in which the work in each field stimulates and orientates that in the other. The experimental work has been mainly aimed at discovering physical or chemical factors responsible for the carcinogenesis, and it has not been aimed at defining a TLV for man by extrapolation from studies in animals. These approaches and their inter-relations are shown in Fig. 1.

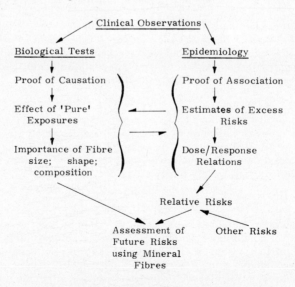

Fig. 1: Research into Asbestos Cancers

Biological tests

The cancers proved in man to be caused by asbestos are also induced in rats and other small animals. For example, intra-pleural injection of all types of asbestos causes a high incidence of mesotheliomas (6), and although by this route part of the defence mechanisms of the lungs are effectively by-passed, the technique is very useful to study the effect of particle shape and composition in the induction of a tumour known to be caused by asbestos in man. Present evidence suggests that to produce mesotheliomas the fibres may have to be $\geq 10 \mu m$ in length and less than about $1 \mu m$ diameter. Larger fibres $> 3 \mu m$ diameter and rounded particles rarely or never cause mesotheliomas by this route. When it is possible to produce narrow sized ranges of fibres for both diameter and length, it should be possible to specify fairly precisely the critical dimensions for the induction of mesotheliomas. Much effort is being put into preparing such test samples of fibres at present because the results of such experiments are likely to influence the way new man-made mineral fibres are produced in the future and also the way fibrous dusts are sampled in air and water.

When the fibres are inhaled it is the diameter rather than the length which controls entry to the periphery of the lung; also curly fibres, such as chrysotile, are more readily arrested in the upper respiratory tract by impaction than the straight fibres of amosite or crocidolite. This may be one of the reasons why the incidence of mesotheliomas in those exposed only to chrysotile is much lower than those exposed to crocidolite or amosite (7).

The composition and chemical structure does not seem very important because all types of asbestos and some very fine glass fibres and ceramic fibres have all produced mesotheliomas by intra-pleural inoculation. The animal experimental results have given no support to hypotheses that trace elements or adsorbed hydrocarbons might be important in the production of the pleural tumours (8). But an interesting recent observation (9) is that fully magnesium leached chrysotile, even though fibrous, produces very few mesotheliomas compared to the untreated mineral. An important result of the experimental work has been a better understanding of the likely mesothelioma risk of man-made mineral, and other fibres. In general the fibres are too large in diameter and the airborne concentrations too low for many to reach deep into the lung. In addition, man-made fibres do not break down within the lung into fine fibrils as occurs with asbestos (10, 11). Thus present evidence indicates the chances of present man-made mineral fibres of a critical size reaching the periphery of the lung in significant amounts must be far less than in the case of the natural fibrous minerals.

Inhalation of asbestos produces in rats a small incidence of bronchial cancers and even fewer mesotheliomas (12). So far nothing is known about the importance of fibre size in the production of the bronchial tumours.

Feeding rats with asbestos, as well as detailed autopsies of rats inhaling the mineral, (during which they will ingest an appreciable amount when cleaning their fur), has shown no excess of gastro-intestinal tumours (13, 14).

Epidemiology

Proof of association between asbestos and cancers of a particular site is based in part on case/control studies of pathological material, and in part on the causes of deaths in cohorts of asbestos-exposed workers. The pathological studies give no measure of the magnitude of the risk and are, therefore, of little predictive value. The simple cohort studies do give a measure of the excess risk for each type of cancer, but due to the long latent period for cancer induction, often 20 or more years, they refer to conditions many years in the past. Such studies, many of which have received wide publicity in the general press, are also of little use for prediction of risks in the future or of extrapolation to the general population.

Dose response studies

For such predictions cohort studies, in which the workers can be sub-divided on the basis of duration and intensity of past exposure, are required. Ideally we need quantitative information on past dustiness, but useful information has been obtained by the detailed historical study of the factory or mill, making the maximum use of the knowledge of long-term employees, and relating these to such dust measurements as there have been. In this approach the men's jobs can often be grouped into those with markedly different dust exposures, so that mortality experience can be related to dose. The advantage of this approach is that groups of employees are identifiable, who almost certainly had exposures several orders of magnitude greater than the general population, and yet show no excess cancer risk or a small one compared to those who have been most heavily exposed.

Men born 1891-1920   Deaths/1000 to 1969

	Dust Exposure (particle/years)					
	<10	10-	100-	200-	400-	800+
All Cancers	58	61	54	41	67	79
*Lung "	10	13	13	16	21	32
Abdominal "	18	14	19	12	26	29
No. of Men	2810	2329	1124	1007	837	585

*Including Mesotheliomas (5)

Table II: Chrysotile Mining and Milling, Quebec

Table II is an example from the chrysotile mining and milling industry in Quebec where about half the world's asbestos is produced. Only in the two highest exposure groups is there an increase of cancer risks relative to the lowest exposed group. For the whole group of workers there was no lung cancer excess over the general population. Thus the least exposed group were not at a significant excess risk despite being exposed to far more asbestos dust than members of the general population. Similar patterns have been reported in the asbestos cement industry, and for lung cancer and mesotheliomas in factories manufacturing asbestos products (15, 16, 17).

Differences within the Industry

Table III summarizes differences in proportional mortality within the industry for lung cancers and mesotheliomas (18). As is often the case in epidemiology, the interpretation of findings is not simple; in this case because the type of asbestos fibre used and the type of industrial process are to some extent confounded. Many manufacturing processes use more than one type of asbestos, often all three of the major types - chrysotile, amosite, and crocidolite. In mining and milling exposures to one type do occur, but unfortunately for the epidemiologist crocidolite and amosite are mined only in South Africa where the opportunities for quantitative epidemiology are slender because of the paucity of records.

	Asbestos	Lung Cancer	Mesoth:	No. of Surveys	Total Men
*	Insulation	18 - 26	5 - 9	6	26,500
*	Factories	8 - 21	1 - 7	5	10,800
+	Mining and Milling	2 - 10	0 - 0.2	3	13,700
	Gen: Popl:				
	E and W	9			
	U.S.A.	5	0.0001		
	Canada	5			

* Mixed fibre exposures

+ Chrysotile (2): Anthophyllite (1):

Table III: Proportion (%) of All Deaths due to Lung Cancer and Mesotheliomas in Cohorts of Asbestos Workers and the General Population

The percentage of lung cancers and mesotheliomas is highest in the insulators; less in the factories making asbestos products; and least in mining and milling, but the mining and milling covers only chrysotile and anthophyllite. In the two large chrysotile-exposed cohorts, one in Quebec and the other in Italy, no general excess of lung cancers was shown, and no or a very few mesotheliomas.

If we add to these quantitative cohort studies the results of 15 years' observations in Southern Africa, where chrysotile, amosite, and crocidolite are mined, it is fairly certain that mining and milling crocidolite is - by a factor of 100 or more - more likely to produce mesotheliomas than amosite or chrysotile. Studies in

Finland suggest anthophyllite almost never causes mesotheliomas (19).

Cofactors

Cigarette smoking is most important in the production of bronchial cancer in asbestos workers. The effects of asbestos and cigarette smoke are multiplicative, not simply additive, Table IV shows this for both sexes (20,21,). Asbestos alone is apparently a weak bronchial carcinogen but it has not been possible to establish this very firmly, because there are few non-smoking asbestos workers with long exposures. It seems likely that stopping cigarette smoking is likely to have a much bigger effect on the incidence of lung cancer in asbestos workers than small changes in exposure to asbestos dust. No cofactors affecting the incidence of mesothelioma have been firmly identified.

	Observed	Expected		
		Non Occup:	Non Occup: and Occup:	
			+	×
Insulation				
Cigarettes	24	2·98	18·9	22·6
Non-Smokers	0	0·05	2·8	0·4
Factories				
Male				
Cigarettes	25·5	9·9	25·1	26·4
Non-Smokers	0	0·0	0·9	0·1
Female				
Cigarettes	15·5	1·4	12·5	15·3
Non-Smokers	1·7	0·2	4·7	1·9

Table IV: Lung Cancer Deaths and Smoking fit with Additive (+) and Multiplicative (×) Hypotheses

Relative risks

These asbestos cancers should be seen in perspective in relation to other occupational cancers and occupational accidents (Table V). The information to d this precisely is not available, but the general pattern is well summarized by Pochin (22,23). All the figures are deaths per million per year. Cancers vary from about 700 for the nasal cancers in woodworkers to 24,000 for the beta-naphthalene manufacturers, with asbestos lung cancers at about 3,000. The occupational fatalities range from 3 in clothing manufacturers up to 11,000 for the professional divers (24). For both occupational cancers and accidents there is a big range of risk for different types of job. In general the cancer risks are the bigger, but affect a selected group of workers.

	Cancers		Accidents	
Wood Workers	Nasal	700	Clothing Mf:	3
Asb: Industry	Lung	2,000 M 4,000 F	Bricks and Cement	80
			Ship Building	160
Rubber Workers	Bladder	7,000	Coal face	600
Nickel Refining (up to 1925)	Lung	15,000	Company Directors	1,800
			Deep sea fishing	3,000
β Naph: Mf:	Bladder	24,000	Prof: Divers	11,000

Table V: Estimated Occupational Mortality/M/year

Finally, compare these occupational risks with risks run by all from accidents, our personal habits, and our age (Table VI). Traffic accidents are about one quarter that of working on the coal face, but cigarette smoking at 20/day is about double that of heavy asbestos exposure. This is, of course, due to cigarette smoking affecting several diseases. The last column shows how rapidly age creeps up on us and perhaps indicates that those who research into occupational hazards should be in the 30's to see things in perspective!

General Accidents; Personal Habits; Your Age (Male)

Accidents		Your Age	
Traffic	150	12	350
Home	130	30	1,000
Suicide	80	42	3,000
All	460	53	10,000
Cigarettes		63	30,000
20/day	5,000	77	100,000

Table VI: Estimated Mortality/M/year in U.K.

SUMMARY

Major differences in excess cancer risks have occurred in the asbestos industry in the past; part of this difference is probably related to dustiness, part to the type of fibre used. In the case of mesotheliomas there is evidence of a major effect of the fibre type in the order of risk, crocidolite > amosite > chrysotile > anthophyllite.

Differences of risk within an industry indicate that there is a dose response relation for both bronchial cancers and mesotheliomas. Also that in some instances it has been possible to identify lightly exposed groups in which no excess risks were detectable. As these least exposed groups are likely to have had several orders of magnitude heavier exposure than the general population, the risks to the general public are likely to be negligible.

Bronchial cancers in absolute numbers are the major excess risk and are highly smoking-related. Stopping cigarette smoking is likely to be of paramount importance in reducing the excess cancer risks in asbestos-exposed individuals.

Cancers in other sites which may possibly be related to asbestos exposure need further study even though the magnitude of the excess risk has been small compared to the lung cancers.

In my view it is now possible to use the epidemiological evidence and experimental results to predict with fair confidence the physical and chemical characters and dose of natural and man-made fibres which will not cause a significant hazard in the future.

References

(1)   Pooley F.D.: Personal communication.

(2)   Elmes P.C. and Simpson M.J.C. : Insulation workers in Belfast. Brit. J. industr. Med., 1971, 28, 3, 226-236.

(3)   Selikoff I.J., Hammond E.C., and Seidman H.: Cancer risk of insulation workers in the United States:in "Biological Effects of Asbestos", IARC Sci. Pub. No. 8, 1973, pp. 209-216.

(4)   McDonald J.C.: Cancer in chrysotile mines and mills. Ibid. pp. 189-194.

(5)   Bogovski P.: Ibid. pp. 189-226.

(6)   Wagner J.C. and Berry G.: Mesotheliomas in rats following inoculation with asbestos. Brit. J. Cancer. 1969, 23,3, 567-581.

(7)   Timbrell V.: Physical factors as etiological mechanisms in "Biological Effects of Asbestos", IARC Sci. Pub. No. 8, 1973, pp. 295-303.

(8)   Wagner J.C., Berry G., and Timbrell V.: Mesotheliomata in rats after inoculation with asbestos and other minerals. Brit. J. Cancer, 1973, 28, 2, 173-185.

(9)   Wagner J.C. : Personal communication.

(10) Corn M. and Sansone E. B.: Determination of total suspended particulate matter and airborne fiber concentrations at three fibrous glass manufacturing facilities. Environm. Res., 1974, 8, 1, 37-52.

(11) Assuncao J. and Corn M.: The effects of milling on diameters and lengths of fibrous glass and chrysotile asbestos fibers. Paper presented at the Annual Meeting of the American Industrial Hygiene Association, Minneapolis, Minnesota, 1975. (To be published).

(12) Wagner J.C., Berry G., Skidmore J.W., and Timbrell V.: The effects of the inhalation of asbestos in rats. Brit. J. Cancer, 1974, 29, 3, 252-269.

(13) Gross P., Harley A.R., Swinburne L.M., Davis J.M.G., and Grane W.B.: Ingested mineral fibres: Do they penetrate tissue or cause cancer? Arch. environm. Hlth., 1974, 29, 6, 341-347.

(14) Wagner J.C.: Personal communication.

(15) McDonald J.C.: Cancer in chrysotile mines and mills: in "Biological Effects of Asbestos". IARC Sci. Pub. No. 8, 1973, pp. 189-194.

(16) Newhouse M.L.: Cancer among workers in the asbestos textile industry. Ibid, pp. 203-208.

(17) Enterline P.E., de Coufle P., and Henderson V.: Respiratory cancer in relation to occupational exposures among retired asbestos workers. Brit. J. industr. Med., 1973, 30, 2, 162-166.

(18) McDonald A. and McDonald J.C.: Paper to International Congress of Occupational Health, Brighton, 1975. (To be published).

(19) Wagner J.C., Gilson J.C., Berry G., and Timbrell V.: Epidemiology of asbestos cancers. Brit. med. Bull., 1971, 27, 1, 71-76.

(20) Doll R.: Practical steps towards the prevention of bronchial carcinoma. Scot. med. J., 1970, 15, 433-447.

(21) Berry G., Newhouse M.L., and Turok M.: Combined effect of asbestos exposure and smoking on mortality from lung cancer in factory workers. Lancet, 1972, 2, 476-479.

(22) Pochin E.E.: Occupational and other fatality rates. Community Health, 1974, 6, 2-13.

(23) Pochin E.E.: The acceptance of risk. Brit. med. Bull., 1975, 31, 3, 184-190.

(24) Crockford G.W. and Dyer D.: Annual Report of the TUC Centenary Institute of Occupational Health, London, 1973-1974, pp. 19-20.

RESUME

De grandes différences de surfréquence du cancer ont été jusqu'ici observées dans l'industrie de l'amiante ; elles sont probablement dues en partie à l'empoussiérage et en partie au type de fibre utilisé. Dans le cas des mésothéliomes a des indices que le type de fibre joue un rôle majeur, l'ordre d'importance du risque étant le suivant : crocidolite > amosite > chrysotile > anthophyllite.

Les différences de risque au sein d'une industrie font apparaître une relation dose/réponse pour les cancers bronchiques et les mésothéliomes. Dans certains cas, on a pu identifier des groupes peu exposés pour lesquels aucun excédent de risque n'était décelable. Comme ces groupes peu exposés ont sans doute été soumis à une exposition plusieurs fois supérieure à celle de la population générale, les risques sont, pour celle-ci, probablement négligeables.

Les cancers bronchiques représentent, en nombres absolus, le principal excédent de risque et ils sont étroitement liés à l'usage du tabac. La suppression de l'usa de la cigarette contribuera sans doute de manière essentielle à la réduction de l'e dent de risque chez les individus exposés à l'amiante.

Les cancers d'autres localisations susceptibles d'être liés à l'exposition à l'amiante appellent une étude complémentaire, même si l'excédent de risque s'est ré lé faible comparativement au cancer du poumon.

A mon sens, on peut maintenant utiliser les données épidémiologiques et les ré tats expérimentaux pour prévoir de façon assez sûre les caractéristiques physiques et chimiques et les doses de fibres naturelles et artificielles qui n'engendreront pas un danger notable dans l'avenir.

FACTORY POPULATIONS EXPOSED
TO CROCIDOLITE ASBESTOS — A CONTINUING SURVEY

J.S.P. JONES (1), F.D. POOLEY (2) and P.G. SMITH (3)

(1) Department of Pathology, City Hospital, Nottingham NG5 1PB, England

(2) Department of Mineral Exploitation
University College, Cardiff CF2 1TA, Wales

(3) Department of Pathology
General Hospital, Nottingham NG1 6HA, England

An important factor in considering the health risk of people exposed to asbestos dust is the type of fibre that is inhaled (Gilson, 1973). In the majority of manufacturing processes a mixture of different types of asbestos is used, and information derived from populations exposed to a single type of fibre has come mainly from the mining areas. The opportunity to collect data from a factory population exposed to a single type of fibre presented itself in Nottingham (Jones, 1968).

In 1965 a woman was diagnosed as having a pleural mesothelioma. Her occupational history indicated that she had spent all her working life in the lace industry except for a period of four years during the Second World War when she worked in a factory engaged in the manufacture of gas masks. Further enquiries revealed that the filter units of the respirators contained pads consisting of merino wool and 15% blue crocidolite asbestos. The asbestos fibre was of a particularly fine diameter, and the entire consignment used for gas mask production had been derived from a mine in Western Australia.

Gas mask production at the Nottingham factory extended from 1939 - 1945. Detailed records of the workers had been destroyed but it was possible to trace via the wages' books that the total number of employees involved in this work amounted to approximately 1,600. (Wignall). Some of the workers were only engaged on this production for a few weeks or a few months, but others worked throughout the war. Because of the known danger at that time of asbestosis, the factory took precautions to minimise the dissemination of dust, and part of the assembly process was conducted in enclosed chambers with extraction ventilation.

Gas mask production, using the same crocidolite fibre was also known to have taken place in two other centres during the war, in Preston (Owen and Miller) and in Birmingham (Kipling and Waterhouse). The size of these exposed populations are not as yet established, but it is known that the numbers at Birmingham were small.

Results:

(i) Nottingham. So far, 26 deaths from mesothelioma have been identified in people who worked in the Nottingham factory where gas mask production was being undertaken (Table I). The tumours have become apparent between 20 and 35 years from the time of possible asbestos exposure. All but one of the cases have occurred in women (Table II). The mesotheliomas have been confirmed histologically in 25 of the 26 cases.

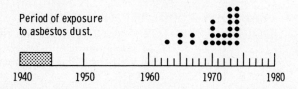

TABLE I. This illustrates clearly the long latent period between exposure to asbestos dust and the appearance of mesothelioma.

GAS MASK WORKERS. MESOTHELIOMAS (1975)

	Pleural	Peritoneal	Total
Male	-	1	1
Female	20	5	25
Total	20	6	26

Table II. A very high proportion of the work force were women.

Lung tissue was available from twelve of the twenty-six cases, and analysis of lung sections by the method of Pooley (1972) has shown that in each case only one type of fibre is present, and that it is of a very fine diameter crocidolite. This peculiarly fine fibre is consistent with the appearance of samples taken from the Western Australian crocidolite mine where the fibre diameter is approximately 800 A^o on average.

The concentration of fibres in the lung tissue has been estimated on a grid area equal to a 3 mm. diameter circle of lung tissue cut 6 μ thick. The results are:

CASE	FIBRES/GRID AVERAGE
1	60
2	400
3	40
4	1,200
5	18,600
6	200
7	860
8	880
9	980
10	1,820
11	280
12	100

It is of interest to note that case No.1 was an inspector who was not involved in the assembly of the cannisters and denied having handled the asbestos pads, or having worked in the same room as the cannister packers. Case No.3 did not work on gas mask production at all, but worked in another part of the same factory.

One woman, who did not have a mesothelioma, died of asbestosis and she is the only case of asbestosis so far detected in this group.

(ii) Preston. A preliminary search of records from 1967 - 1975 in Preston (Owen) shows that autopsies have been performed on at least eleven cases of mesothelioma, and a further seven cases have been diagnosed by biopsy – all 18 having worked on gas mask production during the Second World War. Apart from the mesothelioma cases, only two cases of asbestosis have been identified.

(iii) Birmingham. One case of mesothelioma has so far been identified from the Birmingham group.

The tracing of asbestos-associated diseases and other forms of malignancy in these groups is being continued.

Comment:

At this time interval it is difficult to be certain about the exact concentrations of dust and the periods of exposure of any individual who has subsequently developed pathological conditions associated with asbestos. Certain facts however do clearly emerge from the known data relating to some of the Nottingham cases. The largest numbers of patients developing mesothelioma came from those most heavily exposed to asbestos. However two patients who developed mesothelioma did not work in the gas mask production unit at all, and one of these was engaged in unrelated work in an entirely different part of the factory. Only occasional asbestos fibres were detected in their lungs at autopsy, and it would seem that their dust exposures were relatively trivial. These cases suggest that a dose response relationship is not necessarily applicable.

None of the cases who developed mesothelioma had concurrent asbestosis.

Work is continuing to try and correlate intensity of dust exposure with the development of pathological changes, but at this stage it should be emphasised that mesotheliomas have developed in some workers after only marginal contact with this type of asbestos fibre.

Conclusion:

At this intermediate stage of investigation it is apparent that exposure to the fine-fibre Western Australian crocidolite – even in small concentrations – carries a significant risk to the development of mesothelioma 20 to 40 years plus later. The extent to which the physical properties of the fibre are contributory to tumour formation is still unknown, but these results confirm the dangers of this type of asbestos, and reinforce the need to prohibit its use in the future.

Acknowledgements

It is a pleasure to record our thanks to the many clinicians and pathologists in Nottingham and Mansfield who have contributed to this investigation. Particular thanks are due to Dr. S. Crowther, Dr. D. Davies, Dr. Roderick Smith, Dr. A. McFarlane and Dr. W. K. S. Moore.

For technical assistance we are indebted to Mr. K. Gordon, for photography to Mr. G. B. Gilbert, and for preparing the manuscript to Mrs. Margaret Richardson.

References

(1) GILSON, J.C. (1973) Biological Effects of Asbestos. IARC Publication No.8,
(2) JONES, J.S.P. (1968) Discussion on mesotheliomas. In: Holstein and Anspach, eds. Internationale Konferenz über die biologischen Wirkingen des Asbestes, Dresden.
(3) WIGNALL, B.K. Personal communication.
(4) OWEN, W.G. and MILLER, A. Personal communication.
(5) KIPLING, M.D. and WATERHOUSE, J.A.H. (1975) Personal communication.
(6) POOLEY, F.D. (1972) Brit. J. ind. med. 29, 146.

RESUME

Cette étude porte sur un groupe de personnes qui ont travaillé pendant la guer à la fabrication de masques à gaz, durant laquelle elles étaient exposées à une var. particulière d'amiante : la crocidolite à diamètre fin. Il en est résulté jusqu'ic une forte prévalence de mésothéliomes (45 malades); chez quelques individus, le mése liome n'était pas significatif. Dans ce groupe, personne n'était atteint concurrem d'asbestose et de mésothéliome.

EFFECT OF CHANGED WORKING TECHNIQUES ON ASBESTOS DUST LEVELS IN THE WORKING ENVIRONMENT

A.A. CROSS (*)
114, Park Street, London W1Y 4AB, England

The principal methods of correcting the working environment in those cases where asbestos dust concentrations are found to be too high is tackled in various ways. The principal methods are:-

(i)   Modification of the product so that it is less likely to emit dust.
(ii) Elimination of the dust producing operation or process.
(iii) Suppression of dust, for example, by damping.
(iv) Avoidance of personal exposure by mechanization.
(v)   Enclosure, usually by combination with exhaust ventilation.
(vi) Exhaust ventilation or dust extraction.
(vii) Good factory hygiene.
(viii) Personal protection.

I put this deliberately last since we will all regard this as a line of last resort for those situations where all other measures are impracticable or insufficiently effective. It is certainly not an alternative to other forms of control.

I have not included in this list substitution, although I know this is always quoted as the number one measure to be considered when dealing with any potential hazard. There are certainly cases where asbestos materials are no longer used and where alternative materials have replaced them. In the United Kingdom those asbestos-containing products such as thermal insulation, popularly known as lagging, in the use of which the majority of cases of occupational disease associated with asbestos have occurred in the last twenty years, have almost entirely been replaced by non-asbestos materials. The products which remain are, generally speaking, of an entirely different order of dust producing potential and are susceptable to relatively simple means of control for the avoidance of any occupational risk.

Furthermore, there is growing awareness of the need to ensure that materials substituted should themselves be free from potential risk, as well as adequately meeting technical and safety standards, before they are used in place of other materials where the degree of risk is understood and for which adequate methods of prevention of risk have been evolved.

Special Problems with Asbestos.

In dealing with the control of asbestos dust there are some peculiar difficulties. Most industrial dusts consist of particles - asbestos dust consists of fibres and because of this the dust remains in suspension in the air longer, so increasing the

---

(*) Environmental Control Committee, Asbestosis Research Council and Chairman, International Asbestos Information Conference Standing Committee.

possibility of inhalation. This same fibrous structure also adds to the problem of the engineer in designing methods of automatic handling and conveying, etc. The fibres are inclined to pack or to bridge in the outlets of gravity feeds or cyclones. They tend to cling around any irregularity or angle in the surface. The fibres have a natural inclination to cling together. Pneumatic conveying introduces the problem of controlling the air movement created so that it does not distribute inhalable particles into the working atmosphere.

Our engineers, however, have today certain advantages which were not available to them in former times. By the use of such devices as the Tindal Beam or with more sophisticated dust monitoring equipment, concentrations of hitherto invisible dust can be detected and their point of origin located so that control measures can be designed or modified to cure such situations. The most significant advance, however, is the ability to measure dust concentrations and to relate this to a standard threshold limit value so that the engineer can tell if the equipment he has designed and installed is achieving the desired result.

The standard to which we are working has been clearly shown over an ample number of years to have been effective in controlling risks of asbestos related disease where it has been consistently attained.

When the Asbestos Regulations 1969 were introduced, applying as they did not only an overall improvement in the standard of control, but the application of this standard to every occupation where exposure to harmful quantities of asbestos dust might occur it was obvious that renewed efforts would be called for, particularly among employees who had not previously had any experience of dust control with this unusual material.

United Kingdom asbestos industry realised that it had a duty and a responsibility to make available the experience which it had gained in the last 30 years, working with the Asbestos Industry Regulations 1931, to those who would now for the first time be faced with a similar problem. It was also seen that the industry would have a responsibility to advise its customers in methods of using those products in a manner which would not involve risks to the people who worked with them. To meet this responsibility an Environmental Control Committee was established by the Asbestosis Research Council.

The Committee's first task was to identify the extent of the potential hazard for various types of asbestos-containing products. It was earlier recognized that there were many asbestos-containing materials which by their nature involved no risk in normal use. At the other end of the scale there were situations where control was difficult and where the nature of the product and the traditional methods of use were such that high concentrations of inhalable dust could be produced.

To cover this diversity a number of working groups were set up, each of which was given the task of considering a particular problem area and to formulate advice on practical measures of control which could and should be applied in each case.

I would like to quote you some examples of the problems which had to be tackled.

Certainly the most difficult area for control arises in the demolition of old plant or the stripping of old insulation for replacement. If such operations are carried out without adequate suppression very high concentrations are likely to be produced.

Surgeon Commander Harries reported a few years ago measurements taken during the removal of old crocidolite lagging from the "Ark Royal". He recorded mean concentrations in the range of 150-400 with peak concentrations in some instances with over 3000 fibres per m/l. As a result of this, methods of personal protection of the most sophisticated kind were developed by the Navy for carrying out such work, but have found since then that by thorough wetting out of the material and by strict attention to very high standards of hygiene the production of such enormous quantities of dust can be completely avoided.

Indeed, a survey carried out in recent months during the stripping of old lagging material at a power station demonstrated how effective thorough and conscientious observance of the recommended procedures can be. On this occasion the highest concentration recorded was 0.10 fibres per m/l. This was a measurement, during a period of 16 minutes, of actual removal of pipework lagging and is a record of the atmosphere in the breathing zone of the operative carrying out this work. Counts taken in the general area of the same work in five different locations varied from 0.01 to 0.03 fibres per m/l. These counts were confirmed by the Central Electricity Board monitors. This example demonstrates that even in the most unpromising situations it is possible, with properly trained personnel and determined supervision, for safe levels of working to be established.

Another example of the effectiveness of study of a particular problem was the development of a predamping technique for the application of sprayed asbestos insulation. Formerly such operations gave rise to very high concentrations, usually more than 100 fibres per m/l, and concentrations of even as high as 1500 fibres per m/l have been reported.

By the use of specially developed equipment for predamping the asbestos fibre before application by the spray machine, it was found possible to control the dust levels to between 5 and 10 fibres per m/l. This improvement was sufficient, if not to avoid the need for personal protection for the operatives, to enable them to use the simplest type of respirator, but it was particularly valuable in effectively limiting the area of possible contamination.

Such occupations as these, together with certain asbestos manufacturing industries in the past, are the situations where the vast majority of occupational disease due to asbestos dust have occurred. As far as the asbestos manufacturing industry is concerned, I would like to quote the remark of Mr. Brian Harvey in his report, as HM Chief Inspector of Factories, for 1974. Referring to the representative personal samples taken in workers' breathing zones in connection with the survey of 5000 asbestos workers, he says: "The results were most encouraging, showing that 92.6% of the dust counts taken were below the very stringent hygiene standard of 2 fibres per m/l and reflect the great efforts made by the major firms in the industry to improve standards of control".

Sometimes it is possible to reduce the dustiness of a product not only in manufacture, but in use. Such a case is in the manufacture of asbestos textile products. Dust suppressed qualities have been developed which are virtually dust free in normal handling, and some of the difficulties of dust control in manufacture have been overcome by novel techniques.

The main attention of the Environmental Control Committee, however, has been to the special needs of people working with asbestos materials away from the facilities available in a factory. In collaboration with manufacturers of tools, dust extraction and vacuum equipment, special hoods have been designed for site cutting, drilling, routing and sanding of asbestos board or sheet materials. The use of such tools, coupled to transportable vacuum units fitted with approved standards of filter and collecting bags, enables workers on construction sites or elsewhere to work within the permissible dust levels. The results of effective application of various types of control are brought out very clearly when the concentrations of dust are measured.

However effective the dust extraction equipment of the tools provided for safe working, the elimination of harmful quantities of dust can be frustrated by poor standards of factory hygiene. Considerable attention has, therefore, been given to the matter of vacuum cleaning equipment which can be confidently recommended for use in factories where the waste materials are likely to contain a proportion of asbestos dust. Attention has also been given to working procedures for the creation and maintenance of satisfactory standards of factory cleanliness. The care and maintenance of protective clothing and equipment has also been the subject of detailed advice.

The investigations of the Environmental Control Committee led, in due course, to the publication of a series of Control & Safety Guides which are now widely known throughout the asbestos-using industries. These Guides were prepared with the advice and the benefit of the comments of the Factory Inspectorate and others, and have been in regular and continuous demand since their first publication in a draft form at the first Asbestos & Health Symposium organized jointly by the British Occupational Hygiene Society and the Asbestosis Research Council in 1969.

In addition to these Guides the Environmental Control Committee has applied itself to some problem areas outside the scope of the United Kingdom Asbestos Regulations. It has produced a Code of Practice for the Disposal of Asbestos Waste Materials which preceded the introduction of the Poisonous Waste Act 1973 and of the more recent Control of Pollution Act. This Code of Practice was prepared at the request of and in consultation with the then Ministry of Housing and Local Government, and discussions are currently proceeding with the Department of the Environment in an updating of this Code in so far as this may be necessary after the new legislation.

More recently it has become apparent that while these efforts to inform and assist those on managements concerned with the responsibility for the protection of work people, the ultimate success of these protective measures depends to a large extent on the employees themselves. The Environmental Control Committee has realized it needs to make greater efforts in this aspect of promotion. It has produced a series of illustrated leaflets telling in simple words and pictures the type of fundamental precautions which should be taken. It has made a film designed to make a wide range of the using and professional public aware of the variety of methods of dust control which are appropriate to the use of asbestos.

The availability of advice is being drawn more positively to the attention of employers and employees through meetings, posters and through coupon advertisements in national, trade, professional and safety press.

It is vital that we should continue vigorously to seek and promote new and improved methods of control for the improvement of working methods and evironmental conditions so that all may continue to benefit from the unique properties of asbestos without risk to those people who work with it.

SUMMARY

The methods are presented for correcting the working environment when asbestos dust concentrations are too high. Asbestos dust, being fibrous, remains suspended longer than most industrial dusts, and has a tendency to pack and cling together. Sophisticated monitoring techniques have made possible the attainment of standards that are effective in controlling risks of asbestos related diseases.

The most difficult situation to control is the dust produced in stripping old asbestos lagging, but predamping has greatly reduced the contamination.

Special attention has been paid to the special needs of people working with asbestos on building sites. Special hoods have been designed to protect workers drilling, cutting and sanding asbestos material.

The Environmental Control Committee has prepared a series of Control and Safety Guides and a Code of Practice for the Disposal of Asbestos Waste Material as part of activities to promote new and improved methods of dust control.

RESUME

Les méthodes de correction du milieu de travail, lorsque les concentrations de poussières d'amiante sont trop élevées, sont exposées. La poussière d'amiante étant fibreuse, elle demeure en suspension plus longtemps que la plupart des poussières industrielles, et elle a tendance à s'agglomérer. Des techniques de surveillance perfectionnées ont permis d'atteindre des normes efficaces dans la prévention des maladies liées à l'amiante.

Le problème le plus difficile est celui des poussières produites lors du dégarnissement des anciens revêtements calorifuges à l'amiante, mais l'humidification préalable a grandement réduit la contamination.

Les besoins particuliers des travailleurs exposés à l'amiante sur les chantiers de construction ont spécialement retenu l'attention. Des cagoules ont été conçues pour les ouvriers qui percent, coupent et poncent des matières asbestosiques.

L'Environmental Control Committee a élaboré plusieurs instructions de protection et de sécurité ainsi qu'un code de bonne pratique pour l'évacuation des déchets asbestosiques, dans le cadre des activités tendant à promouvoir des méthodes nouvelles et meilleurs de lutte contre les poussières.

OCCUPATIONAL CHEMICAL CARCINOGENESIS : NEW FACTS, PRIORITIES AND PERSPECTIVES

Cesare MALTONI

Institute of Oncology and Tumour Centre,
Viale Ercolani 412, 40138 Bologna, Italy

The first historical observation of an occupational cancer dates back to the sixteenth century: it was the so-called "mountain disease" in miners of Joachimisthal, which, only three hundred years later was known as pulmonary carcinoma, and which, only in the thirties, was recognized as due to radioactive pollution present in those mines.

Since then, up to a few years ago, the policy has been often the same: that is "let us wait and see" if an agent present in the workplace is carcinogenic or not. Which means, if it produces cancer on man, on the basis of an epidemiological evidence, of course validated by a statistical analysis.

This is not any more the time of the policy of "let us wait and see". It must be rather the starting of a new, more correct, social and scientific approach.

The reasons why occupational oncogenesis, which nowadays is mainly industrial oncogenesis, need a new approach which cannot anymore be delayed, are the following.

First. It has been estimated that from 80% to 90% of the tumours in human beings depends on causes present in the occupational and general environment, and that, therefore, cancer must be largely considered an ecological disease.

Second. The oncogenic risk has been progressively increasing in the last century, in relation to situations depending on industrial trends (table 1). Nowadays the major problem is represented by the multitude of the products of synthetic industry, which are unknown to animal and human protoplasm, and on whose effects we are therefore fully ignorant.

Third. Factories, that is the work-places, represent in the present situation the crucial area: in fact, the most exposed population lives there, and from there the new chemical compounds are spread in the general environment as consumers goods and pollution.

Fourth. There are at present clear-cut facts and knowledge on general oncogenesis whose significance is self-explanatory (table 2).

On the basis of these facts, it emerges that the only real measure to prevent industrial oncogenesis is to identify the potentially oncogenic agents, before workers and the whole mankind are exposed to them. In other words the time has come to predict the oncogenic risk, so preventing the exposure of human beings to it.

Are there nowadays available tools for this purpose? We believe so

It is a fact that the four most important cases of environmental and occupational tumours, discovered in the last years, have been indirectly or directly predicted in some way, experimentally.

The first is the case of the clear-cell-adenocarcinoma of the vagina in adolescence, found in girls born from mothers treated during pregnancy with synthetic nonsteroid estrogen therapy. If we go back, in 1938, Lacassagne (1) already reported that he had produced mammary carcinomas in mice treated with stilbestrol and in the following years it was shown, by several scientists, that the same hormone was producing a variety of tumours in hormone dependent and non-dependent tissues, among different experimental animal species (2).

The second is the case of pulmonary carcinomas among workers exposed to bis(chloromethyl)ether. The discovery of this new type of occupational tumour came together with experimental evidence produced by Van Duuren, in 1968 (3), and by Laskin, in 1971 (4), showing respectively that, when injected subcutaneously into rats, or applied to the skin of mice, the compound was producing subcutaneous fibrosarcomas and skin carcinomas, and when inhaled by rats it induced squamous cell carcinomas of the lung.

The third and most evident example is the history of vinyl chloride carcinogenicity (5, 6, 7, 8, 9, 10).

Experimental bio-assays on the carcinogenic potential of this compound have provided information on:
- its oncogenic effects,
- the target organs and the type of tumours,
- the still effective dose levels.

Following our early observation on the dysplastic changes produced by vinyl chloride on the cells of the respiratory tract in exposed workers, and the early results of Viola (11, 12) that rats exposed to

30,000 ppm of vinyl chloride developed carcinomas of the Zymbal sebaceous gland of the external ear duct, we started in 1971 a project of integrated experiments to study the biological effects of vinyl chloride, in order to predict its cancerogenic risk.

A series of 17 experiments have been started in sequence.

The data of the crucial experiments are now being presented in the tables 3-5.

When administered by inhalation, vinyl chloride produced on rats Zymbal gland carcinomas, nephroblastomas, angiosarcomas of liver and of other sites, subcutaneous angiomas, skin carcinomas, hepatomas, brain neuroblastomas and mammary carcinomas. The compound shows carcinogenic effects down to the dose level of 50 ppm.

In mice vinyl chloride produces pulmonary tumours, mammary carcinomas, liver angiosarcomas, other vascular tumours of different sites and types, and epithelial tumours of the skin, being effective in the species also at the level of 50 ppm.

In hamsters the monomer induces liver angiosarcomas, melanomas, forestomach papillomas, acanthomas and trichoepitheliomas, and it seems to anticipate the onset of lymphomas, being the latency time 48 weeks in the treated animals and 82 weeks in controls.

Vinyl chloride is active also when given by oral administration, producing in rats some of the tumours observed following inhalatory exposure (table 6).

Moreover tumours correlated to vinyl chloride exposure have been observed on offsprings born from mothers exposed during pregnancy to high dose levels of the monomer in the air (table 7).

Before the end of 1972, a short time after it was known that vinyl chloride was inducing in rats not only Zymbal gland tumours, but also nephroblastomas and liver angiosarcomas, these results were also made known to several interested industries, and then were communicated at an international meeting in April 1973. This information promoted the clinical observation which in December 1973, for the first time, identified a liver angiosarcoma in a worker of a U.S.A. factory producing VC-PVC as occupational in origin. Since then epidemiological investigations have led to the discovery of nearly 45 liver angiosarcomas among workers of VC-PVC industries in the U.S.A. and several European countries. Most of these tumours arose before 1973 (the first known case dates from 1961), but in absence of experimental data, they were not linked to the occupational exposure, and as a matter of fact, they were often not properly diagnosed.

Vinyl chloride has caused in one or more of the three animal species used in our experiments, all types of tumours, which at present, on the basis of available epidemiological evidence, are correlated to occupational exposure, i.e. liver angiosarcomas, brain and lung tumours, lymphomas, leukemias and hepatomas (table 8).

There are now going on, in our laboratories, several experiments t better assess the degree of risk from exposure to vinyl chloride in relation to doses and length of treatment.

We are studying the effect of a short exposure (100 hours variousl distributed in time) to high doses of VC on rats (table 9).

The potential oncogenic effects of exposure to 50 ppm of the monomer by inhalation is now furtherly investigated (table 10).

A further experiment studies the effects of vinyl chloride by inhalation at doses below 50 ppm, that is 25, 10, 5 and 1 ppm (tables 11 and 12).

Finally, a few months ago, an experiment started to determine the effects of low doses of vinyl chloride by oral administration (table 13).

The fourth example of the predictive value of experimental bio-assays is represented by the carcinogenicity of chromium pigments.

In 1969 we started a program of investigation on the cancerogenic risk from exposure to several inorganic compounds and pigments. A series of several of the compounds under studies, with the available results, is given in the following table (table 14).

The results dealing with lead-chromate pigments and lead-molybdenum chromate pigments were made known in April 1973 (5-13).

Data published a few months ago (14) and on going epidemiological investigations indicate and excessive incidence of bronchial carcinomas among workers producing chromium pigments.

On the basis of the present evidence it may be concluded that carcinogenicity bio-assays, if properly done, have a high predictive value.

Therefore, they cannot be any more delayed.

Up to the recent past too little importance has been given to predictive bio-assays, for many reasons, among which there were:

1) scepticism on their validity, based upon past results;
2) fatalistic acceptance of the situation because of the huge number of newly produced compounds;
3) complication of elaborate testings.

As regards scepticism, we think that time has come to critically review methods and results of past experiments which, nowadays, may be thought to have been inadequate.

As regards the large number of compounds, it may be true that there are already millions of newly produced chemicals, but the ones which urgently need to be tested, mainly because of their widespread diffusion, actually number in the hundreds.

Concerning complications in performing elaborate testings, it may be true that the more experimental tests on animals can reproduce the conditions of human exposure, the more relevant they are in man. It should be pointed out, however, that agents to be examined should follow a pattern of tests which progress in degree of precision, so as to filter out and expose the most dangerous.

In our opinion the future policy for the prevention of the occupational as well of the environmental oncogenesis, are the following:
1) priority to experimental bio-assays for the scrutiny of compounds already produced and widespread and for the compounds to be still produced on a large scale;
2) basic studies to investigate the potentiality of experimental models in determining the oncogenic risk with particular emphasis on easy and rapid tests which, however, should be carefully and critically evaluated; and
3) the study with proper long term bio-assays already validated, of the most important and widespread compounds, whose effects are still unknown, following a list of priorities determined by all the interested social parties.

On this line in our Institute we are now studying several compounds, among which styrene, acrylonitrile and vinylidene chloride. The plans of these experiments are given in the tables 15, 16, 17.

Ongoing experiments performed in our Institute on the cariinogenic potential of drugs have shown in last weeks that a drug widely used in tumour chemotherapy and also in the prophylaxis of neoplastic recurrency, namely adriamicine, is a potent carcinogen (table 18).

Concluding, occupational oncogenesis should not any more be considered a limited medical problem, but a wider scientific and sociale one, for whose control we look forward to the collaboration of all the interested parties of society: scientific community, unions, industries and governments.

SUMMARY

There is need for a new active approach to the problem of occupational carcinogenesis. It is no longer acceptable to adopt a "wait and see" attitude since:

1. evidence suggests that cancer is an "ecological" disease;
2. the carcinogenic risk has been increasing as industrialisation increased;
3. factory workers are the most exposed population and hazards spread from the work-place to the general population as consumer goods and pollution. It is necessary and possible, to identify potential carcinogens before workers and the general population are exposed.

That predictive experiments could be done, is exemplified by the history of stilboestrol, bis(chloromethyl)ether, vinyl chloride and chromium pigments, all of which were reported carcinogenic in animal tests before any epidemiological results were available. Unfortunately, these results were received with scepticism because the test methods were thought to be inadequate.

It is necessary to develop the future policy of prevention by:

1. establishing a priority for testing compounds already produced and widespread in occurence and those about to be produced on a large scale;
2. establishing experimental models for determining carcinogenic risk, especially rapid screening tests;
3. study of the more important compounds by long-term bioassays.

Full details are given of the testing of vinyl chloride monomer and preliminary results are tabulated for tests on styrene, acrylonitrile and vinylidene chloride.

RESUME

Une nouvelle approche active du problème de la cancérogenèse professionnelle est nécessaire. Adopter une attitude d'expectative n'est plus acceptable pour les raisons suivantes:

1. des risques conduisent à penser que le cancer est une maladie "écologique";
2. le risque cancérogène n'a cessé de croître avec l'industrialisation;
3. les travailleurs des usines sont la population la plus exposée et les risques se propagent du lieu de travail à la population générale comme les biens de consommation et la pollution. Il est nécessaire, et possible, d'identifier les cancérogènes potentiels avant que les travailleurs et la population en général ne soient exposés.

La possibilité d'effectuer des expériences à valeur prédictive est montrée par les exemples du stilboestrol, du bis (chlorométhyl)éther, du chlorure de vinyle et des sels de chrome, substances que les tests effectués sur les animaux ont signalées comme cancérogènes avant toute étude épidémiologique. Malheureusement, ces résultats ont été accueillis avec scepticisme parce qu'on estimait les méthodes d'expérimentation insuffisantes.

Il importe, pour la future politique de prévention, de prendre les mesures suivantes:

1. donner la priorité à l'expérimentation des composés déjà produits et largement répandus et de ceux qui sont sur le point d'être produits à grande échelle;
2. élaborer des modèles expérimentaux permettant de déterminer le risque cancérogène, épreuves de détection rapide notamment ;

3. étudier les composés les plus importants au moyen d'épreuves biologiques de longue durée.

Il est donné des résultats sur l'expérimentation du chlorure de vinyle monomère et les résultats préliminaires sont mis en tableaux pour des tests sur le styrène, l'acrylonitrile et le chlorure de vinylidène.

References

1) Lacassagne, A.: Apparition d'adénocarcinomes mammaires chez des souris mâles traitées par une substance oestrogène synthétique. C. R. Soc. Biol., 1938, 129, 641.

2) Lacassagne, A.: Les cancers produit par des substances chimiques endogènes, 1950. Herman, Paris.

3) Van Duuren, B.L., Sivak, A., Goldschmidt, B.M., Katz, C. and Melchionne, S.: Carcinogenicity of halo-ethers. J. Nat. Cancer Inst., 1969, 43, 481-486.

4) Laskin, S., Kuschner, M., Drew, R.T., Cappiello, V.P. and Nelson, N.: Tumors of the respiratory tract induced by inhalation of bis(chloromethyl)ether. Arch. Environ. Health, 1971, 23, 135-136.

5) Maltoni, C.: Occupational carcinogenesis. "II International Symposium on Cancer Detection and Prevention, Bologna, 1973" in Advances in Tumour Prevention, Detection and Characterization, Maltoni, 1974, vol.II, pp.19-26, Excerpta Medica.

6) Maltoni, C. and Lefemine, G.: Carcinogenicity bio-assays of vinyl chloride. I Research plan and early results. Environmental Res., 1974, 7, 387-405.

7) Maltoni, C. and Lefemine, G.: Le potenzialità dei saggi sperimentali nella predizione dei rischi oncogeni ambientali. Un esempio: il cloruro di vinile. Accademia Nazionale dei Lincei, Rendiconti della Classe di Scienze Fisiche, Matematiche e Naturali, Vol. LVI, Serie VIII, Fasc. 3, March 1974.

8) Maltoni, C. and Lefemine, G.: Carcinogenicity bio-assays of vinyl chloride: current results, in "Toxicity of vinyl chloride-polyvinyl chloride", Selikoff I.J. and Hammond E.C. eds., 1975, pp. 195-218. Annals of the New York Academy of Sciences.

(9) Maltoni, C., Lefemine, G., Chieco, P. and Carretti, D.: La cancerogenesi ambientale e professionale: nuove prospettive alla luce della cancerogenesi da cloruro di vinile. Gli Ospedali della Vita, 1974, 1, 5-6, 4-66.

(10) Maltoni, C., Ciliberti, A., Gianni, L. and Chieco, P.: Insorgenza di angiosarcomi in ratti, in seguito a somministrazione per via orale di cloruro di vinile. Gli Ospedali della Vita, 1975, 2, 1, 65-66.

(11) Viola, P.L.: Cancerogenic effect of vinyl chloride, in "X International Cancer Congress, Houston, 1970, Abstr. vol. 29.

(12) Viola, P.L., Bigotti, A. and Caputo, A.: oncogenic response of rat skin, lungs and bones to vinyl chloride. Cancer Res., 1971, 31, 516-519.

(13) Maltoni, C., Sinibaldi, C. and Chieco, P.: "Subcutaneous sarcomas in rats following local injection of chromium orange and molybdenum orange". V International Symposium on the Biological Characterization of Human Tumours, Bologna, 1973, in Advances in Tumour Prevention, Detection and Characterization, Maltoni, 1974, vol. I, pp. 133-134, Excerpta Medica.

(14) Langard, S. and Norseth, T.: A cohort study of bronchial carcinomas in workers producing chromate pigments. Brit. J. Ind. Med., 1975, 32, 62-65.

OCCUPATIONAL CHEMICAL CARCINOGENESIS

C. MALTONI

Tables 1 to 18

Table 1

CAUSES OF INCREASED ONCOGENIC RISK IN HUMAN ENVIRONMENT

I   - Concentration of oncogenic agents already present in the surface of the Earth.

II - Surfacing of oncogenic agents from the depth of the Earth.

III - Production of new potentially oncogenic compounds by chemical and petrochemical industry.

Table 2

KNOWLEDGE AND FACTS WHICH MAKE URGENT A NEW APPROACH TO THE PROBLEM OF INDUSTRIAL ONCOGENESIS

I   - The potentially oncogenic agents in human environment are progressively increasing.

II - Different oncogenic agents may have additive effects.

III - Changes produced by oncogenic agents are largely irreversible.

IV - Oncogenic agents may exert their effects on different organs and tissues and widely affect the target organs.

V   - A large part of the natural history of tumours take place without any clinically and sometimes otherwise detectable pathological changes.

VI - Cancer is not a reversible disease.

Table 3

Experiments BT1 and BT6 : Exposure by inhalation to VC in air, at 30,000, 10,000, 6,000, 2,500, 500, 250, 50 ppm., 4 hrs. daily, 5 days weekly, for 52 weeks.
Results after 135 weeks (end of the experiments).

GROUPS AND TREATMENT	ANIMALS (SPRAGUE-DAWLEY RATS)		ANIMALS WITH TUMOURS															
	Total	Corrected number (a)	Zymbal gland carcinomas (b)			Nephroblastomas (c)			Angiosarcomas						Subcutaneous angiomas	Skin carcinomas	Hepatomas	Brain neuroblastomas	Mammary carcinomas	Other type and/or site (n)	Total (z)
			No.	%	Average latency time (weeks)	No.	%	Average latency time (weeks)	Liver			Other sites						
									No.	%	Average latency time (weeks)	No.	No.	No.	No.	No.	No.	No.
I VC 30,000 ppm.	60	60	35	58	43	-	-	-	18	30	53	1 (e)	-	-	-	-	27 (o)	51
II VC 10,000 ppm.	69	61	16	26	50	5	8	59	9	15	64	3 (f)	4	3	1	7	3	4 (p)	38
III VC 6,000 ppm.	72	60	7	12	62	4	7	65	13	22	70	3 (g)	3	1	1	3	-	8 (q)	31
IV VC 2,500 ppm.	74	59	2	3	33	6	10	74	13	22	78	3 (h)	3	-	2	5	1	3 (r)	32
V VC 500 ppm.	67	59	4	7	79	4	7	83	7	12	81	2 (i)	1	4	3	-	1	3 (s)	22
VI VC 250 ppm.	67	59	-	-	-	6	10	80	4	7	79	2 (l)	-	4	-	-	1	2 (t)	16
VII VC 50 ppm.	64	59	-	-	-	1	2	135	1	2	135	1 (m)	1	1	-	-	2	7 (u)	10
VIII No treatment	68	58	-	-	-	-	-	-	-	-	-	-	-	-	-	-	-	10 (v)	6
Total	541	475	64	-	-	26	-	-	65	-	-	15	12	11	7	15	8	64	206

a) Alive animals after 24 weeks, when the first tumour (a Zymbal gland carcinoma) was observed. The percentages are referred to the corrected number.
b) Metastases to lung.
c) Metastases to liver, lung, spleen and brain.
d) Metastases to lung.
e) 1 lung angiosarcoma.
f) 1 intrabdominal angiosarcoma (next to liver); 1 angiosarcoma of the lips; 1 angiosarcoma of the nose.
g) 1 angiosarcoma in subcutaneous fibrosing angioma; 1 ossifying parauricular angiosarcoma; 1 intrabdominal angiosarcoma (next to liver).
h) 1 ossifying angiosarcoma of neck; 2 intrabdominal angiosarcomas (1 next to spleen and 1 next to ovary).
i) 1 angiosarcoma of uterus; 1 lung angiosarcoma.
l) 1 intrabdominal angiosarcoma (next to spleen); 1 intrathoracic ossifying angiosarcoma.
m) 1 intrabdominal diffused angiosarcoma.
n) Several cases of breast fibroadenomas, adrenal and pituitary tumours (generally adenomas) have not been considered, since their distribution in the different groups does not vary.
o) 7 Zymbal gland adenomas; 1 skin squamous carcinoma; 1 subcutaneous fibroangioma; 2 mammary carcinomas; 1 hepatic angioma; 1 hepatoma; 1 ovarian angioma; 11 forestomach papillomas; 1 brain neuroblastoma; 1 Harderian gland carcinoma.
p) 2 Zymbal gland adenomas; 1 ovarian cystoadenocarcinoma; 1 neurilemmoma.
q) 4 Zymbal gland adenomas; 2 hepatic and 1 peritoneal angiomas; 1 salivary gland adenocarcinoma.
r) 1 Zymbal gland adenoma; 2 ependymomas.
s) 1 pulmonary fibrosarcoma; 2 lymphomas.
t) 1 Zymbal gland adenoma; 1 lymphoma.
u) 3 Zymbal gland adenomas; 1 subcutaneous angiopericitoma; 3 uterine adenocarcinomas (1 with sarcomatous component).
v) 1 invasive acanthoma of Zymbal gland; 1 subcutaneous fibrosarcoma; 2 peritoneal fibroangiomas; 2 uterine adenocarcinomas (1 with sarcomatous component); 1 uterine leiomyosarcoma; 1 ovarian fibrosarcoma; 1 pulmonary rhabdomyosarcoma; 1 lymphoma.
z) Several animals with 2 or more tumours.

Table 4

Experiment BT4 : Exposure by inhalation to VC in air at 10,000, 6,000, 2,500, 500, 250, 50 ppm., for 4 hrs. daily, 5 days weekly, for 30 weeks. Results after 81 weeks (end of the experiment).

GROUPS AND TREATMENT	ANIMALS (SWISS MICE)											ANIMALS WITH TUMOURS						
	Total			Corrected number (a)			Lung tumours (b)			Mammary carcinomas (c)			Liver an-giosarcoma	Vascular tumours of other type and/or site		Epithelial tumours of the skin	Other type and/or site	Total (z)
	♂	♀	Total	♂	♀	♂♀	No.	%	Average latency time (weeks)	No.	%	Average latency time (weeks)	No.	No.		No.	No.	No.
I VC 10,000 ppm.	30	30	60	22	28	50	35	70	36	13	47	31	8	9 (d)		3 (m)	2 (r)	36
II VC 6,000 ppm.	30	30	60	26	28	54	38	70	38	8	28	33	5	9 (e)		6 (n)	4 (s)	39
III VC 2,500 ppm.	30	30	60	23	30	53	30	57	43	9	30	35	11	12 (f)		3 (o)	2 (t)	31
IV VC 500 ppm.	30	30	60	29	29	58	38	66	41	7	24	37	11	18 (g)		1 (p)	1 (u)	43
V VC 250 ppm.	30	30	60	29	29	58	33	57	45	11	32	39	11	22 (h)		2 (q)	3 (v)	40
VI VC 50 ppm.	30	30	60	27	30	57	2	3,5	51	12	33	43	1	13 (i)		-	1 (w)	18
VII No treatment	80	70	150	74	67	141	8	6	44	-	-	-	-	1 (l)		-	7 (y)	13
Total	260	250	510	230	241	471	184			60			47	84		15	20	220

a) Alive animals after 16 weeks, when the first tumour (a mammary carcinoma) was observed. The percentages are referred to the corrected number.
b) Adenomas, some of which undergoing malignant trasformation.
c) In females, Adenocarcinomas frequently with areas of squamous metaplasia.
d) 3 subcutaneous angiomas; 4 liver fibroangiomas; 1 heart fibroangioma; 1 ossifying interscapular angioma.
e) 1 subcutaneous angiosarcoma; 2 liver fibroangiomas; 4 liver angiomas; 1 renal fibroangioma; 1 thymus angioma.
f) 2 subcutaneous angiosarcomas; 1 subcutaneous angioma; 3 liver angiomas; 2 intrabdominal angiosarcomas; 1 lung angioma.
g) 2 subcutaneous angiosarcomas; 1 subcutaneous fibroangioma; 1 subcutaneous angioma; 4 liver fibroangiomas; 1 liver angioma; 3 intrabdominal angio-sarcomas; 1 intrabdominal fibroangioma; 1 renal angiosarcoma; 1 testicular fibroangioma; 1 angioma of the caecum; 1 lung fibroangioma; 1 lung angioma.
h) 1 subcutaneous angioma; 6 liver fibroangiomas; 10 liver angiomas; 2 intrabdominal angiosarcomas; 1 ovarian angioma; 1 scrotal angioma; 1 lung angioma.
i) 1 subcutaneous angiosarcoma; 1 subcutaneous fibroangioma; 2 subcutaneous angiomas; 2 liver fibroangiomas; 3 liver angiomas; 1 intrabdominal angioma; 1 ovarian angioma; 1 intrathoracic fibroangioma; 1 angioma of the interscapular fat pad.
l) 1 angiosarcoma of uterus.
m) 2 squamous carcinomas; 1 invasive acanthoma.
n) 5 squamous carcinomas; 1 acanthoma.
o) 3 squamous carcinomas; 2 acanthomas.
p) 1 acanthoma.
q) 1 skin adenocarcinoma; 1 basalioma.
r) 1 Zymbal gland adenoma; 1 forestomach papilloma.
s) 1 subcutaneous leiomyosarcoma; 1 forestomach papilloma; 1 Harderian gland adenoma; 1 lymphoma.
t) 1 forestomach papilloma; 1 parotid gland mixed tumour.
u) 1 Zymbal gland adenoma.
v) 1 Zymbal gland adenoma, 1 Leydig cells tumour; 1 lymphoma.
w) 1 parotid gland adenocarcinoma.
y) 1 subcutaneous leiomyosarcoma; 1 adenoma of colon; 1 ovarian carcinoma; 1 leiomyosarcoma of uterus; 3 lymphomas.
z) Several cases with 2 or more tumours.

Table 5

Experiment BT8 : <u>Exposure</u> by inhalation to VC in air at 10,000, 6,000, 2,500, 500, 250, 50 ppm., 4 hrs. daily, 5 days weekly, for 30 weeks. <u>Results</u> after 105 weeks.

GROUPS AND TREATMENT	ANIMALS (GOLDEN HAMSTERS)		ANIMALS WITH TUMOURS						
	Total	Survivors	Liver angiosar- comas	Skin tri- choepithe liomas and basaliomas (a)	Melanomas	Lymphomas	Forestomach epithelial tumours (b)	Other type and/or site	Total (g)
			No.	No.	No.	No.	No.	No.	No.
I VC 10,000 ppm.	35	-	-	6	1	-	4	2 (c)	4
II VC 6,000 ppm.	32	-	1	2	2	2	7	7 (d)	10
III VC 2,500 ppm.	33	-	-	1	-	1	10	3 (e)	12
IV VC 500 ppm.	33	-	2	3	-	1	7	2 (f)	12
V VC 250 ppm.	32	-	-	3	-	1	2	-	6
VI VC 50 ppm.	33	-	-	6	1	1	4	-	10
VII No treatment	70	1	-	2	-	2	-	-	5
Total	268	1	3	23	4	8	34	14	59

a) Several cases with acanthosis and some undergoing malignant trasformation.
b) Papillomas, acanthomas, some of which undergoing malignant trasformation.
c) 1 subcutaneous angioma; 1 gall-bladder adenocarcinoma.
d) 2 hepatomas; 2 liver fibroangiomas; 2 liver angiomas; 1 biliducts adenocarcinoma.
e) 1 hepatoma; 1 liver fibroangioma; 1 liver angioma.
f) 1 subcutaneous angioma; 1 bronchial carcinoma.
g) Several animals with 2 or more tumours.

Table 6

Experiment BT11 : <u>Exposure</u> by ingestion (stomach tube) to VC in olive oil, at 50.00, 16.65, 3.33 mg/Kg body weight, once daily, 4-5 days weekly, for 52 weeks. <u>Results</u> after 84 weeks.

GROUPS AND TREATMENT	ANIMALS (SPRAGUE-DAWLEY RATS)		ANIMALS WITH TUMOURS					
	Total	Survivors	Zymbal gland carcinomas	Nephrobla- stomas	Angiosarcomas		Other type and/or site (b)	Total
					Liver	Other sites		
			No.	No.	No.	No.	No.	No.
I VC 50.00 mg/Kg	80	25	-	-	8	1 (a)	3 (c)	11 (d)
II VC 16.65 mg/Kg	80	33	1	1	5	-	-	7
III VC 3.33 mg/Kg	80	39	-	-	-	-	-	-
IV Control olive oil	80	37	-	-	-	-	-	-
Total	320	134	1	1	13	1	3	18

a) 1 thymus angiosarcoma.
b) Several cases of breast fibroadenomas; adrenal and pituitary tumours (generally adenomas) have not been considered, since their distribution in the different groups does not vary.
c) 1 mammary carcinoma; 2 forestomach papillomas.
d) 1 animal with 2 tumours.

Table 7

Experiment BT5 : Exposure by inhalation to VC in air at 10,000 and 6,000 ppm. of breeders, 4 hrs. daily, for 1 week (from 12th to 18th day of pregnancy).
Results after 143 weeks (end of the experiment).

GROUPS AND TREATMENT	ANIMALS (SPRAGUE-DAWLEY RATS)		ANIMALS WITH TUMOURS					
	Total	Survivors	Zymbal gland carcinomas	Nephroblastomas	Angiosarcomas		Other type and/or site	Total
					Liver	Other sites		
			No.	No.	No.	No.	No.	No.
I VC 10,000 ppm. Breeders	30	-	1	-	-	1 (a)	2 (d)	3
II VC 6,000 ppm. Breeders	30	-	-	-	-	-	-	-
III VC 10,000 ppm. Offsprings	54	-	3	1	-	2 (b)	2 (e)	8
IV VC 6,000 ppm. Offsprings	32	-	1	-	-	2 (c)	4 (f)	6
Total	146	-	5	1	-	5	8	17

a) 1 intrabdominal angiosarcoma.
b) 1 subcutaneous angiosarcoma; 1 angiosarcoma of the leg.
c) 1 subcutaneous angiosarcoma; 1 intrabdominal angiosarcoma.
d) 1 liver fibroangioma; 1 liver angioma.
e) 1 Zymbal gland fibrosarcoma; 1 ovarian leiomyosarcoma.
f) 1 Zymbal gland adenoma; 1 skin carcinoma; 1 subcutaneous fibroangioma; 1 mammary carcinoma.

Table 8

TUMOURS PRESENTLY CORRELATED TO VC EXPOSURE (BY INHALATION), ON EXPERIMENTAL RODENTS AND MAN

Species	TUMOURS										
	Angiosarcomas of liver	Tumours of brain	Tumours of lung	Lymphomas and leukemias	Angiosarcomas and angiomas of other sites	Nephroblastomas	Sebaceous cutaneous carcinomas	Squamous tumours of epidermis	Mammary carcinomas	Hepatomas	Forestomach papillomas and acanthomas
Rat	+	+			+	+	+	+	(+)	+	+
Mouse	+		+		+		+	+	+		+
Hamster	+		(+)	(+)	(+)			+		(+)	+
Man	+	(+)	(+)	(+)					(+)		

Table 9

Experiment BT10 : <u>Exposure</u> by inhalation to VC for 100 hrs., following different schedules. <u>Results</u> after 59 weeks.

GROUPS AND TREATMENT	ANIMALS (SPRAGUE-DAWLEY RATS)		ANIMALS WITH TUMOURS					
	Total	Survivors	Zymbal gland carcinomas	Nephroblastomas	Angiosarcomas		Other type and/or site	Total
					Liver	Other sites		
			No.	No.	No.	No.	No.	No.
I VC 10,000 ppm. 4hd/5dw/5w	120	84	4	-	-	-	-	4
II VC 6,000 ppm. 4hd/5dw/5w	120	85	3	-	-	-	1 (b)	4
III VC 10,000 ppm. 1hd/4dw/25w	120	89	4	-	-	-	-	4
IV VC 6,000 ppm. 1hd/4dw/25w	120	85	-	-	-	1 (a)	-	1
V VC 10,000 ppm. 4hd/1dw/25w	120	96	2	-	-	-	1 (c)	3
VI VC 6,000 ppm. 4hd/1dw/25w	120	97	-	-	-	-	1 (d)	1
VII No treatment	249	180	-	-	-	-	-	-
Total	969	716	13	-	-	1	3	17

a) 1 subcutaneous angiosarcoma.
b) 1 lymphoma.
c) 1 subcutaneous angiopericitosarcoma.
d) 1 lymphoma.

Table 10

Experiment BT9 : <u>Exposure</u> by inhalation to VC in air at 50 ppm., 4 hrs. daily, 5 days weekly, for 52 weeks. <u>Results</u> after 59 weeks.

GROUPS AND TREATMENT	ANIMALS (SPRAGUE-DAWLEY RATS)		ANIMALS WITH TUMOURS					
	Total	Survivors	Zymbal gland carcinomas	Nephroblastomas	Angiosarcomas		Other type and/or site	Total
					Liver	Other sites		
			No.	No.	No.	No.	No.	No.
I VC 50 ppm.	300	216	1	-	-	-	-	1
II No treatment	100	72	-	-	-	-	-	-
Total	400	288	1	-	-	-	-	1

Table 11

Experiment BT15 : <u>Exposure</u> by inhalation to VC in air, at 25, 10, 5, 1 ppm.,
4 hrs. daily, 5 days weekly, for 52 weeks.
<u>Results</u> after 46 weeks.

GROUPS AND TREATMENT	ANIMALS (SPRAGUE-DAWLEY RATS)		ANIMALS WITH TUMOURS					
			Zymbal gland carcinomas	Nephrobla-stomas	Angiosarcomas		Other type and/or site	Total
	Total	Survivors			Liver	Other sites		
			No.	No.	No.	No.	No.	No.
I VC 25 ppm.	120	110	-	-	-	-	-	-
II VC 10 ppm.	120	111	-	-	-	-	-	-
III VC 5 ppm.	120	113	-	-	-	-	-	-
IV VC 1 ppm.	120	112	-	-	-	-	-	-
V No treatment	120	97	-	-	-	-	-	-
Total	600	543	-	-	-	-	-	-

Table 12

Experiment BT17 : <u>Exposure</u> by inhalation to VC in air at 1 ppm., 4 hrs. daily,
5 days weekly, for 52 weeks.
<u>Results</u> after 46 weeks.

GROUPS AND TREATMENT	ANIMALS (WISTAR RATS)		ANIMALS WITH TUMOURS					
			Zymbal gland carcinomas	Nephrobla-stomas	Angiosarcomas		Other type and/or site	Total
	Total	Survivors			Liver	Other sites		
			No.	No.	No.	No.	No.	No.
I VC 1 ppm.	120	96	-	-	-	-	-	-
II No treatment	130	89	-	-	-	-	-	-
Total	250	185	-	-	-	-	-	-

Table 13

Experiment BT27 : Exposure by ingestion (stomach tube) to VC in olive oil, at 1, 0.3, 0.03 mg/Kg body weight, once daily, 4-5 days weekly, for 104 weeks. Results after 21 weeks.

GROUPS AND TREATMENT	ANIMALS (SPRAGUE-DAWLEY RATS)		ANIMALS WITH TUMOURS					
			Zymbal gland carcinomas	Nephroblastomas	Angiosarcomas		Other type and/or site	Total
	Total	Survivors	No.	No.	Liver No.	Other sites No.	No.	No.
VC 1 mg/Kg/bw I	158	147	-	-		-	-	-
VC 0.3 mg/Kg/bw II	159	145	-	-		-	-	-
VC 0.03 mg/Kg/bw III	163	147	-	-		-	-	-
Control: olive oil IV	159	148	-	-		-	-	-
Total	639	587						

Table 14

Experiments BO12 and BT2001 : Treatment: 1 subcutaneous injection of 30 mg of the compound in 1 cc of H_2O.

	Groups and Treatment	Animals (1) (Sprague-Dawley rats)		Tumours (sarcomas at the site of injection)			Ended Experiments length (weeks)	Ongoing experiments Time from start (weeks)
		Total	Survivors	No.of animals with tumours	Average latency time (weeks)	Histology type		
I	Chromite	40	-	-	-	-	133	
II	Neochromium	40	-	9	78	Rhabdomyosarcomas and fibrosarcomas	149	
III	Chromium allumen	40	-	8	72	Rhabdomyosarcomas and fibrosarcomas	137	
IV	Chromium yellow (Lead chromate)	40	-	26	40	Rhabdomyosarcomas and fibrosarcomas	150	
V	Chromium orange (Lead chromate)	40	-	27	47	Rhabdomyosarcomas and fibrosarcomas	150	
VI	Molybdenum orange (Lead chromate, sulphate and molybdate)	40	-	36	32	Rhabdomyosarcomas and fibrosarcomas	117	
VII	Cadmium yellow (Cadmium sulphide)	40	-	16	86	Rhabdomyosarcomas and fibrosarcomas	143	
VIII	Iron yellow (Iron oxide)	40	-	1	77	Rhabdomyosarcomas and fibrosarcomas	125	
IX	Iron red (Iron oxide)	40	-	-	-	-	134	
X	Control	140	-	-	-	-	127	
XI	Titanium oxide	120	116	-	-	-		35
XII	Zinc chromate	80	80	-	-	-		4

(1) 50% males, 50% females

Table 15

PLAN OF THE EXPERIMENTS ON THE BIOLOGICAL EFFECTS OF STYRENE

Experiments	Species and age at start (weeks)	Groups	No. of animals at start			Treatment				Length (weeks)	Weeks from start
			Total	♂	♀	Route	Dose	Stopped	Ongoing		
BT101	Rats (13)	I	60	30	30	Inhalation	300 ppm		+	44	44
		II	60	30	30	"	200 ppm		+	44	44
		III	60	30	30	"	100 ppm		+	44	44
		IV	60	30	30	"	50 ppm		+	44	44
		V	60	30	30	"	25 ppm		+	44	44
		VI	120	60	60	Untreated (controls)	-				44
		Total	420	210	210						
BT102	Rats (13)	I	80	40	40	Ingestion	250 mg/Kg in olive oil		+	37	37
		II	80	40	40	"	50 mg/Kg in olive oil		+	37	37
		III	80	40	40	"	control: olive oil		+	37	37
		Total	240	120	120						
BT103	Rats (13)	I	80	40	40	4 intrabdominal injections	50 mg in olive oil (x 4)	+			38
		II	80	40	40	"	control: olive oil	+			38
		Total	160	80	80						
BT104	Rats (13)	I	80	40	40	1 subcutaneous injection	50 mg in olive oil	+			38
		II	80	40	40	"	control: olive oil	+			38
		Total	160	80	80						
			980	490	490						

146

Table 16

PLAN OF THE EXPERIMENTS ON THE BIOLOGICAL EFFECTS OF ACRYLONITRILE

Experiments	Species and age at start (weeks)	Groups	No. of animals at start			Treatment				Length (weeks)	Weeks from start
			Total	♂	♀	Route	Dose	Stopped	Ongoing		
BT201	Rats (12)	I	62	32	30	Inhalation	40 ppm		+	35	35
		II	60	30	30	"	20 ppm		+	35	35
		III	60	30	30	"	10 ppm		+	35	35
		IV	60	30	30	"	5 ppm		+	35	35
		V	60	30	30	Untreated (controls)	-				35
		Total	302	152	150						
BT202	Rats (12)	I	105	58	47	Ingestion (1)	20 mg/Kg in olive oil	+		6	35
		II	80	40	40	"	control: olive oil	+		6	35
		Total	185	98	87						
BT203	Rats (10)	I	82	42	40	Ingestion (2)	5 mg/Kg in olive oil		+	21	21
		II	158	82	76	"	control: olive oil		+	21	21
		Total	240	124	116						
TOTAL			727	374	353						

(1) 3-5 times weekly
(2) 3 times weekly

Table 17

PLAN OF THE EXPERIMENTS ON THE BIOLOGICAL EFFECTS OF VINYLIDENE CHLORIDE

Experiments	Species and age at start (weeks)	Groups	No. of animals at start			Route	Treatment			Length (weeks)	Weeks from start
			Total	♂	♀		Dose	Stopped	Ongoing		
BT401	Rats (16)	I	120	60	60	Inhalation	150 ppm		+	21	21
		II	60	30	30	"	100 ppm		+	21	21
		III	60	30	30	"	50 ppm		+	21	21
		IV	60	30	30	"	25 ppm		+	21	21
		V	60	30	30	"	10 ppm		+	21	21
		VI	200	100	100	Untreated (controls)	-				21
		Total	560	280	280						
BT402	Mice (16)	I	120	60	60	Inhalation	200 ppm	+		2 days	21
		II	60	30	30	"	100 ppm	+		2 days	21
		III	60	30	30	"	50 ppm	+		4 days	21
		IV	60	30	30	"	25 ppm		+	21	21
	(9)	V	240	120	120	"	25 ppm		+	19	19
		VI	60	30	30	"	10 ppm		+	21	21
		VII	200	100	100	Untreated (controls)	-				21
		Total	800	400	400						
BT403	Rats (9)	I	100	50	50	Ingestion	20 mg/Kg in olive oil		+	32	32
		II	100	50	50	"	10 mg/Kg in olive oil		+	32	32
		III	100	50	50	"	5 mg/Kg in olive oil		+	32	32
		IV	200	100	100	"	control:olive oil		+	32	32
		Total	500	250	250						
BT404	Rats (9)	I	100	50	50	Ingestion	0.5 mg/Kg in olive oil		+	21	21
BT405	Chinese Hamsters (28)	I	60	30	30	Inhalation	25 ppm		+	13	13
		II	35	18	17	Untreated (controls)	-				13
		Total	95	48	47						

148

Table 18

Experiment BT1001 : <u>Treatment</u>: 1 subcutaneous injection of 2 mg of Adriamicine in 1 cc of olive oil.
<u>Results</u> : after 38 weeks.

GROUPS AND TREATMENT	ANIMALS (SPRAGUE-DAWLEY RATS)				TUMOURS (1)		AVERAGE LATENCY TIME (weeks)	
	Sex	Number	Corrected number	Survivors	No.	%	First tumour	Average
I Adriamicine 2 mg	♀	42	42	35	14	35	19	31
	♂	40	40	30	12	30	25	32
II Control: olive oil	♀	40	40	34	-	-	-	-
	♂	40	39	30	-	-	-	-
Total	♀ ♂	162	161	129	26			

(1) At the site of injection.

METABOLISM OF ^{14}C-VINYL CHLORIDE IN VITRO AND IN VIVO

H.M. BOLT, H. KAPPUS, R. KAUFMANN, K.E. APPEL
A. BUCHTER (*) and W. BOLT (*)

Institute of Toxicology, University of Tübingen

(*) Institute of Occupational and Social Medicine, University of Cologne, R.F.A.

INTRODUCTION

Cancerogenic effects of vinyl chloride have been recently reported in man (1-4) and in laboratory animals (5,6). These findings promoted research on metabolism of this compound in the living organism (7-13). The concept has been developed that vinyl chloride is transformed by liver microsomal enzymes to metabolites which show mutagenic effects (14,15), covalent binding to proteins and RNA (12) in vitro, and tissue damage (16) in vivo. According to present theories of chemical carcinogenesis (17,18) pre-carcinogens are metabolised in the organism to ultimate carcinogens which can bind to cellular macromolecules (nucleic acids and proteins).

It has been speculated (7,12,16,18) that the primary intermediate in the metabolism of vinyl chloride, capable of interacting with cellular components, should be the epoxide, i.e. chloroethylene oxide. Other metabolites of vinyl chloride reported to be mutagenic are chloroethanol and chloroacetaldehyde (14). Some recent reports (7-9, 13) deal with the metabolism of vinyl chloride in the rat in vivo, but so

---

Correspondence should be addressed to: Dr.H.M.Bolt, Institut für Toxikologie der Universität, D-74 Tübingen, Wilhelmstr. 56.

far no data are available concerning covalent binding of metabolites of vinyl chloride to cellular macromolecules under in vivo conditions The question, however, is of particular importance to insure that irreversible (or covalent) binding of vinyl chloride to proteins and nucleic acid (RNA) is not a phenomenon occurring solely in vitro. The present report aims at presenting data of irreversible binding of ^{14}C vinyl chloride in vivo. Furthermore, irreversible protein binding of ^{14}C-vinyl chloride in vitro has been assessed using an artificial superoxide generating system. It has been shown that, similar to the rat liver microsomal system, superoxide oxidizes vinyl chloride to an active intermediate which irreversibly binds to protein (12).

EXPERIMENTAL

Radioactive material

1,2-^{14}C-Vinyl chloride was synthesized by the Radiochemical Departmen of Farbwerke Hoechst, Frankfurt/Main. Specific radioactivity of the product was 10.7 mCi/mmol.

Incubations in vitro

Rat liver microsomes with NADPH-regenerating system were incubated in a closed all-glass system under ^{14}C-vinyl chloride containing atmosphere as already described (12). Similar incubations were performed using a superoxide (O_2^-) generating system comprised of phenazine methosulfate and NADH (19). These incubations (1 ml) contained 0.4 m albumin solution (25 mg/ml), 0.4 ml 0.15 M citrate buffer pH 4.5, 0.1 ml 5 mM NADH, and 0.1 ml 0.5 mM phenazine methosulfate (Serva, Heidel berg).

In vivo experiments

Male Wistar rats were used. The rats were exposed to ^{14}C-vinyl chloride in a closed all-glass system which has been described elsewhere (13). After exposure rats were sacrificed and organs were examined for their content of radioactivity (20).

Irreversible binding of radioactivity to microsomal protein of the liver was studied as described previously (21).

Irreversible binding of reactive metabolites of vinyl chloride to protein of tissue homogenates was determined as follows:

Tissue was homogenized with 4 vol. of water by means of an "Ultra-Turrax" tissue homogenizer. One-half milliliter of homogenate, corresponding to 0.1 g tissue wet weight, was added to 1 ml ethanol and re frigerated. After centrifugation the precipitate was washed twice wit 70 % ethanol (0.5 ml), once with boiling 70 % ethanol, and again twic with cold 70 % ethanol. This procedure proved to be sufficient to remove all the extractable metabolites of vinyl chloride and the unmeta bilised vinyl chloride from the tissue proteins. The radioactivity re maining irreversibly attached to the protein was determined after solubilising the protein in hyamine hydroxide by counting with BRAY's solution (22) in a liquid scintillation spectrometer. Protein determinations were done according to LOWRY et al. (23).

Isolation of DNA and RNA from liver of rats exposed to ^{14}C-vinyl chloride followed usual standard procedures (24,25). Radioactivity attached to the nucleic acids was determined by liquid scintillation counting.

RESULTS

Irreversible binding of ^{14}C-vinyl chloride to protein in vitro

It has been already shown that rat liver microsomes in presence of NADPH catalyse irreversible, presumably covalent, binding of ^{14}C-vinyl chloride to the microsomal protein (11,12). If a soluble protein like albumin, or RNA is added to this system, irreversible binding of vinyl chloride metabolites occurs also to these macromolecular compounds (12).

Table 1 shows that irreversible binding of ^{14}C-vinyl chloride to albumin is also achieved by a non-enzymic system comprised of NADH and phenazine methosulfate, which is known (19) to generate the superoxide radical (O_2^-). No significant binding of metabolites of vinyl chloride is observed without microsomes or without phenazine methosulfate. This suggests that activated oxygen converts vinyl chloride to a reactive intermediate capable of binding to protein. Because it is currently considered that activated oxygen, linked to the heme iron of cytochrome P-450, may be involved in microsomal oxidation of xenobiotics, similar mechanisms may be responsible for the microsomal irreversible binding to protein (and RNA) of vinyl chloride.

Irreversible binding of ^{14}C-vinyl chloride in vivo to microsomal protein

That irreversible binding of vinyl chloride is not only an in vitro phenomenon, is shown in Tab.2. Five hours after starting exposure to ^{14}C-vinyl chloride, significant amounts of radioactivity are irreversibly bound to liver microsomal protein, as judged from two different procedures of extraction of microsomes, with ethanol and with trichloroacetic acid. More than half of the microsomal radioactivity immediately after exposure is irreversibly bound to the microsomal protein. However, the bulk of radioactivity of liver at that time is found in cellular compartments other than microsomes, primarily in the cytoplasm. This major part of radioactivity from ^{14}C-vinyl chloride has been found to be readily extractable (see also Tab.3) and consists

TABLE I

Irreversible protein binding of ^{14}C-vinyl chloride in vitro
Uptake of substrate (vinyl chloride): 10 nmol/90 min incubation

a) Microsomes and NADPH-regenerating system:
   Binding to microsomal protein 0.14 ± 0.07 nmoles/mg x)
   Binding to albumin 0.15 nmol/10 mg albumin x)

b) Phenazinmethosulfate and NADH:
   Binding to albumin 12.3 ± 1.5 pmol/10 mg albumin
   Control value, binding to albumin
      without phenazinemethosulfate 0.6 pmol/10 mg albumin

---

x) see ref. (12).

TABLE II

Distribution of metabolites of ^{14}C-vinyl chloride in liver of rats after 5 hrs exposure
Initial concentration of vinyl chloride in atmosphere = 140 ppm
Uptake per rat (250 g) = 15.8 μmol vinyl chloride

a) Total metabolites $\dfrac{\text{nmol radioactive metabolites}}{\text{mg protein}}$

Homogenate of liver	Homogenate after 10.000 g centrifugation	Microsomes (100.000 g pellet)	Cytosol (100.000 g supernatant)
1.00 ± 0.16 (n = 4)	1.55 ± 0.45 (n = 4)	0.48 ± 0.10 (n = 4)	1.87 ± 0.19 (n = 3)

b) Non-extractable radioactivity in microsomes
   after extraction with ethanol: 0.31 ± 0.06 nmol/mg protein (n=4)

   after extraction with trichloroacetic acid: 0.28 ± 0.04 nmol/mg protein (n=4)

Figures show $\bar{x} \pm$ S.D.

TABLE III

Metabolites of ^{14}C-vinyl chloride and irreversible protein binding in different tissues of rats after 5 hrs exposure to ^{14}C-vinyl chloride.

Initial concentration of vinyl chloride in atmosphere = 44 ppm
Uptake per rat (200 g) = 6.5 µmol vinyl chloride

	Radioactive metabolites immediately after exposure		Radioactive metabolites 48 hrs after beginning of exposure		
	nmol metabolites/100 mg tissue		nmol metabolites/100 mg tissue		
	Total metabolites	Irreversibly protein bound metabolites	Total metabolites	Polar, extractable metabolites	Irreversibly protein bound metabolites
Liver	11.45 ± 1.84	1.46 ± 0.47	2.16 ± 0.14	0.38 ± 0.03	1.49 ± 0.06
Spleen	2.25 ± 0.30	0.88 ± 0.02	1.21 ± 0.08	0.33 ± 0.01	0.83 ± 0.06
Kidney	13.15 ± 2.63	1.18 ± 0.14	1.37 ± 0.19	0.40 ± 0.04	0.99 ± 0.07
Lung	n.d.	n.d.	1.01 ± 0.08	0.30 ± 0.11	0.77 ± 0.01
Small intestine	n.d.	n.d.	1.04 ± 0.08	0.33 ± 0.04	0.72 ± 0.07
Brain	1.05 ± 0.05	0.13 ± 0.04	0.22 ± 0.03	n.d.	n.d.
Adipose tissue	1.36 ± 0.17	0.26 ± 0.06	0.30 ± 0.03	n.d.	n.d.
Muscel (M.psoas)	1.97 ± 0.44	0.13 ± 0.01	0.32 ± 0.07	n.d.	n.d.

Figures show $\bar{x}$ ± S.D.; n = 3
n.d. = not determined

mostly of polar metabolites of vinyl chloride (13).

Metabolites of ^{14}C-vinyl chloride and their irreversible protein binding in different tissues

Irreversible binding of metabolites of vinyl chloride to tissue protein is a relatively rapid process (Tab.3). The values of irreversibly bound metabolites obtained immediately after a 5 h period of exposure, and those obtained 48 h after beginning of a similar 5 h exposure, do not differ statistically. On the other hand, this means that the irreversible binding of metabolites of vinyl chloride to cellular proteins is very long lasting. This is consistent with the underlying assumption that the binding is covalent.

Considerable differences are found in the tissues' content of polar extractable metabolites. Whereas the amount of radioactivity, irreversibly bound immediately after exposure, comprises only 10 - 40 % of the total radioactivity in tissues, this percentage rises up to about 70 % after 48 h. This is in agreement with our previous findings (13) that in rats the bulk of radioactivity after exposure to ^{14}C-vinyl chloride is excreted in the urine of the first 24 hours.

The largest amount of radioactivity immediately after exposure is found in liver (i.e., the organ of metabolism), and kidneys (i.e., the organ of excretion of the metabolites). These both organs, especially the liver, also contain the largest amounts of irreversibly protein bound metabolites. Other tissues showing remarkable irreversible binding of metabolites of vinyl chloride are lung, intestine, and spleen. These organs are known to be also active in metabolism of xenobiotics (26,27). Metabolically more inert organs such as brain, fat, and muscle, contain only minor amounts of irreversibly bound metabolites of vinyl chloride.

Irreversible binding of metabolites of vinyl chloride to nucleic acids of liver in vivo

In addition to the irreversible binding of metabolites of ^{14}C-vinyl chloride to protein, some radioactivity is also incorporated into nucleic acids. Table 4 shows the timal course of radioactive labelling of DNA and RNA of rat liver after a 5 h period of exposure to ^{14}C-vinyl chloride.

TABLE IV

Irreversible binding of metabolites of ^{14}C-vinyl chloride to DNA and RNA of liver of rats exposed to ^{14}C-vinyl chloride

Initial concentration of ^{14}C-vinyl chloride in atmosphere = 145 ppm
Time of exposure = 5 h
Uptake per rat (210 g) = 16.4 μmol vinyl chloride

Time after beginning of exposure	pmol metabolites bound to 1 mg		pmol metabolites bound per g liver (wet weight) to	
	DNA	RNA	DNA	RNA
5 h (immediately after exposure)	132	15.9	102	118
12 h	55	19.6	46	164
24 h	61	42.5	43	340
48 h	37	36.6	22	307

Both types of nucleic acids behave different. The peak of incorporation of ^{14}C into DNA is already reached immediately after exposure and is followed by a relatively rapid decline, probabely due to the action of repairing enzymes. Conversely, the specific labelling of RNA increases after exposure to its maximum at 24 h, and decreases much more slowly than that of DNA. The RNA isolated by the applied procedure consists mostly of ribosomal RNA which has a relatively slow turnover.

DISCUSSION

It has been established that in the rat vinyl chloride is rapidly metabolised to polar compounds like chloroethanol, chloroacetaldehyde, and chloroacetic acid (7,10,18). Furthermore, after administration of ^{14}C-vinyl chloride to rats the sulfur containing compounds thiodiglycolic acid and N-acetyl-S-(2-hydroxyethyl)-cysteine (8,9) have been recovered from urine as radioactive metabolites. The occurrence of cysteine derivatives suggests that chloroethylene oxide should be a reactive intermediate, primarily formed from vinyl chloride by action of drug oxidizing enzymes (18). This concept is supported by depletion of hepatic glutathione on exposure of rats to high concentrations of vinyl chloride (7), and by the recent findings (13) that effective inhibitors of microsomal oxidative drug metabolism completely block the uptake of vinyl chloride by rats. The irreversible binding of vinyl chloride to a protein in vitro in the rat liver microsomal system depends on presence of free sulfhydryl groups of the protein (12). This is a further argument for the implication of chloroethylene oxide in the irreversible protein binding of vinyl chloride. The present finding that superoxide radicals effect irreversible binding of vinyl chloride to protein can be explained by assuming that an epoxide intermediate is also formed by action of superoxide.

Interpretation of results dealing with the attachment of radioactivity of ^{14}C-vinyl chloride to macromolecules in vivo should consider that a substantial portion of ^{14}C-vinyl chloride may be degraded in vivo to C_1-fragments. This is indicated by expiration of 9 - 13 % ^{14}C after dosing ^{14}C-vinyl chloride (8). Hence radioactivity may enter the C_1-pool of tetrahydrofolic acid and subsequently may be incorporated into amino acids, particularly methionine and serine, and into purine nucleotides. Intermediates in the generation of $^{14}CO_2$ from ^{14}C-vinyl chloride may be considered to be ^{14}C-formaldehyde and formic acid. It is well known that application of ^{14}C-formaldehyde to rats results in a very rapid formation of ^{14}C-formate (28). Definitely, 82 % of the ^{14}C-radioactivity of ^{14}C-formaldehyde are exhaled as $^{14}CO_2$, and the rest is excreted as urinary metabolites. Urinary ^{14}C-methionine and ^{14}C-serine comprise about 8 % of administered radioactivity of ^{14}C-formaldehyde (29). This means that about 10 % of the excreted $^{14}CO_2$ radioactivity passed the C_1-pool of tetrahydrofolic acid. Accordingly it is to be expected that not more than 1 % of incorporated ^{14}C-vinyl chloride would enter the C_1-pool, and ought possibly be incorporated into proteins and nucleic acids.

However, simple comparison of the tissues' content of total metabolites of vinyl chloride, and the amount of irreversibly bound material immediately after exposure to ^{14}C-vinyl chloride (see Tab.3) demonstrates that as much as 10 - 40 % of the metabolites, depending on the particular tissue, are bound irreversibly. An incorporation of up to 1 % of ^{14}C of vinyl chloride into the C_1-pool, therefore, can not explain this order of magnitude of irreversible protein binding.

On the other hand, the much lesser radioactive labelling of the nucleic acids (Tab.4) ranges within a magnitude where incorporation of radioactive C_1-fragments has to be taken into account: only 0.23 % of radioactive metabolites in liver, immediately after ^{14}C-vinyl chloride exposure, are irreversibly bound to DNA, and maximal labelling of RNA comprises 0.8 % of the initial content of radioactive metabolites in the liver. However, three observations argue against the possibility that all observed radioactivity of nucleic acids may be attributed to C_1-incorporation:

(1) The rapid decline of radioactive labelling of DNA with time is faster than DNA turnover. This suggests that repairing enzymes eliminate radioactivity from the DNA, and elimination by repairing enzymes would not occur if radioactivity would be incorporated by the natural pathway of DNA synthesis.

(2) The totally different timal course of radioactivity in DNA and RNA should not be observed, if all radioactivity entered the nucleic acids <u>via</u> C_1-incorporation, because for synthesis of DNA and RNA the same cellular nucleotide pool is utilised.

(3) The previous finding (12) that rat liver microsomes <u>in vitro</u> catalyse irreversible binding of ^{14}C-vinyl chloride to RNA, is an argument that a similar mechanism under <u>in vivo</u> conditions should be feasible.

Our results show that, from a quantitative point of view, proteins are the major target for irreversible binding of metabolites of vinyl chloride to macromolecules. Coupling of reactive metabolites to cellular glutathione may be viewed as alternative pathway which prevents most of this reactive material from binding to protein. However, for carcinogenic action of vinyl chloride the most important reaction should be irreversible binding to nucleic acids, especially of the liver cell. Future studies, therefore, will answer the question as to the biological role of irreversible protein binding. It may be possible that binding to proteins is not a hazardous process, but has its significance, together with coupling to glutathione, in preventing large amounts of reactive metabolites from reacting with nucleic acids. This view is supported by previous findings on estrogens. Estradiol and ethynylestradiol, which are not at all to be regarded as chemical carcinogens, bind irreversibly to liver protein <u>in vitro</u> and <u>in vivo</u> (21,30,31), but these compounds do not bind to DNA or RNA (31). Similar results have been obtained with a reactive epoxide metabolite, norethisterone-4,5-epoxide, of the progestational drug norethisterone. It turned out that this epoxide, which was not mutagenic, also did not bind to nucleic acids, but did show irreversible binding to proteins (32).

Hence, in future studies particular attention has to be paid to the irreversible binding of possible carcinogens, as vinyl chloride, to nucleic acids, and to its chemical mechanism, both in vitro and in vivo.

ACKNOWLEDGMENTS

We wish to thank the Dynamit-Nobel AG, Troisdorf, for valuable support, and the Radiochemical Department of Farbwerke Hoechst, Frankfurt, for synthesis of the ^{14}C-vinyl chloride. We are indepted to Dr. P.J. Gehring, Midland, Michigan, for his interest in our study and for valuable discussion.

SUMMARY

Rat liver microsomes metabolise ^{14}C-vinyl chloride to intermediates which irreversibly bind to the microsomal protein and to soluble proteins and RNA, when these compounds are added to the incubation. A superoxide (O_2^-) generating system comprised of phenazine methosulfate and NADH also converts ^{14}C-vinyl chloride to metabolites which irreversibly bind to albumin. These data are consistent with the assumption of chloroethylene oxide being the primary reactive metabolite of vinyl chloride.

If rats are exposed to ^{14}C-vinyl chloride, about half of the radioactive metabolites in the liver microsomal fraction is bound irreversibly to microsomal protein, when assessed immediately after exposure Large amounts of polar, extractable, metabolites are present in the cytosol fraction. The amount of radioactivity in tissues of the rats, irreversibly bound immediately after exposure, comprises 10 - 40 % of the total radioactivity in tissues. This percentage rises up to 70 % after 48 hrs. Some radioactivity derived from ^{14}C-vinyl chloride is also incorporated into DNA and RNA of liver. Whereas the peak of incorporation of ^{14}C into DNA is already reached immediately after exposure to ^{14}C-vinyl chloride, specific labelling of RNA increases after exposure until its maximum after 24 hours.

REFERENCES

1) Creech,J.L. and Johnson,M.N.: Angiosarcoma of liver in the manufacture of polyvinyl chloride. J.Occupational Med.,1974, 16, 150-151.
2) Lee,F.J. and Harry,D.S.: Angiosarcoma of the liver in a vinyl chloride worker. Lancet, 1974, I, 1316-1318.
3) Buchter,A., Bolt,W.: Arbeitsmedizinische Aspekte der "Vinylchlorid-Krankheit". 14.Jahrestagung der Deutschen Gesellschaft für Arbeitsmedizin, Hamburg 1974. A.W. Gentner-Verlag, Stuttgart, 1974, pp. 307-312.
4) Thomas,L.B., Popper,H., Berk,P.D., Selikoff,I. and Falk,H.: Vinyl chloride induced liver disease. New. England J.Med. 1975, 292, 17-22.
5) Maltoni,C. and Lefemine,G.: Carcinogenicity bioassays of vinyl chloride. Env.Res. 1974, 7, 387-405.
6) Maltoni,C., Lefemine,G., Chieco,P. and Carretti,D.: La cancerogenesi ambientale e professionale: Nuove prospettive a luce della cancerogenesi da cloruro di vinile. Gli Ospedali della Vita, 1974, 1, 7-66.
7) Hefner,R.E., Watanabe,P.G. and Gehring,P.J.: Preliminary studies of the fate of inhaled vinyl chloride monomer in rats. Ann.N.Y. Acad.Sci. 1975, 246, 135-148.
8) Watanabe,P.G., McGowan,G.R. and Gehring,P.J.: Fate of ^{14}C-vinyl chloride after single oral administration in rats. Toxicol.Appl. Pharmacol.(in press)
9) McGowan,G.R., Watanabe,P.G. and Gehring,P.J.: Vinyl chloride urinary metabolites: Isolation and identification. Toxicol.Appl. Pharmacol. (in press)
10) Radwan,Z., and Henschler,D.: Uptake and metabolism of vinyl chloride in the isolated perfused liver preparation. Naunyn-Schmiedeberg's Arch.Pharmacol., Suppl., 1975, 287, R 100.
11) Bolt,H.M., Kappus, H., Buchter, A., Bolt, W.: Metabolism of vinyl chloride. Lancet 1975, I, 1425
12) Kappus, H., Bolt, H.M., Buchter, A., Bolt, W., Rat liver microsomes catalyse covalent binding of ^{14}C-vinyl chloride to macromolecules. Nature 1975, 257, 134-135
13) Bolt, H.M., Kappus, H., Buchter, A., Bolt, W., Disposition of 1,2-^{14}C-vinyl chloride in the rat. Arch.Toxicol. (in press)

(14) Malaveille, C., Bartsch, H., Barbin, A., Camus, A.M., Montesano, R., Croisy, A., Jacquignon, P.: Mutagenicity of vinyl chloride, chloroethyleneoxide, chloroacetaldehyde and chloroethanol. Biochem. Biophys. Res. Commun. 1975, 63, 363-370.

(15) Bartsch, H., Malaveille, C., Montesano, R.: Human, rat and mouse liver-mediated mutagenicity of vinyl chloride in S. Typhimurium strains. Int.J.Cancer 1975, 15, 429-437.

(16) Jaeger, R.J., Reynolds, E.S., Conolly, R.B., Moslen, M.T., Szabo S., Murphy, S.D.: Acute hepatic injury by vinyl chloride in rats pretreated with phenobarbital. Nature 1974, 252, 724-726.

(17) Miller, J.A.: Carcinogenesis by chemicals. Cancer Res. 1970, 30, 559-576.

(18) Van Duuren, B.L.: On the possible mechanism of carcinogen actic of vinyl chloride. Ann. New York Acad. Sci. 1975, 246, 259-267.

(19) Prema Kumar, R., Ravindranath, S.D., Vaidyanathan, C.S., Appaji Rao, N.: Mechanism of hydroxylation of aromatic compounds. Biochem.Biophys.Res.Commun. 1972, 48, 1049-1054.

(20) Bolt, H.M., Remmer, H.: The Accumulation of mestranol and ethyny oestradiol metabolites in the organism. Xenobiotica, 1972, 2, 489-498.

(21) Kappus, H., Bolt, H.M., Remmer, H.: Irreversible protein binding of metabolites of ethynylestradiol in vivo and in vitro. Steroid 1973, 22, 203-225.

(22) Bray, G.A.: A simple efficient liquid scintillator for counting aqueous solutions in a liquid scintillation counter. Anal.Biochem. 1960, 1, 279-285.

(23) Lowry, O.H., Rosebrough, N.J., Farr, A.L., Randal, R.J.: Protein measurement with the folin phenol reagent. J.biol.Chem. 1951, 193, 265-275.

(24) Georgiev, G.P., Mantieva, V.L.: The isolation of DNA-like RNA and ribosomal RNA from the nucleo-chromosomal apparatus of mammalian cells. Biochim.Biophys.Acta 1962, 61, 153-154.

(25) Scherrer, K., Darnell, J.E.: Sedimentation characteristics of partly labelled RNA from Hela cells. Biochem.Biophys.Res.Commun 1962, 7, 486-490.

(26) Orrenius, S., Ellin, A., Jakobsson, S.V., Thor, H., Cinti, D.L. Schenkmann, J.B., Estabrook, R.W.: The cytochrome P-450-containing mono-oxygenase system of rat kidney cortex microsomes. Drug Metab.Disp. 1973, 1, 350-356.

27) Litterst, C.L., Mimnaugh, E.G., Reagan, R.L., Gram, T.E.: Comparison of in vitro drug metabolism by lung, liver and kidney of several common laboratory species. Drug Metab. Disp. 1975, 3, 259-265.
28) Malorny, G., Rietbrock, N., Schneider, M.: Die Oxydation des Formaldehyds zu Ameisensäure im Blut, ein Beitrag zum Stoffwechsel des Formaldehyds. Naunyn-Schmiedeberg's Arch.Pharmacol. 1965, 250, 419-436.
29) Brock-Neely, W.: The metabolic fate of formaldehyde-^{14}C intraperitoneally administered to the rat. Biochem.Pharmacol. 1964, 13, 1137-1142.
30) Marks, F., Hecker, E.: Metabolism and mechanism of action of oestrogens. Biochim.Biophys.Acta 1969, 187, 250-265.
31) Bolt, H.M., Kappus, H.: Irreversible binding of ethynyl-estradiol metabolites to protein and nucleic acids as catalyzed by rat liver microsomes and mushroom tyrosinase. J.Steroid Biochem. 1974, 5, 179-184.
32) Kappus, H., Bolt, H.M.: Irreversible protein binding of norethisterone (norethindrone) epoxide. Steroids (in press).

RESUME

Les microsomes de foie de rat métabolisent le chlorure de vinyle marqué au C^{14} en produits intermédiaires qui se lient irréversiblement à la protéine microsomique et aux protéines solubles et à l'ARN, lorsque ces composés sont ajoutés à l'incubation. Un système générateur de superoxyde (O_2^-) composé de méthosulfate de phénazine et de NADH convertit également le ^{14}C - chlorure de vinyle en métabolites qui se lient irréversiblement à l'albumine. Ces données corroborent l'hypothèse selon laquelle l'oxyde de chloréthylène serait le métabolite réactif primaire du chlorure de vinyle.

Si des rats sont exposés au ^{14}C - chlorure de vinyle, la moitié environ des métabolites radioactifs de la fraction microsomique de foie apparaissent liés irréversiblement à la protéine microsomique, immédiatement après l'exposition. De grandes quantités de métabolites polaires extractibles sont présentes dans la fraction cytosole. La radioactivité dans les tissus de rats, liée irréversiblement tout de suite après l'exposition, représente 10 à 40% de la radioactivité totale dans les tissus. Cette proportion s'élève jusqu'à 70% après 48 heures. Une certaine radioactivité issue du ^{14}C - chlorure de vinyle est aussi incorporée à l'ADN et à l'ARN du foie. Si le pic d'incorporation du ^{14}C à l'ADN est déjà atteint immédiatement après l'exposition au ^{14}C - chlorure de vinyle, le marquage spécifique de l'ARN augmente après l'exposition jusqu'à un maximum après 24 heures.

HYDROCARBURES POLYCYCLIQUES AROMATIQUES CANCEROGENES DANS LES PRODUITS PETROLIERS PREVENTIONS POSSIBLES DU CANCER DES HUILES MINERALES

C. THONY (1), J. THONY (1),
M. LAFONTAINE (2) et J.C. LIMASSET (2)

(1) Centre de Médecine du Travail de Cluses
B.P. n°113, 74302 Cluses, France

(2) Institut National de Recherche et de Sécurité, Service Chimie — Toxicologie
B.P. n°27, 54500 Vandoeuvre-les-Nancy, France

Les produits d'origine pétrolière sont suffisamment largement dispersés dans le monde industriel pour qu'on puisse considérer qu'ils font partie de l'environnement de l'homme moderne.

Si tout le monde, à la suite d'études historiques bien connues, a conscience que les résidus de combustions incomplètes diverses ou les produits de pyrogénation de la houille véhiculent des agents cancérogènes notoires, on n'a, par contre, pas l'habitude d'associer un risque cancérogène au contact avec les produits pétroliers.

Dans le travail que nous souhaitons résumer ici*, c'est le milieu professionnel qui sert de révélateur. Ce travail comprend une enquête clinique chez des personnes exposées aux huiles d'usinage des métaux, associée à une étude chimique des produits mis en cause. La possibilité d'introduire des critères toxicologiques dans les cahiers des charges spécifiant les qualités des produits pétroliers, et par là, la possibilité d'éviter la dispersion de produits présentant des risques cancérogènes sont ensuite discutées.

1ère Partie : RISQUE CANCEROGENE POUR L'HOMME DES HUILES D'USINAGE

L'enquête a porté sur 650 entreprises de mécanique spécialisées dans le décolletage (usinage au tour automatique) employant 6.500 personnes dont 5.000 sont exposées à l'huile par contact cutané direct, par macération avec les vêtements de travail imprégnés et par le séjour dans l'aérosol produit par les machines.

Un suivi médical a été effectué par le Centre de Médecine du Travail de CLUSES (Haute-Savoie) depuis 15 ans. Parmi les affections cutanées constatées, les plus graves sont : le papillome, très répandu et considéré comme la lésion pré-cancéreuse typique, le kérato-acanthome dont la réputation de bénignité paraît douteuse et l'épithélioma spinocellulaire dont 133 cas ont été enregistrés en 14 ans (1960-1974); les lésions malignes siègent au niveau du scrotum (63 % des cas), aux avant-bras et au mains (30 %) et plus rarement à la face et au cou.

---

* Pour plus de détails, voir : THONY, LAFONTAINE, LIMASSET [1] Arch. Mal. Prof. 36, 37-52 (1975).

Tableau I : Huiles de coupe utilisées pendant la période d'après-guerre (origine distillation de la houille)

Nombre de cycles polycondensés	H.P.A. identifiés
3 cycles	anthracène phénanthrène
4 cycles	pyrène fluoranthène chrysène benzo[a]fluorène benzo[a]anthracène
5 cycles	pérylène benzo[a]pyrène benzo[b]fluoranthène benzo[k]fluoranthène
6 cycles	benzo[g,h,i]pérylène indéno-1,2,3[c,d]pyrène ou o-phénylène pyrène

Le tableau II donne la liste des H.P.A. identifiés dans une huile de coupe d'origine pétrolière du type de celles qui sont utilisées depuis 1950.

Tableau II : Huile de coupe actuelle (origine pétrolière)

Nombre de cycles polycondensés	H.P.A. identifiés
4 cycles	pyrène méthylpyrène chrysène alkylchrysène fluoranthène benzo[a]anthracène
5 cycles	benzo[a]pyrène benzo[b]fluoranthène benzo[k]fluoranthène

- L'étude des facteurs du risque (17 paramètres) a porté sur 3.000 personnes choisies au hasard parmi les travailleurs de toutes professions de la même agglomération. Il apparaît que certains facteurs jouent un rôle : la biotypologie, les prédispositions, les données psychosociologiques propres à la région et les conditions locales de travail.

  Mais les facteurs favorisants les plus importants sont :
  - l'exposition à l'huile au cours du travail,
  - l'hygiène individuelle et professionnelle (la fréquence des lésions est en moyenne 3 fois plus élevée dans les ateliers de type artisanal beaucoup plus pollués).

- L'importance du risque est facile à apprécier étant donné la rareté de la localisation scrotale chez l'homme (0,1 pour 100.000 dans la population générale d'après une enquête effectuée dans 3 départements français). Ce taux monte à 25 pour 100.000 dans la région de CLUSES (population de 40 à 45.000 personnes) et atteint 330 pour 100.000 pour la catégorie " décolleteurs ", la plus exposée à l'huile.

- Pendant la durée de l'enquête, une flambée épidémique s'est manifestée de 1955 à 1965 correspondant assez bien à l'introduction dans les ateliers, quinze ans auparavant, d'huiles "anthracèniques" (dérivées de la houille) utilisées comme succédané des huiles de pétrole pendant la période de guerre et d'après-guerre. Ces huiles ont disparu officiellement du marché en 1947 (pratiquement en 1950) et à partir de 1965, sont apparus des cas de cancer chez des personnes n'ayant jamais été en contact avec ces huiles "de guerre", mais uniquement avec des huiles d'origine pétrolière du type de celles qui sont encore utilisées aujourd'hui.

2ème Partie : CONCENTRATION EN SUBSTANCES CANCEROGENES DE LA FAMILLE DES HYDROCARBURES POLYCYCLIQUES AROMATIQUES (H.P.A.) DES HUILES UTILISEES.

Le Centre de Recherche de l'I.N.R.S. a développé une méthode d'extraction et de dosage des H.P.A. à partir d'échantillons huileux**, qui a été appliquée à l'analyse d'huiles de coupe anciennes (huiles "de guerre") et actuelles (huiles "entières" d'origine pétrolière) utilisées dans l'industrie du décolletage de la région de CLUSES.

a) - ASPECT QUALITATIF

Le tableau I donne la liste des H.P.A. identifiés dans une huile "de guerre". Il s'agit d'une base "anthracènique" qui avait été épurée et formulée à l'époque pour usage en tant qu'huile de coupe.

** Le protocole d'analyse comprend : une extraction liquide-liquide (cyclohexane/nitrométhane), une première séparation chromatographique sur colonne d'alumine activée (élution avec un gradient cyclohexane-benzène), une seconde sur plaque de cellulose acétylée à 30 % (éluants : EtOH/CH_2Cl_2/H_2O-20:10:1), puis une identification en spectrophotométrie de fluorescence par comparaison avec des échantillons authentiques (voir [1]).

TABLEAU III : CONCENTRATIONS EN BENZO[a]PYRENE DE DIVERSES HUILES MINERALES :

NATURE OU UTILISATION DE L'HUILE	ECHANTILLON	RENSEIGNEMENTS DIVERS	Conc. en B[a]P :	
			H. neuve	H. usagée
Huiles "anthracéniques" (distillation de la houille) Utilisées pour la coupe pendant la guerre et l'après-guerre		1 - Huile anthracénique (déphénolée et désaminée)	1 g/kg	
		2 - Spindle anthracénique (Procédé GERLAND-1947)	10 mg/kg	
		3 - H. anthracénique épurée (autre procédé-1942-45)	0,4 mg/kg	
Huiles de coupe : (origine pétrolière) (collectées en juillet et septembre 1973)		Type d'usinage et métal travaillé		
	1	Décolletage (acier ord.)	0,6 µg/l	0,6 µg/l
	2	Reprise (acier et laiton)	0,9 µg/l	1 µg/l
	3	Décolletage (acier ord.)	3 µg/l	9 µg/l
	4	Décolletage	5 µg/l	5 µg/l
	5	Décolletage et usinage général (tous métaux, sauf laiton)	6 µg/l	15 µg/l
	6	Décolletage (acier ord.)	10 µg/l	20 µg/l
	7	Décolletage (inox)	13 µg/l	24 µg/l
	8	Taillage-tournage (acier, laiton et duralumin)	25 µg/l	45 µg/l
	9	Décolletage et usinage général (tous métaux)	45 µg/l	60 µg/l
	10	Usinage général (aciers spéciaux)	95 µg/l	105 µg/l
	11	Décolletage et usinage général (laiton)	110 µg/l	120 µg/l
	12	Décolletage	133 µg/l	141 µg/l
	13	Décolletage et usinage général (acier ord.)	160 µg/l	200 µg/l
Huiles de trempe : (origine pétrolière) (collectées en 1974)		Température des pièces trempées		
	1	1100°C	2 µg/l	20 µg/l
	2	875-900°C	7 µg/l	100 µg/l
	3	950°C	8 µg/l	500 µg/l
	4	875-900°C	15 µg/l	110 µg/l
	5	900°C	15 µg/l	550 µg/l
	6	1050°C	40 µg/l	90 µg/l
Huile lubrifiante pour moteur			30 µg/l	9 mg/l

b) - ASPECT QUANTITATIF

Le tableau III rassemble des résultats de dosage des quantités de benzo[a]pyrène présentes dans les huiles.

- les huiles "anthracèniques" contiennent des quantités de l'ordre de plusieurs mg/kg de B[a]P (associé à d'autres H.P.A.) et leur pouvoir cancérogène évident est à rapprocher de celui des goudrons et brais de houille.

- les huiles pétrolières neuves (telles qu'elles sont livrées actuellement aux utilisateurs) contiennent assez généralement des H.P.A. avec des concentrations de B[a]P variables (de 0,5 à 150 µg/l).

- sur des prélèvements d'huiles usagées (ayant travaillé sur machine) effectués parallèlement, on constate une tendance à l'enrichissement en B[a]P. Ce phénomène, qui peut s'expliquer par des réactions de polycondensation aromatique par élévation de température, apparaît de façon beaucoup plus nette sur une série d'huile de trempe ayant subi des échauffements plus importants (trempe de pièces d'acier à température voisine de 1.000° C).

3ème Partie : PREVENTIONS POSSIBLES DU RISQUE CANCEROGENE DES HUILES MINERALES

Une réflexion commune sur les actions possibles est engagée, en FRANCE, entre l'I.N.R.S. et les Professions du raffinage du pétrole et de l'industrie des lubrifiants.

En dehors de la nécessité évidente d'informer les utilisateurs des risques graves présentés par les huiles minérales actuelles, il est souhaitable d'agir sur les produits eux-mêmes pour les assainir.

Il semble que les procédés modernes de raffinage, en particulier l' "extraction au solvant" (qui élimine la fraction aromatique) et l'hydrogénation catalytique (qui peut parvenir à transformer chimiquement les cycles aromatiques) puissent permettre d'abaisser notablement le taux d'hydrocarbures polyaromatiques des bases pétrolières servant à l'élaboration des huiles industrielles.

Le "groupe d'étude sur la toxicité des huiles de coupe " qui s'est constitué, cherche actuellement, par le biais d'essais circulaires inter-laboratoires, à mettre au point une méthode de dosage simple et rapide des H.P.A. totaux d'une fraction pétrolière, qui permette de contrôler les performances des installations de raffinage dans ce sens.

Le problème que pose le "vieillissement" des huiles en cours d'utilisation pourrait être réglé par la définition de rythmes de vidange adaptés au travail effectué. L'industrie de la régénération pourrait ensuite intervenir pour rénover les huiles non seulement en fonction de critères technologiques comme cela se pratique actuellement, mais aussi de critères toxicologiques.

L'établissement de telles normes toxicologiques dans les spécifications des produits pétroliers qui devra s'appuyer sur des données expérimentales établies sur animaux, (qui existent déjà partiellement), sera sans doute le point le plus délicat.

Summary.

Carcinogenic Polycyclic aromatic hydrocarbons in Petroleum products.
Possible Prevention of Mineral Oil Cancer

A real epidemy of cutaneous cancer has appeared from 1954 in a french area specialized in the metal machining industry (valley of the Arve, Haute-Savoie).

Two investigations have been carried out :

- one, of clinical and epidemiological nature over a period of 15 years, has recorded 133 cases of spinocellular epitheliomas often located in the scrotum and has allowed to estimate the hazard for the population of this region at 25 per 100 000 (thats is to say 36 times more than for the general population).

- the other, of chemical nature, has revealed on one hand that the new cutting oils, such as delivered to the utilizers, rather frequently contain benzo pyrene (concentration : 0,5 to 150 µg/ℓ), and on the other hand that the used oils are inclined to grow richer in carcinogenic hydrocarbons.

The prevention must be directed towards a supplementary effort in the refining of industrial oils and proposals for toxicological threshold value, which would allow supply machine shops with oils no longer presenting a carcinogenic hazard.

Environmental Pollution and Carcinogenic Risks
Pollution de l'environnement et risques cancérogènes

CARCINOGENIC POTENTIAL OF CHLORINATED ETHYLENES TENTATIVE MOLECULAR RULES

D. HENSCHLER (*), G. BONSE (*) and H. GREIM (**)

(*) Institute of Toxicology, Universität of Würzburg, D-8700 Würzburg, Versbacher Landstrasse 9.

(**) Department of Toxicology, Gesellschaft für Strahlen- und Umweltforschung, D-8042 München-Neuherberg, R.F.A.

The carcinogenic action of vinyl chloride (VCM) is now well established in man (1) and in experimental animals (2). Malignant liver tumors in mice have also been produced with trichlorethylene (TRI), though after aministration of rather high daily oral doses (0.5 and 1.0 g/kg) in vegetable oil (3). Strong mutagenic activity in vitro after enzymatic activation has been demonstrated with VCM (4,5), vinylidene chloride (VDC), and chloroprene (6). From these findings the possibility has been discussed by consumers and producers but also by the scientific community that all chlorinated ethylenes might carry a mutagenic and/or carcinogenic risk. We therefore have carried out a systematic study on the mechanisms of bioactivation and on the mutagenic activity of the whole series of chlorinated ethylenes with a short term in vitro system which makes use of the metabolic activating liver microsomal enzymes.

Details of the methods used have been described elsewhere (7,8). All chlorinated ethylenes are metabolized to oxiranes (epoxides), as a first step. The amount metabolized varies widely (7). The oxiranes, as strong electrophilic molecules, may react directly with nucleophiles of the cell, or may undergo intramolecular rearrangement as follows:

Cl\C=C/Cl → Cl\C—C/Cl → CCl₃—C(=O)Cl
Cl/   \Cl   Cl/ O \Cl
TETRACHLOROETHYLENE

Cl\C=C/H → Cl\C—C/H → CHCl₂—C(=O)Cl
Cl/   \Cl   Cl/ O \Cl
TRICHLOROETHYLENE

H\C=C/H → H\C—C/H
Cl/ \Cl   Cl/ O \Cl
cis-1.2-DICHLOROETHYLENE
                                    → CHCl₂—C(=O)H
H\C=C/Cl → H\C—C/Cl
Cl/ \H   Cl/ O \H
trans-1.2-DICHLOROETHYLENE

Cl\C=C/H → [Cl\C—C/H]* → CH₂Cl—C(=O)Cl
Cl/ \H   Cl/ O \H
1.1-DICHLOROETHYLENE

Cl\C=C/H → Cl\C—C/H → CH₂Cl—C(=O)H
H/   \H   H/ O \H
VINYL CHLORIDE

The thermal rearrangement results, according to a series of in vitro studies (9-11), either in acyl chlorides or in chlorinated acetaldehydes. Asymmetric chlorine substitution of the ethylenes and oxiranes, respectively, as with TRI, VCM and VDC, renders instability to the molecules (12), whereas the compounds with symmetric substitution are rather stable (tetrachloroethylene, cis- and trans-1.2-dichloroethylene); the oxirane of the most polar compound, VDC, which is expected to possess the lowest stability has up to now resisted to synthesis (7).

Mutagenicity experiments (8) gave the results presented in fig.1. The most active compound was VCM; TRI and VDC were far less but still significantly mutagenic. These findings are indicative of the following rule. as outlined in fig.2:

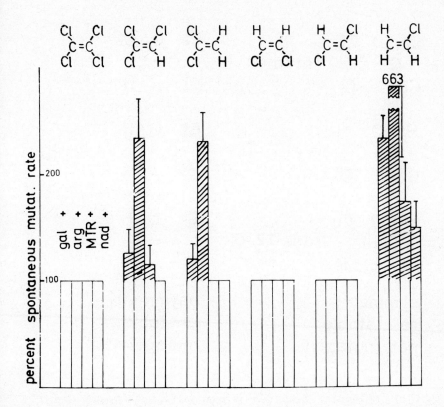

Fig.1. Mutagenicity of chlorinated ethylenes after incubation in a metabolic activating microsomal system (8). Results as per cent of spontaneous mutation rate in different operons (galactose, arginine, methyltryptophane, NAD^+) of E.coli K_{12}.

Chlorinated ethylenes which are metabolized via symmetric oxiranes which in turn are relatively stable (tetrachloroethylene, cis- and trans-1.2-dichloroethylene) are not mutagenic whereas those forming asymmetric and thus unstable oxiranes exert mutagenic activity.

No interrelationship exists between the type of product of
thermal rearrangement (aldehyde or acyl chloride). This again is
suspected of a direct reaction of oxiranes with susceptible targets
forming the biochemical lesion for mutagenesis. In the case of VCM,
this suspicion has been validated in that VCM oxirane is far more
effective than chloroacetaldehyde in an *in vitro* bacterial test
system (5).

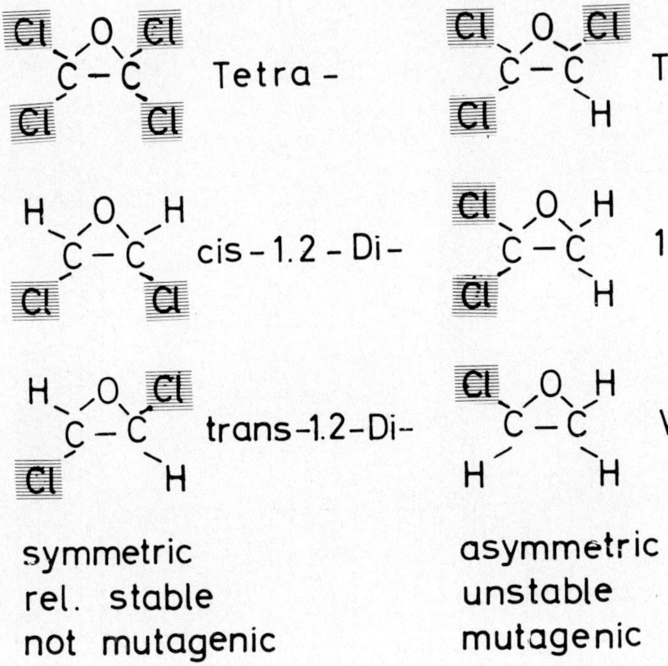

```
   Cl   O   Cl                          Cl   O   Cl
    \  / \  /                            \  / \  /
     C - C      Tetra-                    C - C      Tri-
    /     \                              /     \
   Cl      Cl                           Cl      H


   H    O    H                          Cl   O   H
    \  / \  /                            \  / \  /
     C - C      cis-1.2-Di-               C - C      1.1.-Di-
    /     \                              /     \
   Cl      Cl                           Cl      H


   H    O   Cl                          Cl   O   H
    \  / \  /                            \  / \  /
     C - C      trans-1.2-Di-            C - C      VC
    /     \                              /     \
   Cl      H                            H       H
```

symmetric   asymmetric
rel. stable   unstable
not mutagenic   mutagenic

Fig.2. Molecular features of oxidative metabolic inter-
       mediates (oxiranes) of chlorinated ethylenes in
       relation to their mutagenic potential.

With trichloroethylene, the reaction mechanism is equivacol.
In vitro the thermal rearrangement of the oxirane entirely goes to
dichloroacetylchloride, as outlined above; only under catalysis of
Lewis acids chloral hydrate is formed (7,12). *In vivo*, however
essentially no dichloroacetic acid is formed metabolically (7,13);
the only metabolites detected are chloral hydrate and its reduction
or oxidation products, respectively, trichloroethanol and trichloroacetic acid. This in connection with the above considerations of the
chemical reactivity of the oxiranes and their products of rearrangement again supports the view of a direct attack of the oxirane
as (an) essential reaction for mutagenesis.

Summary. Chlorinated ethylenes are activated in mammalian metabolism to oxiranes. Only those with asymmetric chlorine substitution are mutagenic (trichloroethylene, vinylidene chloride and vinyl chloride) whilst the symmetric molecules (tetrachloroethylene, cis- and trans-1.2-dichloroethylene) are inactive in this respect. Thus, stability of the oxiranes (higher in the symmetric molecules, lower in the asymmetric ones) seems to be the important feature for the mutagenic, and possibly carcinogenic potential.

Literature

(1) Lloyd, J.W.: J.occup.Med. 16, 809 (1974. - 17, 333 (1975).
(2) Viola, P.L., A.Bigotti and A.Caputo: Cancer Res. 31,516 (1971). Maltoni, C., G.Lefemine: Environm.Res. 7, 387 (1974).
(3) Memorandum, Dept.of Health, Education & Welfare, Washington, 20.3.1975.
(4) Rannug, U., A.Johansson, C.Ramel and C.A.Wachtmeister: Ambio 3, 194 (1974).
(5) Bartsch, C.Malaveille and R.Montesano: Int.J.Cancer 15,429 (1975).
(6) Bartsch, H., C.Malaveille, R.Montesano and L.Tomatis: Nature 255, 641 (1975).
(7) Bonse, G., Th.Urban, D.Reichert and D.Henschler: Biochem.Pharmacol. (1975) in press.
(8) Greim, H., G.Bonse, Z.Radwan, D.Reichert and D.Henschler: Biochem.Pharmacol. (1975) in press.
(9) Gross, H., and J.Freiberg: J.für praktische Chemie 311, 506 (1969).
(10) Griesbaum, K., R.Kibar and B.Pfeffer: Liebigs Ann.Chem. 1975 214.
(11) Frankel, D.M., C.E.Johnson and H.M.Pitt: J.org.Chem. 22, 1119 (1957).
(12) Bonse, G. and D.Henschler: in preparation.
(13) Leibman, K.C. and E.Ortiz: 6th Int.Congress of Pharmacol. Helsinki, 1975. Abstracts 257, Nr.608.

Résumé. Les éthylènes chlorés sont activés en oxiranes dans le métabolisme mammalien. Seuls ceux de substitution chlorée asymétrique sont mutagènes (trichloro-éthylène, chlorure de vinylidène et chlorure de vinyle) alors que les molécules symétriques (tétrachloro-éthylène, cis- et trans dichloro -1,2 éthylène) sont, de ce point de vue, inactives. La stabilité des oxiranes (plus forte dans les molécules symétriques et plus faible dans les asymétriques) semble donc la caractéristique importante du pouvoir mutagène, et, peut-être aussi, cancérogène.

IV

RADIATION POLLUTION
POLLUTION PAR LES RADIATIONS

Chairman / *Président* : Dr. E. MASTROMATTEO

LES RISQUES CANCERIGENES
LIES A LA POLLUTION PAR LES RADIATIONS

Raymond LATARJET

Fondation Curie-Institut du Radium, 26, rue d'Ulm, 75005 Paris, France

INTRODUCTION

Les rayons UV d'une part, les radiations ionisantes d'autre part, sont cancérigènes. La participation de l'irradiation naturelle dans la cancérogénèse dite spontanée chez l'homme, est encore incertaine. On a de bonnes raisons de penser qu'elle est faible.

On a pu calculer que la dose de radiations ionisantes qui double la fréquence des mutations survenant chez l'homme est en moyenne de l'ordre de 60 rems (variable selon l'étalement de la dose), soit environ 30 fois la radiation naturelle intégrée en 30 ans, durée moyenne de la vie des gamètes humains mis en jeu dans la reproduction. Ceci prouve que le rayonnement ionisant naturel ne peut jouer qu'un rôle très faible dans la mutagénèse dite spontanée chez l'homme. Cette conclusion est certainement valable pour la cancérogénèse dite spontanée.

En revanche, le rayonnement UV solaire est responsable de la plupart des cancers cutanés épidermiques chez l'homme, et de ceux-là seuls, puisque ce rayonnement n'atteint aucun autre tissu, et n'agit pas à distance par substance photoformée diffusible.

Quoi qu'il en soit, toute pollution qui se traduit par un apport supplémentaire en radiations de ces types suggère un risque cancérigène. Le problème est d'apprécier ce risque en termes quantitatifs de façon à définir quand il est acceptable et quand il ne l'est pas. En outre la connaissance des mécanismes de la cancérogénèse par les radiations devrait nous fournir des moyens de diminuer ce risque.

Dans un premier paragraphe on va montrer que certaines pollutions chimiques en s'attaquant à l'ozone atmosphérique, se transforment en pollution par un excès de rayonnement UV solaire. Ce problème est accessible à un traitement quantitatif.

Dans un second paragraphe consacré aux radiations ionisantes, on discutera des divers moyens susceptibles de fournir des données quantitatives sur la radiocancérogénèse chez l'homme. On discutera ensuite du problème fondamental du "seuil".

Dans un troisième paragraphe, on discutera d'une nouvelle théorie de la radiocancérogénèse, laquelle serait applicable à la cancérogénèse chimique.

Enfin, dans un dernier paragraphe, forts de cette similitude possible, nous envisagerons comment se situe la pollution par les radiations par rapport à la pollution chimique cancérigène.

Au cours d'une telle étude il ne faut pas oublier que la cancérogénèse dite spontanée affecte environ un homme sur cinq, proportion considérable, et que tout effet de pollution s'exerce au-dessus de cette base dont le niveau est déjà très élevé.

RADIATIONS NON IONISANTES : ULTRAVIOLET SOLAIRE ET OZONE

Le rayonnement UV solaire qui parvient au sol est limité du côté des courtes longueurs d'onde vers 3000 Å. Cette limitation est due à l'absorption sélective par l'ozone atmosphérique. Le rayonnement UV solaire de très courtes longueurs d'onde ($\lambda < 2500$ Å), lorsqu'il atteint les couches supérieures de l'atmosphère terrestre, réagit sur l'oxygène, et produit de l'ozone. Inversement, les métabolismes azotés des êtres vivants, et notamment des bactéries de la mer et du sol, dégagent des composés oxygénés de l'azote très stables, qui diffusent jusqu'à la stratosphère, et réagissent sur l'ozone pour redonner de l'oxygène. L'ensemble de ces réactions peut être schématisé comme suit :

$$\begin{cases} O_2 + h\nu \longrightarrow O + O \\ O + O_2 \longrightarrow O_3 \end{cases}$$

$$\begin{cases} O_3 + NO \longrightarrow NO_2 + O_2 \\ O + NO_2 \longrightarrow NO + O_2 \end{cases}$$

Ainsi s'établit un équilibre $2 O_2 \rightleftarrows O_3 + O$ autour d'une valeur moyenne relativement stable correspondant à une couche unique de 3 mm d'épaisseur à la pression de 76 cm de mercure et à 0° C. La majeure partie de cet ozone se trouve dans la stratosphère entre 16 et 28 km d'altitude.

L'ozone absorbe formidablement les radiations UV au-dessous de 3100 Å. C'est pourquoi le rayonnement au sol se trouve coupé vers 3000 m ce qui permet à la vie terrestre de se dérouler à l'abri des radiations abiotiques.

Les activités biologiques du rayonnement UV solaire qui dépendent de l'absorption par les nucléoprotéines (effets mutagène et cancérigène notamment) dépendent étroitement des radiations de longueurs d'ondes les plus courtes qui nous parviennent. Ces activités dépendent donc de la quantité d'ozone et de ses variations. Si la quantité d'ozone diminue, ces effets augmentent rapidement, et inversement. C'est ainsi que les effets solaires augmentent beaucoup en altitude et que les cancers de la peau chez l'homme se produisent de préférence chez les individus soumis à de fortes insolations[*].

On peut alors penser établir une relation entre la quantité d'ozone et ce qu'on peut appeler "l'activité cancérigène" du rayonnement solaire au sol. A cet effet on dispose des données quantitatives nécessaires, à savoir :

a) la distribution spectrale du rayonnement solaire au dehors de l'atmosphère

b) la courbe spectrale d'absorption de l'ozone;

c) l'absorption atmosphérique par d'autres facteurs que l'ozone, notamment par la diffusion moléculaire (Rayleigh);

---

[*] Lorsqu'en 1894 le dermatologiste allemand Unna étudia ces cancers, il leur donna le nom de "Seemanshautkarzinom", c'est à dire "cancer de la peau des marins".

d) le spectre d'action cancérigène, c'est à dire la courbe spectrale d'efficacité cancérigène des radiations UV monochromatiques (ce spectre est mal connu mais on peut, sans commettre d'erreur importante, l'assimiler au spectre d'action érythémale, c'est à dire la courbe d'érythème - coup de soleil - qui est assez bien connue).

En 1935, alors que l'héliothérapie de la tuberculose, de l'ostéomyélite, du rachitisme, et d'autres maladies, conservait sa valeur, j'eus l'idée, pour la première fois, d'établir cette relation entre la quantité d'ozone et les effets du rayonnement UV solaire au sol (1). Ce calcul était particulièrement destiné à établir la variation d'activité au cours de la journée. Lorsque le soleil s'abaisse sur l'horizon, la quantité d'ozone traversée augmente (comme le cosinus de la distance zénitale du soleil). Le problème est ensuite tombé en sommeil, d'autant plus que l'héliothérapie était progressivement remplacée par la pharmacothérapie.

Le problème est sorti de ce sommeil en 1972 par suite du développement prévu pour les vols supersoniques. Les avions du type CONCORDE volent dans la stratosphère à des altitudes où la quantité d'ozone est importante. Les réacteurs dégagent de grandes quantités d'oxyde d'azote, NO_x, qui contribuent, comme on vient de voir, à transformer l'ozone en oxygène, donc à augmenter l'activité du rayonnement solaire au sol. Dès lors, les calculs de 1935 furent repris un peu partout sur des bases plus modernes. Ils ne posaient aucune difficulté; aussi des résultats quasi superposables furent-ils obtenus presque en même temps en France (2), aux Etats-Unis (3), et aussi en Angleterre et en Allemagne.

En ce qui concerne l'action cancérigène qui nous intéresse, sa relation avec la quantité d'ozone apparait simple. On peut dire sommairement que si la quantité d'ozone diminue de n pour 100, l'activité cancérigène augmente de 2n pour 100. Par exemple, si la quantité d'ozone diminue de 5 %, il y aura 10 % de cancers cutanés en plus, soit environ 1800 par an en France et 8500 par an aux Etats-Unis. Parmi ces cancers, 1 % sont des cancers mélaniques dont le pronostic est très sombre.

Connaissant la quantité d'oxyde d'azote rejetée par les réacteurs et la relation entre cette quantité d'oxyde et la quantité d'ozone décomposée, on peut établir la relation entre "heures de vol et quantité d'ozone détruite", et décider ce qui parait admissible et ce qui ne l'est pas*.

Le problème des vols supersoniques n'est pas le seul qui soulève la question de la pollution de l'ultraviolet solaire par l'intermédiaire de l'ozone. En effet la décomposition de ce gaz n'est pas le privilège des oxydes d'azote. Elle peut être produite également par l'anhydride sulfureux, SO_2, produit de combustion du charbon et du pétrole. Ce gaz, très stable, monte jusqu'à la stratosphère où il se décompose en SO + O. Dès lors SO agit sur l'ozone comme NO.

---

* La tendance actuelle consiste à considérer que 500 supersoniques du type CONCORDE, volant 7 heures par jour en moyenne, devraient constituer un plafond mondial. Cette conclusion peut évidemment être modifiée si les caractéristiques de combustion des réacteurs sont elles-mêmes modifiées. D'importants travaux sont en cours pour diminuer le dégagement d'oxydes d'azote.

Une autre source de pollution, qui procède du même mécanisme, vient d'être mise en évidence. Il s'agit du fréon, fluorochlorure double de carbone, dont la production industrielle et le dégagement dans l'atmosphère croissent continuellement. Ce produit constitue en effet la base des aérosols utilisés en cosmétique. C'est également le produit utilisé dans les réfrigérateurs. De formules CF_2Cl_2 et $CFCl_3$, il est extrêmement stable et diffusible. On estime qu'une fois libéré dans l'air il peut y demeurer de 30 à 150 ans. Il diffuse donc dans la stratosphère et, vers 30.000 m, se trouve exposé aux rayons UV de courtes longueurs d'ondes de 1750 à 2200 A. Il est alors décomposé en libérant du chlore à une altitude où la concentration d'ozone est élevée. Se produisent alors les réactions suivantes (4) :

$$Cl + O_3 \longrightarrow ClO + O_2$$

$$ClO + O \longrightarrow Cl + O_2$$

On se trouve donc en présence d'un phénomène de décomposition catalytique comparable à celui des oxydes d'azote. Quantitativement, on estime que la production mondiale de fréon a atteint 400.000 tonnes en 1972. Cette quantité dégagée dans l'atmosphère, année après année, devrait aboutir à une production d'atomes de chlore dans la stratosphère de 5.10^7 cm^{-2} sec^{-1} qui est du même ordre que les 5.10^7 atomes de NO induits par une large flotte d'avions supersoniques (5). Ce problème est donc comparable, sinon plus aigu que l'autre.

II - RADIATIONS IONISANTES

On peut distinguer à priori deux grands types de pollution, soit par les radiations électromagnétiques d'origine externe qui entrainent une irradiation générale de l'organisme, soit par des corps radioactifs inhalés ou ingérés et qui obéissent à des tropismes organiques (par exemple le Strontium 90 qui s'accumule dans les os, l'Iode 131 dans la thyroïde, etc...).

Les risques cancérigènes des fortes doses appliquées à quelques individus sont connus et étudiés depuis le début de ce siècle. Citons par exemple les accidents produits par les rayons X chez les radiologues et les ingénieurs de "l'époque héroïque", ou les accidents anciens du radium et du thorium chez les malades traités. Plus récemment, citons les leucémies consécutives à des irradiations thérapeutiques loco-régionales pour les hypertrophies du thymus chez l'enfant ou pour la spondylite ankylosante chez l'adulte, lymphoïdes pour les premières, myéloïdes pour les secondes.

Mais ceci n'est pas notre propos. La pollution, c'est l'administration de faibles doses à un grand nombre d'individus. Ce problème est d'actualité puisqu' il est lié au développement des centrales nucléaires.

On doit alors considérer :

a) Les effluents gazeux radioactifs qui passent dans l'atmosphère.

♦ Ce sont principalement le Xénon 133 de période 5,27 jours et le Krypton 85 de période 10,74 ans (le rejet de l'Iode 131 est négligeable). C'est surtout le Xénon qui compte puisqu'il y en a 500 fois plus que de Krypton. Notons que cette pollution a été beaucoup diminuée au cours des récentes années par nos progrès dans les processus de rétention. C'est ainsi que le réacteur franco-belge de Shooz, dont la puissance est de 1000 Mw thermiques et 300 Mw électriques dégageait dans l'atmosphère 20.000 curies en 1972 et n'en a dégagé que 1500 en 1974. D'ailleurs ces gaz, qui ne sont pas absorbés par l'organisme, n'agissent que comme source extérieure de rayonnement.

b) Les effluents radioactifs très divers qui passent dans les eaux de refroidissement, et qui atteignent ensuite la faune et la flore où ils peuvent d'ailleurs s'accumuler.

c) Les dépôts de déchets radioactifs.

d) Enfin la centrale elle-même qui peut laisser fuir des rayonnements électromagnétiques.

Quels que soient les types de rayonnement et leurs origines, le problème de pollution se ramène ici (comme pour la pollution chimique d'ailleurs) à celui des doses admissibles, avec un aspect particulier très important qui est la question du "seuil".

+ + +

La cancérogénèse par les faibles doses n'est pas directement accessible à une étude chez l'homme, ni même, il faut le reconnaître, chez les mammifères de laboratoire. Les seuls renseignements directs nous viennent de statistiques effectuées dans des circonstances particulières. Par exemple le rayonnement cosmique qui règne sur les hauts plateaux des Andes est environ trois fois celui que reçoivent les habitants des plaines. Les statistiques de cancer n'ont pas permis de détecter de différence entre les deux pollutions, mais notons ici, comme je l'ai déjà fait dans l'introduction, qu'il est pratiquement impossible de détecter quelques cas significatifs parmi les centaines de milliers de cas naturels.

Les statistiques d'Hiroshima illustrent cette remarque. Il s'agit là, non pas de faibles doses, mais des doses maximum compatibles avec la survie. Pourtant il a fallu vingt ans de travaux effectués par un institut spécialement mis en place dans ce but, pour démontrer que la fréquence des leucémies chez les survivants a atteint environ sept fois la fréquence normale (passant de 2,5 à 17,4 pour 100.000).

Les voies indirectes sont malaisées. La voie théorique se heurte encore à l'insuffisance de nos connaissances sur la cancérogénèse, et à cet égard je dois attirer l'attention sur une erreur de raisonnement fréquemment commise et sur laquelle sont fondées (à tort) des décisions récentes.

Ce raisonnement est le suivant : la transformation d'une cellule normale en cellule cancéreuse est, par définition, une mutation. Ce qui est vrai des mutations l'est donc aussi de la cancérisation. En particulier, puisqu'il n'y a pas de seuil pour les radiomutations, il ne doit pas y en avoir pour les radiocancers. La fréquence de base des mutations humaines peut être estimée par interpolation selon le schéma de Abrahamson et coll. (6). D'après ce schéma il existe une relation linéaire entre l'efficacité de l'induction des radiomutations, c'est à dire le nombre de mutations par locus génétique et par rad, et la quantité d'ADN contenue dans le génome haploïde. Nous connaissons cette quantité pour le génome humain. On peut donc situer l'homme sur la courbe (entre la souris et l'orge), et en déduire sur l'axe des ordonnées que la fréquence des radiomutations chez l'homme serait d'environ $3,5 \cdot 10^{-7}$ par locus et par rad.

Sur cette base, on calcule que la radiation naturelle doit être responsable d'une fraction non négligeable des cancers dits spontanés, et que si on lui ajoute une radiation artificielle, même en faible quantité, on produit de nouveaux cancers dont la fréquence est calculable en fonction de la quantité de radiation ajoutée. C'est sur la base de ce raisonnement que des mesures draconiennes de protection ont été récemment préconisées aux Etats-Unis.

En fait il y a là une grave confusion entre la transformation cellulaire maligne, qui est bien une mutation, et qui constitue la première étape du processus cancérigène, et le développement clônal de cette cellule transformée pour donner une tumeur.

La présence d'une cellule transformée dans l'organisme n'est pas un cancer, et c'est fort heureux car nous sommes probablement tous porteurs de très nombreuses cellules cancéreuses. Le cancer n'apparaîtra que si une cellule transformée trouve les conditions favorables à son développement dans la compétition avec les cellules normales. Il lui faudra notamment vaincre certaines réactions de défense de l'organisme. C'est au niveau de cette deuxième étape que des différences importantes se manifestent vis à vis des mutations. Par exemple, j'ai signalé dans l'introduction que la radiation naturelle ne peut rendre compte que d'un nombre très faible de cancers dits spontanés.

Ainsi cette voie théorique n'est-elle pas applicable de façon satisfaisante, puisque nous restons dans l'ignorance de la deuxième étape du processus cancérigène

On peut s'adresser alors à l'expérimentation animale, mais celle-ci n'apporte pas grand chose en ce qui concerne les doses faibles. En effet les "méga expérience à faibles doses sur les mammifères sont frappées d'un défaut fondamental qui en supprime pratiquement toute signification. Supposons en effet qu'on administre à des souris une dose faible qui produise environ un cancer sur 10^5 animaux. Si l'on veut relever 10 résultats positifs, ce qui est un minimum, pour un résultat significatif, il faut irradier un million d'animaux ne présentant pratiquement pas de cancers "spontanés". Si l'on y joint les témoins, on est obligé d'opérer de telle manière que les paramètres d'expérimentation varieront au sein de l'expérience : nourriture, âge, logement, techniciens, etc.. etc.. facteurs qui suffiront pour introduire eux-mêmes des variations supérieures à 10^{-5}.

On est ainsi obligé d'utiliser des doses relativement fortes, de les diminuer progressivement autant qu'il est possible, et d'extrapoler par la suite aux doses plus faibles sans connaître les aléas de cette extrapolation.

+ + +

Notion de seuil

Lorsqu'on fait agir de faibles doses d'un polluant sur un système biologique, on dit que l'effet produit comporte un "seuil" lorsque la courbe dose-effet ne commence pas à l'origine, c'est à dire lorsqu'il existe une dose seuil au-dessous de laquelle l'agent ne produit aucun effet. Inversement, on dit qu'il n'y a pas de seuil lorsqu'une dose si faible soit-elle, a une probabilité non nulle de produire l'effet. Cette question est très importante car elle gouverne les notions de risque et de responsabilité.

Les mutations géniques et chromosomiques produites par les radiations répondent à un mécanisme sans seuil. C'est pourquoi, compte-tenu du raisonnement fallacieux présenté au paragraphe précédent, beaucoup considèrent que la radio-cancérogénèse doit être également un processus sans seuil. De nombreuses expériences vont à l'encontre de cette conception simpliste. Ainsi par exemple, dans la radioleucémogénèse chez la souris, les résultats suivant ont été obtenus par Duplan et Latarjet (7). Il s'agissait de souris C57BL âgées de 10 semaines, et irradiées en totalité par les rayons X :

- 4 X 175 r à une semaine d'intervalle ont donné 81 % de leucémies lymphoïdes, la plupart thymiques.

- 2 X 175 r ont donné 6 % de leucémies lymphoïdes, aucune thymique.

- 1 X 175 r a donné moins de 1 %, et probablement aucune leucémie.

Il est clair que ces résultats ne sont pas conformes à une relation de proportionnalité entre l'effet et la dose. Ils suggèrent plutôt l'existence d'un seuil élevé au-dessus duquel l'incidence croît plus vite que la proportionnalité avec la dose.

On obtient des résultats comparables chez l'homme. Ainsi, lors des leucémies induites par les traitements dont il a été question ci-dessus (hypertrophie thymique de l'enfant, spondylite ankylosante de l'adulte) on n'a jamais observé de leucémie lorsque la dose administrée était inférieure à 100 rads. La même conclusion a été obtenue pour les ostéosarcomes induits par le strontium 90. Même conclusion pour les cancers cutanés humains produits par les rayons UV solaires.

Mais ceci ne semble pas vrai dans tous les cas. La production de cancers mammaires chez le rat par les rayons X s'approche beaucoup d'une relation linéaire avec la dose.

Puisqu'on ne peut pas atteindre directement par l'expérience le seuil éventuel, et puisqu'on ne peut pas décider rigoureusement de sa présence ou de son absence par voie indirecte, je crois important d'introduire la notion de "seuil pratique". J'ai déjà, dans l'introduction, souligné le fait que le "bruit de fond" du cancer chez l'homme est très élevé puisqu'il atteint 20 % de la population. Je définis le "seuil pratique" comme la dose de radiations au-dessous de laquelle aucune augmentation significative de cette fréquence de base ne se manifeste. Ainsi, par exemple, j'ai précisé que la cancérogénèse chez les populations des hauts plateaux Andins, où le rayonnement cosmique est de l'ordre de 90 à 130 mrads par an, n'est pas supérieure à celle qu'on observe chez les populations des plaines où le rayonnement cosmique est trois fois moindre (35 à 40 mrads par an).

De même, la cancérogénèse n'est pas supérieure dans les régions granitiques où la dose annuelle reçue par les gonades de l'homme est de 180 à 350 mrems, que chez les populations des plaines alluviales et calcaires où l'irradiation annuelle est quatre fois moindre, à savoir de 45 à 90 mrems.

On en conclut qu'il existe pour la cancérogénèse chez l'homme un "seuil pratique" assurémment supérieure à cette zone de fluctuation de la dose, c'est à dire à 350 mrems par an. La conséquence de cette importante notion est qu'il est inutile de pousser la protection au-dessous de cette valeur, certainement minimum, du seuil pratique.

III - UN MECANISME POUR LA RADIOCANCEROGENESE

La cancérisation, quels qu'en soient l'origine et le mécanisme, comporte deux étapes. D'abord une mutation d'une cellule somatique qu'on peut appeler "la transformation cellulaire maligne". La cellule normale devient cancéreuse par une mutation particulière qui affecte d'une certaine manière la régulation de sa croissance. Mais une cellule maligne ne constitue pas un cancer. Un cancer est une tumeur maligne, c'est à dire un clône de cellules provenant de l'émergence de la cellule transformée. Il faut donc, ainsi qu'il a été déjà mentionné au paragraphe 2, que celle-ci trouve les conditions pour se développer malgré la compétition des cellules normales environnantes.

La transformation cellulaire maligne est sans doute fréquente, et chaque individu porte sans doute de nombreuses cellules transformées malignes. Pourtant un homme sur cinq seulement, en moyenne, développe un cancer.

Considérons la première étape c'est à dire la mutation. Depuis qu'on s'intéresse à la radiocancérogénèse, c'est à dire depuis 1910, on a admis implicitement jusqu'en 1970, c'est à dire pendant 60 ans, que la cancérisation cellulaire était le résultat de l'évolution chimique des radiolésions, c'est à dire des lésions produites par les rayonnement dans certaines structures cellulaires, et plus probablement dans le matériel génétique, et encore plus probablement dans le DNA. Cette conception était admise également pour la cancérogénèse chimique, la radiolésion étant alors remplacée par une lésion chimique du DNA.

Cette manière de voir s'est toujours heurtée à deux difficultés fondamentales :

a) Les radiations, comme les cancérigènes chimiques, sont des agents cytotoxiques. Lorsqu'ils lèsent une cellule, en général ils la stérilisent. Une cellule stérilisée ne peut pas se multiplier pour donner une tumeur.

b) Les divers processus cancérigènes aboutissant dans tous les cas à la même situation finale, c'est à dire à une tumeur maligne. Ces processus, initiés par des agents différents (radiations, produits chimiques, virus) doivent converger quelque part. En quoi consiste ce point de convergence et où se situe-t-il dans le processus de radio-cancérisation ?

Aux environs de 1970 nos conceptions sur la mutagénèse par rayonnement UV ont beaucoup évolué, particulièrement à la suite des expériences et des hypothèses de Witkin (8). La réplication normale du DNA lors du cycle mitotique, s'effectue dans de telles conditions que de nombreuses erreurs paraissent inévitables. La continuité génétique ne peut être assurée que si la cellule est dotée de puissants mécanismes capables de réparer ces erreurs. Ces mécanismes sont sans doute ceux qui ont été mis en évidence dans la réparation de la plupart des radiolésions. Witkin a suggéré que de nombreuses mutations produites par les rayons UV résultent non pas de la photolésion du DNA elle-même, mais d'une erreur dans sa réparation. Erreur minime qui permettrait à la viabilité d'être rétablie, mais suffisante pour entrainer l'apparition d'un caractère héréditaire nouveau. Cette conception n'est pas limitée aux radiomutations, mais serait valable également pour les mutations spontanées et les mutations chimiques. On a pu démontrer en effet que, s'il existe des réparations non mutagènes qui restituent intégralement la structure initiale du DNA, il en existe d'autres dont la mise en action entraine la production de nombreuses mutations (9, 10, 11). Cette théorie a pour conséquence que la mutagénèse en général, et, par suite, la cancérogénèse en particulier, devraient être favorisées par le déroulement des mécanismes de réparation au lieu d'en être entravées (en effet, selon la conception classique où la cancérisation résulte de la lésion, si on répare la lésion, on supprime la cancérisation; si au contraire il faut une erreur de réparation, donc une réparation, on favorise la cancérisation

Des expériences récentes suggèrent qu'il pourrait bien en être ainsi :

a) Borek et Hall (12) ont réussi des transformations malignes cellulaires in vitro par rayons X et neutrons. Ils ont observé que la fréquence des transformations augmente si l'on fractionne la dose (de même que la fréquence des mutations augmente). Or, le fractionnement de la dose favorise les réparations.

b) Zajdela et Latarjet (13) ont produit des cancers cutanés par rayons UV chez la souris selon la méthode classique d'irradiations quotidiennes, et ont constaté que si l'on badigeonne une oreille avec la caféine ou la théobromine (l'autre oreille servant de témoin), on diminue de moitié, c'est à dire significativement, l'incidence des tumeurs. Or ces bases inhibent certains processus de réparation mutagène.

c) Setlow (14) a mis en évidence la production de cancers chez les poissons par les dimères de pyrimidine (photolésions principales du DNA). Si l'on supprime ces dimères par photorestauration, la cancérogénèse diminue. L'interprétation la plus plausible des expériences de Setlow, vu les conditions expérimentales, est que la cancérogénèse résulte d'erreurs dans la réparation mutagène des dimères (par recombinaison ou par système S.O.S.).

On remarquera que cette conception de la cancérogénèse, ou plus exactement de sa première phase, résoud la première difficulté mentionnée, c'est à dire la létalité consécutive à l'action de l'agent. S'il faut réparation, c'est que la viabilité est rétablie. En outre la deuxième difficulté, celle d'un point de convergence avec les autres agents cancérigènes, se trouve levée, du moins avec les cancérogènes chimiques. En effet ceux-ci produisent des lésions du DNA qui, elles aussi, sont réparables selon des processus dont certains paraissent mutagènes. La convergence avec la cancérogénèse virale reste mystérieuse et n'est pas résolue ici.

On remarquera que cette conception permet d'interpréter la cancérogénèse dite spontanée. En effet, la réplication normale de l'ADN ne devant se faire qu'au prix de multiples erreurs, les réparations semblent inéluctables. S'il en est ainsi, le cancer comporte une part importante de fatalité. La division cellulaire est nécessaire; les erreurs sont inévitables; leur réparation est nécessaire; les erreurs de cette réparation sont inévitables. Ainsi, à mon sens c'est une erreur de considérer, comme certains font, que la grande majorité des cancers sont provoqués de l'extérieur, donc évitables.

Cette conception suggère-t-elle des interventions possibles susceptibles de diminuer la cancérogénèse par pollution ? Il est trop tôt pour se prononcer.

IV - POLLUTION CHIMIQUE CANCERIGENE et RADS EQUIVALENTS

A l'heure actuelle la pollution chimique mutagène et cancérigène est beaucoup plus importante que la pollution par radiations. Les centrales thermiques à charbon ou à pétrole, et de nombreuses industries qui procèdent à des combustions massives, les automobiles, le tabac dégagent dans l'atmosphère des produits qui attaquant le matériel génétique.

Pourtant cette pollution n'est assujettie à aucune règle de limitation et, on doit l'avouer, n'a fait l'objet d'aucune étude traduisant une préoccupation générale, alors que la pollution par radiations, depuis 20 ans, connait des doses internationalement fixées comme dangereuses ou, inversement, comme admissibles.

Cette anomalie apparente est due à l'horreur d'Hiroshima et de Nagasaki. L'émotion internationale a entraîné la création d'un organisme au sein des Nations Unies, lui-même promoteur de recherches qui ont permis d'aboutir à des règles raisonnables et généralement observées. En revanche, jusqu'ici, la pollution chimique est demeurée diffuse et sournoise. Aucun accident spectaculaire de grande dimension n'a alerté une opinion internationale peu imaginative. Ainsi, bien que quantitativement plus importante, la pollution chimique a au moins vingt ans de retard sur la pollution par radiations en ce qui concerne la prévention et la sécurité. Il est temps de rattraper ce retard.

Il serait très long et très laborieux de procéder, pour chaque substance cancérigène, aux étapes par lesquelles il a fallu passer pour aboutir aux règles adoptées pour les radiations. Mais il semble exister un autre chemin plus général et plus court. En voici le principe. Les lésions produites dans le matériel génétique par les substances mutagènes d'une part, et par les radiations d'autre part, sont très semblables. Dans les deux cas il s'agit soit de ruptures de chaîne, soit de la formation de composés d'addition sur les bases qui entraînent une déformation de la double hélice, soit de pontages inter-chaînes au sein de la double hélice, soit de pontages entre l'ADN et les protéines. Ces similitudes sont telles que les lésions, dans les deux cas, mettent souvent en oeuvre les mêmes systèmes de réparation (lesquels procèdent soit par soudure des ruptures, soit par ouverture des ponts, soit par excision des bases lésées, soit par recombinaison génétique, soit enfin par un système de réparation inductible). On peut s'en assurer en vérifiant (généralement sur des microorganismes, mais également parfois sur des cellules de mammifères) que telle souche qui résiste aux radiations parce qu'elle est douée de certains systèmes de réparation, est également résistante aux mutagènes chimiques.

Cette similitude entre les lésions suggère qu'on puisse établir expérimentalement une équivalence, sur un système donné, entre la quantité d'un certain produit et une quantité de radiation. Sur ce système, on pourrait tester la létalité, la mutagénicité, et enfin dénombrer les lésions produites dans le matériel génétique par les deux agents. On pourrait alors définir sur ce système la dose de radiation qui produit quantitativement le même effet qu'une certaine concentration du produit chimique étudié, c'est à dire définir les rads équivalents à une concentration donnée de ce produit. Une fois cette équivalence établie, on pourrait extrapoler les règles de la radio-sécurité à la sécurité vis à vis de ce produit.

Ce principe est relativement simple, mais son application ne l'est évidemment pas. L'équivalence recherchée n'est pas nécessairement la même, et on peut dire à priori n'est assurément pas la même, selon le système expérimental utilisé; et ceci pour plusieurs raisons, dont la plus évidente est que l'absorption du composé par ce système dépend des conditions dans lesquelles il se trouve. Par exemple, pour une simple cellule, elle varie avec les conditions de croissance (croissance exponentielle, quiescence, etc..). Si l'on choisit un certain composé, par exemple le formaldéhyde d'une part et les rayons X d'autre part, on pourra établir toute une collection d'équivalences :

a) selon le système étudié: bactéries, levures, cellules de mammifères, etc..

b) selon les conditions expérimentales : quiescence, croissance, ploïdie, etc..

c) selon l'effet considéré: létalité, mutagénicité, dénombrement des lésions chimiques de l'ADN, etc...

De tous ces paramètres, c'est à dire de toutes les équivalences observées dans tous ces cas, on pourra déduire un "centre de gravité" du spectre de dispersion des résultats, et définir l'équivalence correspondant à ce centre de gravité. Je suis convaincu que la signification de cette valeur vis à vis du problème général de la pollution chimique sera grande, et qu'on pourra alors extrapoler à cette valeur, pour le composé considéré (qui pourrait être un mélange complexe relativement bien défini), les règles de protection et de sécurité peu à peu élaborées pour les radiations.

Rien n'empêchera par la suite si nos connaissances se précisent de modifier les règles admises, comme on l'a fait d'ailleurs périodiquement pour les radiations au cours des vingt dernières années.

Résumé :

Cet exposé considère quatre aspects du problème :

1. La pollution cancérigène par les radiations non ionisantes se limite au cas du rayonnement UV solaire dont l'activité au niveau du sol peut être accrue par une diminution de l'ozone stratosphérique, elle-même provoquée par certaines pollutions chimiques : oxydes d'azote des avions supersoniques, fréon.

2. En ce qui concerne les radiations ionisantes, on discute du problème fondamental du "seuil" et des moyens dont on dispose pour obtenir des données quantitatives sur la cancérogénèse par les faibles doses chez l'homme. Un nouveau concept, celui du "seuil pratique" est proposé.

3. On discute une théorie qui associe la radiocancérogénèse et la cancérogénèse chimique à des erreurs commises dans la réparation du DNA.

4. On discute du projet des "rads-équivalents" pour les mutagènes et cancérigènes chimiques.

Summary

CARCINOGENIC RISQS ASSOCIATED TO POLLUTION BY RADIATIONS

In this paper, four aspects of the problem are taken into consideration :

1. The cancerogenic pollution by non-ionizing radiations is limited to the case of solar ultraviolet, whose activity at ground level may be increased as a consequence of the stratospheric depletion of ozone, itself produced by certain chemical pollutants : nitrogen oxydes from supersonic aircrafts, freon.

2. As regards ionizing radiations, the discussion is focused on the fundamental problem of the "threshold", and on the means by which one may obtain some quantitative data related to carcinogenesis by small radiation doses in Man. A new concept, that of a "practical threshold" is proposed.

3. One discusses a theory which links radiocancerogenesis, as well as chemical cancerogenesis, to errors produced in the repair of lesions in the DNA.

4. One presents and discusses the "rads-equivalent" project for chemical mutagens and cancerogens.

REFERENCES

1. Latarjet R. Revue d'Optique 1935, 14, 398
2. Chavaudra J. et Latarjet R. Comptes-Rendus Acad. Sc. Paris 1973, 276, 3481
3. Impacts of Climatic Change on the Biosphère, C.I.A.P. monograph n° 5, 1974
4. Clyde M.A. Nature 1974, 249, 796
5. Molina M.J. et Rowland F.S. Nature 1974, 249, 810
6. Abrahamson S., Bender M.A., Conger A.D., Wolf S. Nature 1973, 245, 460
7. Duplan J.F. et Latarjet R. Cancer Res. 1966, 26, 395
8. Witkin E.M. Ann. Rev. Microbiol. 1969, 23, 487
9. Weigle J.J. Proc. Nat. Acad. Sc. U.S.A. 1953, 39, 628
10. Radman M. et Dévoret R. Virology 1971, 43, 504
11. George J., Dévoret R. et Radman M. Proc. Nat. Acad. Sc. U.S.A. 1974, 71, 144
12. Borek C. et Hall E.J. 5th Int. Congress Rad. Res., Seattle 1974
13. Zajdela F. et Latarjet R. Comptes Rendus Acad. Sc. Paris 1973, 277, 1073
14. Setlow R.B. 5th Int. Congress Rad. Res. Seattle 1974

RADIATION CANCER, SAFETY STANDARDS AND CURRENT LEVELS OF EXPOSURE

R.H. MOLE

MRC Radiobiology Unit, Harwell, Oxon OX11 ORD, England

It sometimes seems characteristic of radiation-induced tumours that they are naturally somewhat uncommon, for example leukaemia, osteosarcoma, thyroid carcinoma. This is partly because there is a wide range of radiosensitivity between different tissues - bone marrow happens to be about the most sensitive to malignant transformation by ionizing radiation - and partly because radionuclides of practical significance happen to have particular physical and metabolic properties which ensure that irradiation of cells is highly localised - the endosteum of bone in which tumours originate is specifically irradiated by radium, because it is an alkaline earth as calcium is and, therefore, a bone seeker, and also by plutonium which deposits in the endosteum on bone surfaces; the colloid of the thyroid gland selectively concentrates the radio-iodines which are thus enabled to irradiate the adjoining thyroid cells. However, naturally common tumours can also be induced by ionizing radiation, e.g. cancer of the female breast and of the lung and there is now a mass of information which supports the generalisation that cancer can be induced by radiation in any tissue where cancer occurs naturally (Table I). I will concentrate hereafter on presenting a personal view and interpretation.

Apart from a few exceptional situations, such as in experiments involving gross physiological disturbances, radiation-induced cancer originates only in directly irradiated tissues. Its frequency depends on the amount of locally absorbed energy, that is on radiation dose. At the cellular level energy absorption is commonly non-uniform both in space and time for a combination of metabolic and physical reasons, for instance when radio-active materials are incorporated into the body. Incorporation into a tissue is rarely uniform on a cellular scale and only the most energetic β-emitters with β-tracks each traversing many cells can compensate for this. An α-particle travels only some 40 µm in soft tissue (less in calcified tissue which is denser) so that unless uptake of an α-emitter were to be exceptionally uniform, tissue irradiation must be very non-uniform. This is especially clear when α-emitters are present in particulate form (Figure 1). The statistical character of all radio-active decay inevitably causes marked temporal non-homogeneity of irradiation of individual cells at the levels of exposure which concern radiological protection. If ten µCi of radio-active material were distributed uniformly within 1 kg of tissue containing 10^{12} cells each of volume = 1000 µm^3 then on average one radio-active atom would disintegrate in each cell each month. But in fact in any given month no disintegration would occur in 37 per cent of the cells and in one of the 10^{12} cells there would be a disintegration every other day. The unit of radiation dose used to be defined in terms of energy absorbed per gram of tissue. For the unwary such a definition may conceal the statistical character of the interaction of radiation and matter. Absorption of radiation energy is quantized and always highly non-uniform at a sub-cellular level.

TABLE I   Sources of quantitative information on induction of cancer in man by ionizing radiation (1) (2)

	Information on dose	Kinds of malignant disease with confirmed increase in frequency
Occupational exposure		
Uranium and hard rock mining	+	Lung
Hospitals, X-ray set manufacture	?	Skin, leukaemia, cancer overall
Luminizing with Radium	++	Bone, air-containing cavities in the head
Medical exposure of patients		
X-ray diagnosis	+/−	All kinds of cancer and leukaemia in childhood
X-ray fluoroscopy	+	Breast
X-ray with Thorotrast	+/−	Liver, connective tissue, haemangiosarcoma
Treatment with X-rays, Radium, etc.	++	Leukaemia, lung, bone, connective tissue, thyroid, salivary gland, larynx, pharynx, skin, large bowel (?)
Treatment with Radionuclides	++	Bone
Atomic warfare		
Japanese bomb survivors	++	Leukaemia, lung, breast thyroid, cancer overall
Marshall islanders	?	Thyroid (?)

TABLE II   Comparative sensitivities of different radiobiological responses (3)

	Probability of effect per cell after 200 rads of low LET radiation at dose-rates of 1-1,000 rads/minute
Clonogenic sterilization	0.1-0.9
Visible chromosomal aberrations	0.2-0.4
Genetic effects:	
Dominant lethals in spermatogonia and adult dictyate oocytes	0.01-0.1
Specific mutations per locus	10^{-4}-10^{-6}
Myeloid leukaemia:	
Mouse	10^{-9}
Man	10^{-14}

In Japanese bomb survivors a whole-body dose of about 200 rads was followed by 0.6-0.7 per cent of leukaemia. Thus to get one case of leukaemia the bone marrow of 150 persons, about 225 kg, had to be irradiated, or 10^{14} cells assuming that the average volume of each marrow cell is 2250 μm^3.

It should be noted that radiation dose in radiological protection is always thought of in terms of dose commitment, that is the dose which a particular exposure will give to an individual over the rest of his life. This convention is immaterial where X-rays are concerned but is clearly important when considering radio-active materials which have long half-lives and are well retained by the body.

Figure 1. Schematic representation of a mass of tissue containing two deposits of an insoluble α-emitting radionuclide of equal activity

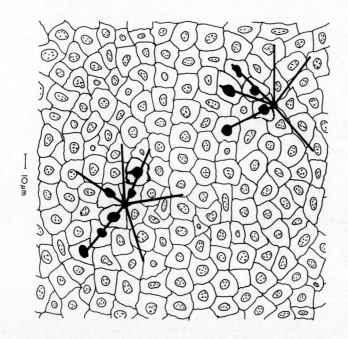

The track of each α-particle is represented by a thick black line. When the nucleus of a cell has been traversed by an α-particle the cell cytoplasm is shown dark and mottled to indicate that the cell has been killed or rendered incapable of further cell division.

Linearity Between Radiation Dose and Cancer Frequency

The Linear Hypothesis

It is non-homogeneity, not knowledge of the intracellular mechanism of radia[tion] action, which is the real scientific basis for the working hypothesis underlying current radiation safety standards intended to limit cancer induction. This is [the] so-called linear hypothesis that cancer induction is simply proportional to radi[ation] dose. Because tissue dose is a kind of Poissonian average, an increase or decrea[se] in dose increases or decreases the proportion of cells absorbing a given amount [of] energy. When α-emitters are present in particulate form, some cells, those trave[rsed] by an α-particle track, absorb a substantial concentration of energy; the majori[ty of] cells, those escaping traversal, get no α-dose at all however small or large the [dose] might be averaged throughout the tissue (Figure 1). Thus, if a radiation-induce[d] tumour originates in a single cell "transformed" by absorption of radiation energ[y,] the frequency of cancer observed in a population of individuals all receiving th[e] dose would be expected to depend simply on the total number of cells transformed (allowance, of course, being made for the Poissonian statistics of "transformati[on of] two or more cells within one individual who, nevertheless, will be recorded as a [case] as showing just one tumour). Any biological factors which may modify quantitati[vely] the process of growth and development from the single "transformed" cell to the [state] of overt clinical cancer will merely be scaling factors affecting the constant o[f] proportionality relating dose and observed frequency of cancer.

A great deal is known about the quantitative aspects of some cellular actio[ns of] ionizing radiation, the production of chromosome aberrations and gene mutations [and] the inactivation or sterilisation of cells which often leaves them alive and fun[ctional] but no longer able to divide. In terms of probability per cell per dose, the car[cino]genic transformation is less probable by many orders of magnitude (Table II). M[y] somewhat iconoclastic conclusion is that nothing that is known about cellular rad[iation] biology has necessarily any bearing on the mechanism of carcinogenesis by ionizi[ng] radiation (3). For example, if cancer induction depended not simply on changes within a single cell ("transformation") but on the co-operation of two (or more) [cells,] the working hypothesis of linearity would be profoundly modified. It seems to m[e] empiricism will be our guide for some time to come.

Linearity as Part of a System of Radiological Protection

It is important to distinguish two quite different uses of the linear hypot[hesis.] First, any system of radiological protection, which includes the setting of safe[ty] standards which should not be exceeded, must provide for some means of adding up [the] risks from different kinds of radiation exposure so as to keep the combined risk [below] the specified maximum. The International Commission for Radiological Protection [does] this by weighting the radiation dose in different tissues according to the inver[se of] their radiosensitivity and then adding the products thus obtained in simple line[ar] fashion. Linearity seems the only practicable basis for such a convention and, a[s] already noted, it may have a radiobiological justification. Conventions of this [kind] which are required for practical reasons are nonetheless valuable even if to the thoughtful biologist they may seem sometimes to be merely crude approximations.

Linearity as a Realistic Means of Estimating Risk

The second and quite different use of linearity is in deciding the level at [which] to set a safety standard in terms of radiation dose. Basic standards are intend[ed to] limit risk and it is therefore necessary to know the risk to be attached to expo[sure] at the maximum permissible level. Epidemiological data of various kinds are available on the frequencies of a variety of malignant diseases and the corresponding radiation doses and it is the current orthodoxy to fit a linear dose-response relation to these data. The slope of the straight line through the point (zero dose, zero induced cancer) can then be used as the risk coefficient to calculate the risk after exposure to the maximum permissible dose and thus to verify that the basic safety standards of dose correspond to whatever may have been decided upon as an acceptable level of risk. Whatever the standard there will always be some risk since using the linear hypothesis also presupposes that there is no threshold of dose and that, however small the dose, there will be some effect.

Linear risk coefficients certainly make risk calculations appear very easy. Average dose per person is all that is required to determine the number of induced cancers in an exposed population: the distribution of dose is immaterial. When making such calculations those who may have wished to make radiation exposure appear dangerous have often inflated the risk coefficient employed on the grounds that there must be statistical uncertainties about the slope of the straight line fitted to the observations. On the other hand the Establishment's view has usually been that the risk coefficients are not best estimates but almost certainly overstate the risk and that the true risk could be anywhere between that calculated and zero.

One reason for this view is that the risk coefficients refer to brief exposures to large doses but the safety standards deal with exposures protracted over years. The general argument is that biological effects are always reduced by protraction in time. This is certainly a commonplace for chemical actions, whether of drugs or of traditional beverages, and for radiation effects like acute mortality, burns and radiotherapeutic reactions. Perhaps surprisingly, since the question is practically important and highly amenable to experimental investigation, there is as yet remarkably little information about the importance of these factors in carcinogenesis. What information there is (cf. Table III) may be interpreted as suggesting that neither protraction nor magnitude of dose are the significant variables.

It should be emphasized that, whatever the true dose-response relationship might be, a succession of small exposures is bound to give a simple proportional response between frequency of induced cancer and numbers of exposures, i.e. total dose, if each exposure acts independently of all the others. This is exemplified in the only quantitative data on the matter, breast cancer following repeated fluoroscopy for monitoring pneumothorax treatment of pulmonary tuberculosis (2). Figure 1 is a schematic representation of tissue containing two localised deposits of Plutonium-239 sufficiently far apart for there to be no overlap of the emitted α-particles. Whatever the cancer risk related to a single deposit, clearly the risk from $\underline{n}$ deposits is n x risk from a single deposit whatever that may be. The major scientific question is how the frequency of cancer induction at the site of a deposit might vary with the quantity of radio-activity in it.

A radiobiological flaw in the assumption of linearity for estimation of risk

This assumption takes it for granted that linearity is radiobiologically justified at radiation doses very much higher than the basic safety standards. With one exception (considered below) it is only at these larger doses that induced cancer frequency has been measurable, for instance in Japanese bomb survivors or in patients receiving radiotherapy. In fact the simplest kind of empirical data on cancer frequency, those after single brief exposures, often do not fit a simple uncomplicated linear regression on dose and there is a simple radiobiological reason why this should be so, even if the true induction process was indeed linear with dose. Ionizing

TABLE III Variation with experimental conditions in incidence of radiation-induced myeloid leukaemia in RF mice (data from references 4 and 5)

Radiation Physical dose-rate	Manner and overall duration of exposure		Myeloid leukaemia (per cent) Total dose	
			150 rad	450 r
X-rays 85 rad/minute	Brief	2-5 minutes	32	27
	Two or three fractions	2-4 days	18	18
		6-10 days	24	20
Gamma 0.2-3 rad/hour	Continuously protracted	2 days	6	-
		8-10 days	6	9
		30 days	4	-
Unirradiated controls			3-4	

TABLE IV Radiation dose to the population of Great Britain averaged out per individual (millirem)

Source of Radiation	Annual	Total Over the Year 1945-1999 AD (Estimat
Natural background - geographical variation in UK more than 2 x	100	5,500
Occupational i.e. exposure of radiation workers in industry, hospitals, etc. (7)	less than 1 ╪	30 ╪ *
Miscellaneous (TV sets, watch dials, etc.)	less than 1 ╪	40 ╪
Medical Practice: Diagnosis 3/4 Therapy 1/4 (8)	19	1,100
From Fall-out from nuclear weapons and devices 1945 onwards (1)	-	170

╪ These values should be regarded merely as approximations. They are larger by order of magnitude than those given in (6) because of different attitudes to the inevitable uncertainties. Many radiation workers who wear film badges or other individual monitoring devices have no measurable dose: for purposes of record it common practice to assume that their exposure gave a dose equal to the minimum detectible by the particular kind of monitor concerned. Other radiation workers not individually monitored and their dose is estimated from environmental measure ments. For members of the public also individual doses are not measured but are calculated from more or less fallible data and assumptions.

* <1 is attributable to irradiation of the public by disposal of radio-active wa into the environment.

Doses to individuals from particular activities cover a range of 6+ orders of mag from c 10 million millirem to the thyroid gland in the treatment of thyroid disea by radio-iodine to 2 millirem whole body exposure for a transatlantic flight in a subsonic commerical aeroplane.

radiation sterilises cells in increasing proportion the larger the dose, overt clinical cancer cannot develop from "transformed" cells unless they divide, and the observed cancer yield must therefore fall increasingly below that expected as dose increases. Allowance for this complicating effect can be made by setting tumour yield = $aD^n.f(S)$ where D is dose and $f(S)$ is the survival function for retention of reproductive integrity. For high LET radiation, like α-particles and neutrons, $f(S) = e^{-\lambda D}$ but for low LET radiation, like X- and γ-rays and the more energetic β-particles, $f(S)$ is more complex, cell survival becoming exponential with dose only after dose exceeds a value commonly in the region of 100 rad or more. Making due allowance for cell sterilisation seems to rationalise much of the experimental and clinical data on dose-response relationships for cancer induction by radiation but does not by any means always lead to a simple induction model with n = 1. Sometimes a $(dose)^2$ model with n = 2 seems more likely (3). These considerations are not merely theoretical but have implications for risk estimation. Some risk coefficients may need to be reduced, others to be raised and it may become necessary to consider individual types of malignant disease on their own merits rather than to expect all types to show the same dependency on dose, LET and protraction.

Overall cancer risk at low doses

A defensible overall linear coefficient for risk estimation given the present state of knowledge could be 100 cases of cancer of all types per million persons for uniform exposure of all the tissues of the body to one rad of low LET radiation, i.e. X- or γ-rays, without any allowance for protraction as such, but open to modification if variation in physical (instantaneous) dose rate is confirmed as important for carcinogenesis (cf. Table III). An influence of dose rate at low dose would be denied by biophysical theory although there is still argument as to whether "low" in this context means less than 10 rad or less than 100 rad. The question is important when comparing the level of natural background and of exposures resulting from human activities (Table IV) since background exposure is uniformly protracted over time whereas some human activities involve low doses, but at dose rates 10^6 x higher than background.

The Threshold Question

Professor R. Latarjet has emphasized the significance of the concept of "threshold" when considering large populations exposed to tiny doses. Certainly the existence of a threshold dose which must be exceeded if any induced cancer at all is to occur could make nonsense of many calculations of risk. However, no physicist would expect an instrument for measuring radiation of a particular kind to have a genuine threshold: there would certainly be a minimum measurable dose but that is something different. I do not see any reason for expecting animate matter to behave differently from inanimate matter. Of course, in principle neither the existence nor the absence of a threshold can be proved by observations (9): it is the quantitative formulation of a scientific hypothesis intended to explain observations which determines whether or not a threshold is to be expected. But observations on cancer in the human species after a quite small dose of radiation are now available, on a far larger scale than any conceivable experiment, and they seem to me to provide as good observational evidence as could ever be expected against the existence of a threshold dose for carcinogenesis. Moreover, the extremely low probability of carcinogenic transformation per cell even at quite high levels of dose and at substantial frequencies of induced malignancy (Table II) may of itself suggest that a threshold is extremely unlikely.

TABLE V Deaths from malignant disease at 0-9 years of age after antenatal radiography (3)

		Singleton births	Dizygotic twins	Monozygotic twins
Live born 1945-1964		14.8 million	254,000	99,000
Proportion X-rayed (per cent)		10	55	55
Deaths from malignant disease per 100,000				
Not X-rayed	Leukaemia	23	16	nil
	Solid cancers	27	21	11
X-rayed	Leukaemia	35	24	30
	Solid cancers	39	31	30
Excess in X-rayed	Leukaemia	12	8	30
	Solid cancers	12	10	19

Based on published data from Oxford Childhood Cancer Survey (10)

TABLE VI Fatal accidents in England and Wales (1961) according to occupation, excluding motor vehicle accident and accidents in the home (data from 12)

	per million per year
Aircraft crew, company directors, barge and boatmen	1000+
Constructional work, railway labourers	900
Coalface workers, armed forces, agriculture	600
Metal manufacture, shipbuilding, marine engineering	100-200
Timber, furniture, bricks, pottery, glass, chemical and allied	60-90
Electrical goods, textiles, vehicles, printing, food, drink, tobacco, leather goods	20-40

The Carcinogenic Effect of Antenatal Radiography and Some Implications

The human foetus in utero is exposed to whole body irradiation when X-ray pictures are taken of its mother's abdomen. Observations on children born in UK over the 20 years 1945-1964 show the same additional cancer risk in those X-rayed in this manner whether the radiography rate was 10 per cent as in singletons or 55 per cent as in non-identical twins (Table V). (Monozygotic, that is identical, twins may possibly be a special case). This result makes it very difficult to deny causation of cancer by the radiation dose involved in antenatal radiography, that is ~ a few rads in 1945-1964. Technical improvements should now be keeping the dose lower than it used to be.

From the point of view of this meeting on pollution it is worth stressing that "the direct demonstration that low-level radiation exposure can cause a low frequency of cancer depends on very special circumstances. It is only because malignant disease in childhood is normally so uncommon ~ 1/2,000, because exposure to antenatal radiography is brief and memorable and can be recorded, because the whole body is irradiated so maximizing the chances of producing cancer, and because a population numbered in hundreds of thousands or millions could be examined over a period of years that antenatal radiography giving a dose of a few rads has been shown to cause cancer with a probability of ~ 1/4,000." (3) Perhaps an even more important consideration is that all the other circumstances of life, both pre-natal and post-natal, would have been broadly similar for all subjects so that any difference in cancer frequency between the irradiated and the unirradiated could be attributed with some confidence to the irradiation itself. When pollutants are widely distributed in the atmosphere, in drinking water or in food every member of a population is exposed. It is then much more of a problem to decide whether any differences from a control non-polluted population are to be attributed to the pollutant or to other, perhaps unrecognised, differences between the two populations. As a direct demonstration of a low level of carcinogenesis by a small dose of "pollutant" antenatal X-radiography may well remain a unique example (3).

One further comment may help to introduce what is said later on non-scientific aspects of acceptability of risk. It can be deeply disturbing to be told that radiography of the unborn child causes cancer. But that qualitative statement may have a different impact if expressed quantitatively as follows. Each practising obstetrician in UK by ordering antenatal radiography at the current frequency of about 10 per cent of all pregnancies will cause about 1 additional case of childhood cancer in the whole of his working life-time. Given the obstetrician's responsibility for so many potentially serious problems in a life-time of work, for example the birth of about 1000 seriously handicapped children to women in his care, is this level of cancer risk sufficient to expect him to take particular thought as to whether an X-ray is really necessary on each of the 4000 occasions he orders one? There must be some level of risk to other people sufficiently low for it to be regarded as negligible.

Current Levels of Exposure to Radiation

Table IV gives in round terms the average radiation dose to the gonads in the UK population in recent years classified according to its source. As already noted, if the linear hypothesis is accepted, cancer induction in a population will be proportional to average dose and for the purposes of this discussion the uniform whole body dose can be regarded as similar to this gonad dose. Thus in terms of numbers of cases of cancer in the UK population the natural background of the environment appears to be 5 x more dangerous as a carcinogen than the routine practice of medicine is to patients and the routine practice of medicine at least 25 x more dangerous than all kinds of occupational exposure taken together (including medical occupational exposure). These data cover the time when as much as 10 per cent of all electricity was being generated in nuclear power plants in UK. Comments are needed on two quite separate issues: first, the difference between average dose in a population and dose to the individual, that is between risk to the population and to the individual, and second the potential risks from uncontrolled releases of radio-active pollutants by terrorists or in accidents.

The Acceptability of Risks to Individuals and for Populations

On the linear hypothesis a given average dose to a population would cause the same number of cancers whether a large number of individuals each received a small dose or a small number of individuals each received a large dose and every one else no dose at all. If the distribution of radiation dose amongst the population was not known the average risk would be all that could be discussed but as soon as a d to an individual is known the risk to him can be assessed separately. Thus it cou be decided to set occupational dose limits for radiation workers, technically call maximum permissible doses, so that the cancer risk to individual radiation workers did not exceed whatever might be regarded as an appropriate maximum occupational d rate for a well-run industry. If a level of accidental death per year at about 10 per million is regarded as the dividing line between sufficiently safe and not suf ficiently safe industry (Table VI) (11), then 50 cases of induced cancer per milli per year might be acceptable for designated radiation workers who in practice have an exceptionally low death rate from accident. Given a risk coefficient of 100 can per million per rem this corresponds to an average exposure of 0.5 rems per year. The maximum permissible dose over the last decade for whole body exposure per year been 5 rem per year and in UK the average exposure to external penetrating radiati for radiation workers has been about one-tenth or somewhat less, i.e. just about wh these considerations suggest it should be. But if it is permissible for anyone to exposed for a working life-time close to the recommended maximum permissible level then important questions about acceptability of risk are raised as soon as it is recognised that an occupational average is derived from a range of exposures, some which are likely considerably to exceed the average, certainly by 10-20 x in the example given. The argument from "average dose" also depends on how "radiation wor whose dose is individually recorded, are defined. The average dose in the French nuclear industry is considerably lower than in the British, presumably at least in because administrative staff carry film badges in French nuclear power stations bu in British.

The ICRP system of Recommendations on Radiological Protection also includes d limits for individual members of the general public and at present these are set (one exception) at one-tenth of the corresponding occupational dose-limits. Since limits are set in terms of annual doses, and since the average life span at 70 year is nearly twice the working life time of, say, 40 years, the permitted total dose fo individual members of the general public is appreciably higher than that for a lif time's work at the actually observed UK average occupational exposure. But member of the public, unlike designated radiation workers, do not know (except exceptiona that they are being exposed at all to radiation in excess of background. Thus the additional risk to which they are being exposed is being accepted for them by prox by those who make regulations.

In practice the ICRP system involves defining what is called a critical group persons, critical not because of the nature of any harm which might come to them b critical because their radiation exposure will be the ultimately decisive factor i determining the amount of controlled pollution which is allowable in a given case.

A critical group is usually very small in numbers and the assumptions involved in determining the radiation dose are always chosen to maximise the assessment of risk. Two examples may be given. The effluent from one nuclear power station in UK runs into a lake which is fished for trout by literally a handful of individuals. The allowable effluent is determined on the assumption that the members of the critical group of fishermen will eat 100 grams of trout flesh from this lake every day of the year. The second example refers to fishermen near the site where effluent is discharged into the sea from the Dounreay Fast Breeder Reactor Site. A certain part of the radio-activity discharged by pipelines out at sea accumulates on the sea-shore. The maximum allowable discharge of some radionuclides (much greater in fact than the currently authorised maximum) would depend on the theoretically calculated dose received by the hands of 4-5 salmon fishermen whose fishing gear becomes contaminated with silt, using assumptions maximising the time they might spend handling their gear.

The purpose of these examples is not to consider the great variety of critical groups or to consider the details of how radiation dose is calculated but to emphasize the distinction between acceptability of risk to an individual and to a population. That a risk should be acceptable to the individual is fundamental to occupational medicine and this requires that he be given a full explanation of the nature and magnitude of the risk. But when pollution is controlled and the permitted level of environmental pollution carries with it some risk - as must always be the case when a non-threshold hypothesis for cancer induction is accepted, however low the level of pollution - the decisions about acceptability of risk are being taken not by those directly exposed but by others on their behalf, sometimes perhaps their representatives, but sometimes also anonymous and faceless bureaucrats. It seems to me that both representatives and bureaucrats need to know if, in circumstances where an individual has no power to lessen risk by an action of his own, there is a level of risk which would be regarded by the individual as negligible, that is, to be wholly ignored. That there is such a risk and that its quantitative value is reasonably established is suggested by every day behaviour in a variety of contexts (13, 14). The numerical value of the negligible risk, about 10^{-5} or less in a life time for fatal accidents, could be a guide to the appropriate dose-limit for individual members of the public for radiation and other pollutants.

Acceptability of risk to a population requires a different set of criteria, including the familar, if somewhat inadequate, concept of cost-benefit analysis and may often be much more restrictive in terms of average dose per person than the limit for an individual. Perhaps there should be much more open discussion of the principles to be employed when considering the distribution of levels of controlled pollution across a large population.

Uncontrolled Pollution: Accidents and Sabotage

If dose-limits and everything which depends on them, such as authorizations of discharge of radioactivity into the environment, are determined on the basis of a linear hypothesis relating dose and cancer induction, combined with the assumption of life-long exposure, then transient over-exposures may have little or no biological significance even if their magnitude is substantial and public concern acute. If the basic protection standards are linked to a low enough frequency of effect, and especially if the assumptions used in arriving at them have tended to maximise pessimism, all of which is consciously true of ICRP Recommendations, then an uncontrolled release of radio-activity, even on a large scale, may well not be intolerable. Plutonium-239, for example, can be a most efficient carcinogen when it is lodged in the tissues. But Plutonium spilt on the ground enters food chains only with the greatest difficulty. When eaten the fraction f crossing the intestinal wall into the blood is tiny. The conventional assumption is $f = 10^{-5}$ but as may be imagined such a low value is not likely to be very accurately determined. The cancer risk is no to the skin or the gut: Plutonium α-particles will rarely penetrate as far as the living basal layer of the epidermis of the skin or as the self-renewing cells of intestinal crypts. The major hazard is inhalation and deposition in the lungs but this requires some means by which the Plutonium is converted into the appropriate sized particulates and then suspended in the air to be breathed.

The Medical Research Council (15) has recently reiterated its advice that counter measures in the event of an accidental release of radio-activity, such as banning of food (milk) or evacuation of people, should not be implemented unless predicted additional exposure of members of the general public exceeds the total which would be received in 20 years continuous exposure at their annual dose limi and not even then if the hazards associated with the counter measures (traffic accidents for example) seem likely to exceed the expected reduction in radiation risks.

Sabotage is effective only if it is damaging. At the present time sabotage involving release of radio-active material into the environment would be effective principally because of the fear and anxiety aroused in the minds of the technical uninformed. It seems likely that this would be out of all proportion to the like level of harm. One reason may be that recommended dose limits are too often thou of as an upper limit just short of disaster, the nature of the linear hypothesis which recommendations are based being altogether overlooked. Thus release of rad activity which gave an individual member of the general public 1000 x more dose t. his annual dose limit would give him double the maximum permissible life-time's exposure of a designated radiation worker which itself is intended to keep work w radiation amongst the safer kinds of industrial work. I do not regard this argum as water-tight or complacent, merely a statement of fact. It will be the accurac our assessment of cancer induction per unit of radiation dose which determines it validity.

Summary

Cancer can be induced by radiation in any tissue where cancer occurs naturally. The observation that antenatal diagnostic radiography causes a small but definite increase in childhood cancer is as good evidence as could be expected in support of the scientific expectation that there would be no threshold of dose for carcinogenesis.

A linear relation between radiation dose and frequency of induced cancer is a necessary assumption for a system of radiological protection but is not necessarily a reasonable basis for realistic assessments of cancer risk. Indeed there are radiobiological and epidemiological reasons to the contrary.

If the linear hypothesis is accepted then at the present time in the UK the routine practice of medicine is of about 2 orders of magnitude more important in causing cancer than environmental pollution by discharge of radio-activity. The acceptability of radiation safety standards for occupational exposure may be justified by comparison of radiation cancer risks with risks from fatal accidents in the safer industries. The acceptability of the corresponding standards for members of the public seems to require more public discussion of the concept of negligible risk. Emotional reactions to uncontrolled releases of radio-activity are based at least in part on a failure to appreciate the hypothesis of linearity.

RESUME

Les rayonnements ionisants peuvent induire un cancer dans tout tissu où le cancer apparaît spontanément. L'observation selon laquelle la radiographie diagnostique prénatale entraîne une augmentation faible, mais certaine, du taux de cancers de l'enfance, constitue une preuve aussi bonne que possible à l'appui de l'hypothèse scientifique d'une absence de dose liminale pour la cancérogènèse.

L'existence d'une relation linéaire entre la dose de rayonnements et la fréquence du cancer induit est une supposition nécessaire pour organiser un système de protection radiologique, mais elle ne constitue pas obligatoirement une base de travail raisonnable pour une évaluation réaliste du risque cancérogène. Il existe en fait des raisons de croire le contraire, d'ordre radiobiologique et épidémiologique.

Si l'hypothèse linéaire est admise, la pratique courante de la médecine est aujourd'hui au Royaume-Uni environ deux fois plus importante en tant que cause de cancer que la pollution de l'environnement par les décharges radioactives. La définition de normes admissibles de sécurité radiologique pour l'exposition professionnelle peut se justifier lorsqu'on compare les risques de cancer dus aux rayonnements à ceux d'accident mortel dans les industries les moins dangereuses. Accepter les normes correspondantes pour des membres de la population nécessite, semble-t-il, un examen public plus approfondi de la notion de risque négligeable. Les réactions émotionnelles aux décharges sans contrôle de radioactivité s'expliquent en partie du moins par une imcompréhension de l'hypothèse de la linéarité.

References

(1) UNSCEAR: Ionizing radiation levels and effects Vol. II: Effects. 1972, United Nations, New York.

(2) BEIR: The effects on populations of exposure to low levels of ionizing radiation. 1972, National Academy of Sciences - National Research Council, Washington, U.S.A.

(3) Mole, R.H.: Ionizing radiation as a carcinogen: practical questions and academic pursuits. British Journal of Radiology, 1975, 48, 157-169.

(4) Upton, A.C., Wolff, F.F., Furth, J. and Kimball, A.W.: A comparison of the induction of myeloid and lymphatic leukaemias in X-radiated RF mice. Cancer Research, 18, 842-848.

(5) Upton, A.C., Randolph, M.L., Conklin, J.W. and others: Late effects of fast neutrons and gamma-rays in mice as influenced by the dose rate of irradiation: Induction of neoplasia. Radiation Research, 1970, 41, 467-491.

(6) Mole, R.H. and Ardran, G.M.: Environmental hazards: Hazards in every day use of radiation and radio-active materials in diagnosis and therapy. Modern Medicine, 1971, May, 334-339.

(7) Webb, G.A.M.: Radiation exposure of the public - the current levels in the United Kingdon. 1974 National Radiological Protection Board, Harwell, England.

(8) Committee on Radiological Hazards: Radiological Hazards to patients. Second Report 1960, Her Majesty's Stationery Office, London.

(9) Mole, R.H.: Radiation as a toxic agent. Lectures on the Scientific Basis of Medicine 1958-59, 1960, 8, 65-86, Athlone Press, London.

(10) Mole, R.H.: Antenatal irradiation and childhood cancer. British Journal of Cancer, 1974, 30, 199-208.

(11) Pochin, E.E.: Occupational and other fatality rates. Community Health, 1974, 6, 2-13.

(12) Registrar General Decennial Supplement Occupational Mortality Tables, 1971, Her Majesty's Stationery Office, London.

(13) Knox, E.G.: Negligible risks to health. Community Health, 1975, 6, 244-251.

(14) Mole, R.H.: The risks of everyday life: Accepting risks for other people. Proceedings of the Royal Society of Medicine, 1976, in the press.

(15) Medical Research Council: Criteria for controlling radiation doses to the public after accidental escape of radio-active material, 1975, Her Majesty's Stationery Office, London.

V

IDENTIFICATION OF CARCINOGENS
IDENTIFICATION DES CANCÉROGÈNES

Chairman / *Président* : Dr. F. E. ZAJDELA

CANCEROGENESE CHIMIQUE EN CULTURE DE TISSUS : CRITERES ET TESTS DE TRANSFORMATION

I. CHOUROULINKOV et C. LASNE

Institut de Recherches Scientifiques sur le Cancer — CNRS
7, rue Guy Mocquet, B.P. n° 8, 94800 Villejuif, France

La culture cellulaire par la relative simplicité de la méthode et de la cible comparée à l'animal promettait un modèle expérimental accessible pour les études sur le mécanisme de la cancérogénèse chimique et l'élaboration de tests rapides et fiables pour une évaluation de l'activité cancérogène des agents chimiques, comme cela a pu être fait pour les virus.

Les travaux sur la cancérogénèse chimique faits en culture de tissus sont considérables. Il n'est pas dans nos intentions d'en faire état. Notre objectif est plutôt de voir les critères importants de la transformation cellulaire et de présenter les schémas expérimentaux actuellement utilisés.

CRITERES DE LA TRANSFORMATION CELLULAIRE EN CULTURE ET SCHEMAS EXPERIMENTAUX.-

Les études sur la transformation cellulaire en cultures par des agents chimiques ont débuté en même temps que les études sur la physiologie de la cellule, alors que les méthodes de culture laissaient à désirer. Tout était donc à découvrir et, à priori, rien ne s'opposait à l'idée que la cellule normale ne se transforme pas spontanément. C'est dans ces conditions et dans cet état d'esprit qu'EARLE et Coll. (1, 2) ont commencé leurs expériences sur l'action cancérogène en culture cellulaire du 3-méthylcholanthrène (MCA). Ce fut un demi-succès. Les cellules témoins se sont transformées en même temps que les cellules traitées.

Actuellement nous disposons de beaucoup d'éléments permettant de distinguer la cellule transformée de la cellule normale. Au cours de la transformation en culture la cellule subit des modifications morphologiques, un changement de comportement. et des modifications biochimiques (revue Réf. 3). En faisant brièvement allusion aux modifications morphologiques (cytologiques) nous nous arrêterons aux changements du comportement de la cellule transformée, <u>formation de colonies atypiques, de foyers de transformation</u>.et <u>de colonies en milieu semi-liquide</u>, et à <u>l'acquisition de la potentialité tumorigène</u>, seuls critères donnant lieu à des tests d'évaluation quantitative de la transformation par les agents chimiques.

I - <u>MODIFICATIONS CYTOLOGIQUES</u>.-

Il est bien connu que les cellules néoplasiques se différencient morphologiquement des cellules normales respectives par le volume et l'aspect général de la cellule, du noyau, du nucléole et du cytoplasme. EARLE et Coll. déjà en 1943 (1), au cours de leurs expériences ont constaté "des altérations morphologiques" plus accentuées chez les cellules traitées avec le MCA. BARKER et SANFORD en 1970 (4) ont repris les études sur la possibilité de diagnostiquer la transformation des cellules de souris en culture par leurs modifications cytologiques. Cependant le diagnostic cytologique de transformation était au mieux posé au moment et

surtout après que les cellules inoculées sur animal aient formé
une tumeur. Les auteurs ont conclu qu'un bon cytologiste peut
reconnaitre la transformation par les modifications cytologiques.
Actuellement les critères cytologiques, d'une valeur certaine pour
les chercheurs, ne sont pas utilisables pour une étude quantitative
de la transformation cellulaire.

II - FORMATION DE COLONIES ET FOYERS DE TRANSFORMATION.-

PUCK et MARCUS (5) ont mis au point une technique qui
permet l'obtention de colonies à partir d'une cellule. Il s'agit
de mettre un petit nombre de cellules (10 à 100 cell/cm^2) en culture. Certaines de ces cellules après avoir adhéré au fond de la
boite se divisent et forment des colonies. Ces colonies d'une
structure régulière sont constituées de cellules fusiformes du
type fibroblastique ou arrondies du type épithélial (Fig. 1).

Avec l'objectif précis de suivre la transformation des
cellules individuellement BERWALD et SACHS (6,7) ont eu recours
à la méthode de clonage de PUCK et MARCUS en traitant les cellules avec des cancérogènes connus, avant, pendant ou après le moment du clonage. Les auteurs ont observé qu'un petit nombre de
colonies traitées par le BaP ou le MCA n'avait pas l'aspect de
colonies normales. Elles avaient une structure désorganisée et les
cellules qui les constituaient étaient entrecroisées et souvent
empilées (Fig. 1). Les auteurs ont considéré la morphologie de ces
colonies comme la conséquence de l'action des cancérogènes et comme
l'extériorisation de la transformation d'où le terme de colonies
transformées.

Sur le plan physiologique la formation des colonies atypiques s'explique par un bouleversement du "comportement social"
des cellules (3) et par l'abolition de l'inhibition de contact.
Normalement les cellules obeissent à un "contrôle mutuel", se ran-
gent de façon régulière, comme dans l'organisme, et forment des
colonies structurées et caractéristiques du type cellulaire. Quand
ce contrôle mutuel est aboli elle s'entrecroisent ou s'orientent
de façon désordonnée (Fig. 1). Le phénomène de l'inhibition de
contact (8) contraint les cellules arrivant en confluence à arrêter

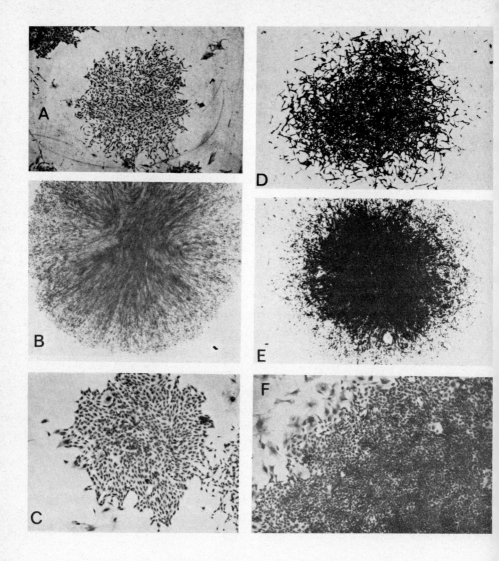

FIGURE 1 : A - Colonie normale de fibroblastes de Hamster.
B - Colonie normale de fibroblastes de Rat.
C - Colonie normale de cell. épithélioïdes de Hamster.
D et E - Colonie transformée-fibroblastes de Hamster.
F - Colonie transformée-cell. épithélioïdes de Hamster.

leur croissance. Or les cellules transformées continuent à se diviser, à se superposer, à s'empiler. CHEN et HEIDELBERGER ont observé (9) que des cellules de prostate de souris restaient à l'état quiescent quand elles étaient parvenues à former une couche monocellulaire. Par contre le traitement avec le MCA entrainait la formation de foyers de cellules empilées (Fig. 2).

Les colonies transformées et les foyers de transformation, morphologiquement visibles ont été considérés comme critères de la transformation cellulaire en culture.

En fonction des cellules utilisées et du délai d'observation après le traitement trois procédés ou schémas expérimentaux se dégagent : le schéma de BERWALD et SACHS (7) ou test à court terme, le schéma de CHEN et HEIDELBERGER (9) et le schéma de l'équipe HUEBNER (10) ou tests à moyen terme.

1°) - <u>Schéma expérimental de BERWALD et SACHS (7) ou test à court terme.</u> (Schéma I)

SCHEMA I. TEST DE TRANSFORMATION CELLULAIRE DE BERWALD ET SACHS (6,7) OU TEST A COURT TERME

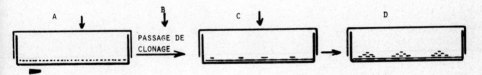

- CELLULES EMBRYONNAIRES DE HAMSTER (SURTOUT UTILISEES)
- COUCHE NOURRICIERE - CELLULES EMBRYONNAIRES DE RAT IRRADIEES (B ET C LIGNE CONTINUE)
- TRAITEMENT (↓) - A - AVANT LE PASSAGE DE CLONAGE - CELLULES CIBLES EN NAPPE.
    B - PENDANT LE PASSAGE DE CLONAGE - AVEC LA SUSPENSION CELLULAIRE
    C - APRÈS LE PASSAGE DE CLONAGE - GÉNÉRALEMENT 24 HEURES APRÈS
- FIXATION, COLORATION 9 À 14 JOURS APRÈS LE PASSAGE DE CLONAGE (D) ET ÉVENTUELLEMENT SUBCULTURE POUR INOCULATION SUR ANIMAUX.

Ce schéma est basé sur le clonage de cellules sur couche nourricière, procédé de PUCK et MARCUS (5). Un petit nombre (200 à 5×10^3 par boite de 6 cm de diamètre) de cellules embryonnaires de Hamster du premier ou du deuxième passage, est cloné dans des

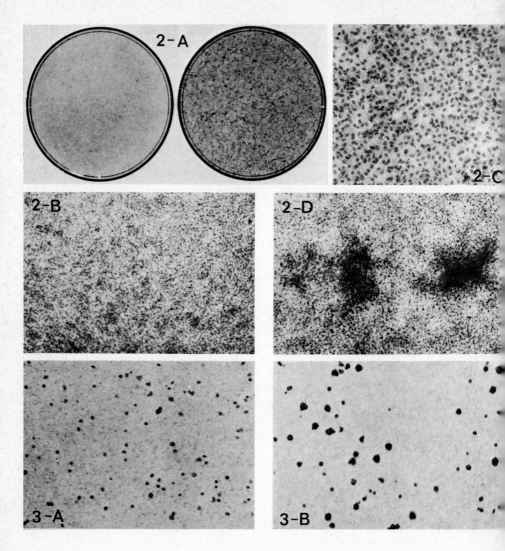

FIGURE 2 : Cellules humaines transformées (?)-culture de 28 jours. A : Aspect macroscopique : multiples foyers de transformation à droite. B et C - témoins : cellules épithélioïdes en arrêt de croissance. D - aspect histologique des foyers de transformation avec nombreuses mitoses.

FIGURE 3 : Clonage en milieu semi-liquide (15 jours de culture). A - Témoins-nombreuses cellules isolées ou à deux (50 000 cell/boite). B - Cellules traitées-nombreuses colonies tridimensionnelles constituées de plusieurs cellules. (25 000 cell/boite).

boites de Pétri sur une couche nourricière. Cette dernière est préparée à partir d'une culture primaire ou du premier passage de cellules embryonnaires de Rat irradiées à 4-5000 r. La répartition des cellules nourricières dans les boites est faite avant le clonage des cellules de Hamster. Les deux manipulations peuvent cependant être faites simultanément en mélangeant les populations cellulaires. Les cellules embryonnaires de Hamster sont traitées avant, pendant ou et surtout 24 heures après le passage de clonage. Selon le milieu utilisé les cellules sont incubées dans une étuve humide sous une atmosphère d'air contenant 5 ou 10 % de CO_2. La durée du traitement peut varier de 1 à plusieurs jours, voire la durée totale de l'incubation (9 à 14 jours) avec ou sans changement de milieu. A la fin les colonies sont fixées au méthanol et colorées au May-Grunwald-Giemsa.

Le nombre de colonies par rapport au nombre de cellules mises en culture détermine l'efficacité de clonage (EC), exprimée en pourcentage. Ce paramètre indique l'effet toxique des substances étudiées et peut être d'une grande utilité.

Le nombre de colonies atypiques détermine l'activité transformante de la substance étudiée. Les résultats sont exprimés en nombre absolu ou pourcentage par rapport à la totalité des colonies dénombrées.

B(a)P µg/ml	Nbre boites	Nbre colonies	Efficacité clonage	Toxicité %	colonies transformées	
					Nombre	%
0,1	10	493	16,4	11,4	21	4,3
1,0	10	438	14,6	21,1	18	4,1
10,0	9	339	12,4	33,0	15	4,4
acétone	10	556	18,5	0	5	0,9

TABLEAU I - Effet du BaP sur des cellules embryonnaires de Hamster (300 cellules/boite) ensemencées sur couche nourricière de cellules embryonnaires de Rat (10^5/boite) irradiées (4500 r). BaP additionné 24 heures après le clonage. - Durée 9 jours.
Réf. C. LASNE : Le benzo(a)pyrène : ses effets biologiques et sa pénétration in vitro dans la cellule - Thèse (Université de PARIS VI 1973).

Le tableau I donne un exemple de présentation et les résultats obtenus avec des cellules de Hamster traitées par le benzo(a)pyrène (BaP). Il est observé une transformation et un effet-dose pour la cytotoxicité, celle-ci étant indépendante de la transformation (11).

Di-PAOLO et Coll. en particulier ont travaillé avec ce modèle expérimental tel qu'il a été décrit (12,13) ou en prétraitant les cellules avec des agents chimiques (14), agents physiques - rayonx X (15) ou en utilisant des cellules de souris qui se clonent sans couche nourricière (16). Les auteurs ont constaté que la 7,8-benzoflavone et le benz(α)anthracène diminuent l'effet toxique des hydrocarbures cancérogènes et augmentent la transformation, tandis que la 5,6-benzoflavone n'a aucun effet. De même l'utilisation de cellules de Hamster irradiées dans des conditions bien précises facilite considérablement l'effet transformant du BaP.

Nous présentons la liste non exhaustive des substances cancérogènes ou non éprouvées par ce procédé (tableau II). Sur les 11 cancérogènes connus 8 ont induit des colonies transformées. Trois, l'uréthane (6), le N-nitrosodiethylamine (NDEA) (16) et N-2-fluorenylacetamide (FAA) (17) se sont montrés inactifs. Ces faux-négatifs s'expliquent par la nécessité d'un métabolisme autre que celui effectué par les cellules. Effectivement Di-PAOLO par traitement transplacentaire et la mise en culture des cellules des embryons ainsi traités a obtenu une transformation par les deux premiers composés (19).

Dans notre Laboratoire nous avons reproduit ces résultats avec le benzo(a)pyrène (BaP), le 7,12-diméthylbenz(a)anthracène et la N-méthyl-N-nitro-N-nitrosoguanidine (MNNG). Cependant un de nos problèmes a été d'élargir l'application de ce procédé à d'autres substances, plus complexes, tels que les condensats de fumée de cigarettes. Nous avons éprouvé plus de 30 substances complexes. Aucune transformation n'a été obtenue. Cependant certaines de ces substances se sont montrées cancérogènes pour la peau de souris. "In vitro" nous n'avions pas la possibilité d'augmenter la dose à cause de l'effet toxique. La prédominance de la toxicité empêchait donc l'activité transformante.

SUBSTANCES	Activité en culture	Activité in vivo
Benzo(a)pyrène (BaP)	+	+
7-12 diméthylbenz(a)anthracène (DMBA)	+	+
3-méthylcholanthrène (MCA)	+	+
10-méthylbenz(a)anthracène	+	+
Dibenz(a,h)anthracène	+	+
Benzo(e)pyrène (BeP)	−	−
Benz(a)anthracène	−	−
Dibenz(a,c)anthracène	−	−
3-méthylbenz(a)anthracène	−	−
Anthracène	−	−
Pyrène	−	−
Phénanthrène	−	−
Chrysène	−	−
N-2-fluorenylacétamide (FAA)	−	+
N-hydroxy - FAA	+	+
N-acétoxy - FAA	+	+
N-méthyl-N'-nitro-N-nitrosoguanidine (MNNG)	+	+
N-nitrosodiéthylamine (NDEA)	−	+
Aflatoxine B1	+	+
Uréthane	−	+

TABLEAU II - Liste (non exhaustive) des substances éprouvées par le test à court terme (de BERWALD et SACHS) et la corrélation avec les résultats in vivo (Réf. 6, 7, 12, 16, 17, 18).

Ces observations nous ont amenés à considérer que ce test, utile pour des substances pures, ne l'est pas pour des mélanges complexes. Toutefois cette technique de clonage trouve toute son utilité au cours des tests à long terme comme nous le verrons par la suite.

2°) - <u>Schéma expérimental de CHEN et HEIDELBERGER (test à moyen terme)</u>.

Le deuxième schéma expérimental basé sur l'induction de foyers transformés a été mis au point par CHEN et HEIDELBERGER (9) modifié ensuite par MARQUARDT et Coll. (20). Il s'agit de mettre en culture 10^3 cellules de prostate de souris C_3H sensibles à l'inhibition de contact. Le lendemain la substance à étudier est ajoutée pour 7 jours avec un renouvellement. Ensuite le milieu normal est renouvelé une ou deux fois par semaine. Après huit semaines environ (50-54 jours) de culture les cellules sont fixées au méthanol et colorées pour examen.

SCHEMA II. TEST DE TRANSFORMATION CELLULAIRE DE CHEN ET HEIDELBERGER OU TEST A MOYEN TERME.

CELLULES DE PROSTATE DE SOURIS C_3H - 10^3 cell/boite.

A - TRAITEMENT (↓)-24 HEURES APRES LA MISE EN CULTURE. DUREE 7 JOURS.

B - 54 JOURS APRES : - FOYERS DE TRANSFORMATIONS
- SUBCLONAGE POSSIBLE POUR INOCULATION SUR ANIMAL.

La même équipe a isolé (21) une lignée de cellules embryonnaires de souris C_3H sensibles à l'inhibition de contact et qui ne montrent pas une transformation spontanée. Par contre sous l'effet du MCA ou du DMBA il se forme 3 types de colonies dont deux isolées provoquent des tumeurs après 2 à 4 passages seulement quand elles sont greffées sur l'animal (22). Ce dernier fait est très important car il confirme sans équivoque la nature transformée des cellules constituant les foyers induits par le MCA ou DMBA.

Ce test a été utilisé presqu'exclusivement pour connaître l'action transformante des hydrocarbures polycycliques et certains dérivés (23,24). Les résultats sont satisfaisants. Il faut regretter qu'il n'y ait pas eu, à ma connaissance des investigations de routine, car sur le plan évolutif les foyers observés 50 à 54 jours après le début du traitement sont plus proches de la transformation maligne que les colonies transformées de 9 à 14 jours. De plus, sur le plan pratique l'aspect subjectif de l'interprétation morphologique de colonies transformées n'existe presque pas pour les foyers de transformation.

3°) - <u>Schéma expérimental de l'équipe HUEBNER (10) (test à moyen terme)</u>.

Un troisième procédé basé sur la formation de foyers de transformation par des cellules sensibles à l'inhibition de contact est proposé par l'équipe de HUEBNER (10, 25, 26, 27, 28). La différence par rapport au test précédent est que les cellules cibles sont plus susceptibles à la transformation. Cette susceptibilité est liée à "l'âge avancé" des cellules (phase II - d'héréroploidie stable - voir Réf. 10) et à la surinfection par le RLV (Rauscher Leukemia Virus).

La procédé de routine est relativement simple (10). Des cellules de Rat de lignée F 1706 (passage 88) sont infectées avec le RLV. Une semaine après les cellules sont subdivisées dans deux flacons et traitées pendant 8 jours avec la substance à étudier. Un des flacons est conservé en renouvelant périodiquement le milieu pendant les 60 jours, de la durée du test. Le deuxième flacon est subdivisé deux semaines plus tard. Le subclonage se répète toutes les deux semaines en conservant toujours un flacon et ceci 4 fois. De cette manière si l'effet est fort les foyers de transformation apparaitront dans le flacon de départ. Si l'effet est faible le subclonage augmente la possibilité pour les cellules transformées de développer ces foyers.

Famille	In Vivo	Nbre substances actives/Nbre total testées
Colorants azoïques	Actifs	3/3
	Inactifs	0/1
Hydrocarbures polycycliques	Actifs	8/8
	Inactifs	0/4
Amines aromatiques	Actifs	6/7
	Faiblement actifs	1/2
	Inactifs	0/3
Divers	Actifs	4/4
	Faiblement actifs	1/1
	Inactifs	0/2

TABLEAU III - Activité transformante des cancérogènes chimiques et analogues - (FREEMAN et Col., In Vitro, 11, 107-116, 1975).

Avec ce modèle expérimental les auteurs ont étudié l'action transformante de 12 fractions de condensat de fumée de cigarette (28) et 35 substances de nature différentes (10). Parmi les fractions de condensat 4 ont été trouvées actives en culture tandis que sur l'animal trois seulement se sont montrées cancérogènes. Par contre la corrélation entre l'activité "in vivo" et l'activité transformante des 35 substances est bien meilleure (tableau 3). Sur les 22 substances fortement cancérogènes 21 se sont montrées actives en culture (à l'exception du N-2-fluorenylacetamide). Sur les 3 faibles cancérogènes 2 ont induit la formation de foyers de transformation. Enfin toutes les substances non cancérogènes sont restées inactives.

Ces résultats obtenus à l'aveugle sont largement satisfaisants. Il n'y a aucune erreur de faux-positifs et l'erreur de faux-négatifs est dans des limites acceptables, d'autant plus que cette erreur a des chances d'être détectée par d'autres tests auxquels la substance sera probablement soumise par exemple des tests de mutagénèse.

III - LA CROISSANCE SELECTIVE DE LA CELLULE TRANSFORMEE EN MILIEU SEMI-LIQUIDE (FORMATION DE COLONIES TRIDIMENSIONNELLES).

La cellule normale n'a pas la capacité de se diviser, donc de proliférer quand elle est en suspension dans un milieu semi-liquide à base d'agar (0,3-0,5 %). Or McPHERSON et MONTAGNIER (29) et MONTAGNIER (30) ont montré que les cellules transformées par un virus oncogène prolifèrent dans un tel milieu et forment des colonies tridimensionnelles (Fig. 3), dont il est possible de déterminer le nombre. Cette propriété de la cellule transformée a été appliquée à la recherche sur la cancérogénèse chimique pour contrôler la transformation.

Le clonage des cellules (2×10^4 ou 10^5 cell/boite de 6 cm de diamètre) en milieu semi-liquide s'effectue soit après le traitement, soit périodiquement au cours d'une expérience à long terme (voir schéma III). Le développement des colonies tridimensionnelles est suivi le plus longtemps possible.

Nous avons effectué de nombreux essais de contrôle de la transformation immédiatement après le traitement des cellules en nappe. Aucun résultat positif n'a été obtenu, même avec des cancérogènes puissants. Par contre au cours de tests à long terme nous avons pu suivre et détecter, dans nos conditions expérimentales, la transformation par la formation de colonies sur agar avant l'obtention des tumeurs sur l'animal (tableau VI).

IV - POTENTIALITE TUMORIGENE DE LA CELLULE TRANSFORMEE.-

Les cellules qui ont subi une transformation maligne spontanée ou induite forment une tumeur quand elles sont inoculées sur un animal isologue ou tolérant. C'est la potentialité tumorigène, critère absolu de la transformation maligne des cellules en culture. C'est par cette donnée que la signification des critères précédents a été établie. Des colonies transformées ou des foyers de transformation ont été subcultivés plus ou moins longtemps jusqu'à l'obtention de tumeurs sur animal, les subcultures des colonies ou cellules normales restant négatives.

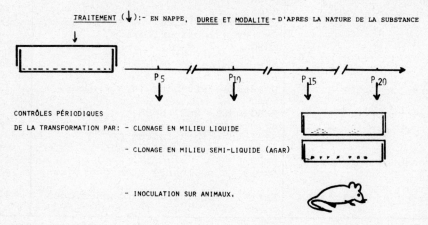

SCHÉMA III. TEST DE TRANSFORMATION À LONG TERME D'EARLE (COMBINE)

Le schéma général du test (Schéma III) est simple : traitement des cellules plus ou moins longtemps, entretien de la culture et inoculation périodique sur animal jusqu'au développement de tumeurs. Le test peut ainsi durer plusieurs mois d'où le nom de test à long terme ou schéma d'EARLE, EARLE ayant réalisé avec ses collaborateurs le premier essai avec l'obtention de tumeurs sur animal (2).

De nombreux chercheurs ont utilisé ce schéma expérimental. Le tableau IV présente les types cellulaires qui ont servi de cible et les substances qui ont entrainé la transformation maligne.

Cellules	Substances	Réf.
Souris : -conjonctives	Benzo(a)pyrène	7,16,31
	7,12-diméthylbenz(a)anthracène (DMBA)	16
	3-méthylcholanthrène (MCA)	7
	N-méthyl-N'-nitro-N-nitrosoguanine (MNNG)	16
	métylnitroso-urée	32
- épith. peau	DMBA	33
Rat : -conjonctives	DMBA	34
	MCA	34
	Nitrosométhylurée	35
	4-nitroquinoline-1)oxide	36
	MNNG	37
- épith.-vessie	Nitrosamine	38
Hamster : -conjonctive	Diméthyl-7,10-benzo(a)acridine	39
Cobaye : -conjonctives	MNNG, MCA, BaP, DMBA	)
	Aflatoxine B1	) 40
	N-nitrosodiéthylamine	)
Rat : -conjonctives+RLV^x	DMBA	27
	Extraits polluants atmosphériques	41
Hamster : -conjonctive+RLV^x	Extraits polluants atmosphériques	41

TABLEAU IV - Résultats (liste non exhaustive) de transformation à long terme. Types cellulaires et cancérogènes étudiés. (x - Rauschert Leukemia Virus).

Les cellules de Souris et de Rat sont les plus utilisées mais leur utilisation n'exclue pas l'emploi d'autres espèces. Les cellules de Souris se transforment spontanément assez vite entre 3 et plusieurs mois selon les conditions expérimentales et la lignée. La réponse à un traitement peut donc être très rapide. Cependant la faible marge entre la transformation induite et la transformation spontanée, et la présence fréquente de virus sont des inconvénients pour l'interprétation des résultats en valeur absolue et pour la détermination du rôle effectivement joué par les agents chimiques étudiés. Pour ces raisons les cellules de Rat sont souvent préférées, mais, étant donné leur transformation spontanée plus tardive le test risque de durer plusieurs mois voire un an. Ceci évidemment parait long. Pour palier à cet inconvénient il est possible d'introduire des contrôles périodiques de la transformation par clonage des cellules en milieu liquide et en milieu semi-liquide (agar) (Schéma III), et en modulant le traitement (la dose et la durée).

La transformation par clonage en milieu liquide est évaluée par le pourcentage des colonies transformées ou par les foyers de transformation s'il s'agit de cellules sensibles à l'inhibition de contact. En milieu semi-liquide le nombre des colonies tridimensionnelles détermine le degré de la transformation.

Nous avons adopté ce schéma expérimental en utilisant les cellules embryonnaires de Rat. Les tableaux V et VI présentent les résultats observés au cours de deux expériences. Le contrôle de la transformation a été effectué par clonage en milieu liquide pour la première expérience (tableau V) et en milieu semi-liquide pour la seconde (tableau VI).

Dans tous les cas la transformation cellulaire a été décelée avant que la potentialité tumorigène se manifeste pour nos conditions expérimentales. De plus il faut noter que l'induction de tumeurs avec des cellules transformées en culture relève de l'immunité de la transplantation. Pour obtenir les tumeurs plus tôt il faut d'une part inoculer soit beaucoup plus de cellules de la culture globale, soit subcultiver des colonies transformées et obtenir une population de cellules transformées homogènes, et d'autre part irradier les animaux pour diminuer, voire abolir

Passage	Trait.	E.C(%)	T (%)	Inoculation sur animal x	
				an. avec tum/total	temps latence (semaines)
P 32	1	3,7	5,4	0/10	-
	2	6,0	20,0	0/11	-
	3	53,0	4,7	0/11	-
	t^x	6,3	0,0	0/9	-
P 36	1	5,3	23,2	0/8	-
	2	21,3	35,6	1/8	6
	3	62,5	32,3	0/12	-
	t	7,8	11,5	0/12	-
P 40	1	35,7	52,1	0/12	-
	2	33,5	60,3	13/13	4
	3	73,0	57,1	11/11	6
	t	8,0	7,6	0/10	-
P 46	1	63,5	95,0	1/10	8
	2	38,8	94,4	11/11	4
	3	73,3	92,2	10/10	4
	t	7,6	11,5	0/10	-
P 50	1	100,0	100,0	5/9	7
	2	62,0	100,0	8/8	4
	3	94,4	100,0	9/9	4
	t	7,5	14,5	0/8	-

TABLEAU V - Cellules embryonnaires de Rat traitées avec des agents chimiques - test à long terme. Relation entre l'efficacité de clonage (EC la transformation (T) et la formation de tumeurs sur animal. (x = inoculation de 3×10^6 cell/nouveaux-nés pour P32, P36 et P40, et 2×10^6 pour P46 et P50. t = témoins). (Résultats non publiés).

Passage	Trait.	formation de colonies sur agar	Inoculation sur animal	
			an. avec tumeurs sur total	temps de latence en semaines
P 28	1	++	0/12	-
	2	+	0/12	-
	T	-	0/10	-
P 37	1	+++	10/10	3
	2	+	1/10	4
	T	-	0/10	-
P 41	1	++++	11/11	3
	2	++	0/11	-
	T	-	0/11	-
P 48	1	++++	11/11	2
	2	++	1/11	8
	T	-	0/10	-

TABLEAU VI - Cellules embryonnaires de Rat traitées avec BaP (tr.2), BaP + TPA (tr.1) -(Test à long terme). Relation entre la formation de colonies sur agar et de tumeurs sur animal. (P29 et P37 - inoculation de 2×10^6 cellules ; P41 et P48 - inoculation 1×10^6 cellules. T - témoins solvant). (Réf. 31).

leurs réactions immunitaires de rejet, comme cela a été fait par EVANS et Col. (40). Dans tous les cas des tumeurs seront obtenues avec des cellules qui ont montré un certain pourcentage de colonies transformées. Dans ces conditions on peut se demander s'il est nécessaire de recourir à l'inoculation sur animal ce qui ajoute un délai supplémentaire, alors que les tests de clonage ont déjà montré que la transformation s'était produite.

Le contrôle périodique de la transformation par clonage en milieu liquide permet de classer les traitements selon leur activité transformante. Le traitement 2 est le plus actif, suivi de près par le traitement 3 et beaucoup plus actif que le traitement 1, exactement comme la potentialité tumorigène (tableau V). Il en est de même pour la clonage en agar (tableau VI).

On peut trouver long les 28-30 passages (environ 150 jours) pour déceler la transformation. L'objectif de ces expériences était de déceler une différence entre les traitements et non pas une performance de vitesse. Cet objectif a été atteint. Toutefois, une transformation accélérée peut être obtenue soit en augmentant la dose, dans les limites de solubilité possible et de toxicité acceptable, soit en prolongeant le traitement au cours des passages, soit enfin en utilisant des cellules rendues susceptibles comme les cellules utilisées par l'équipe HUEBNER (10).

A notre avis le test de transformation à long terme, combiné avec des contrôles périodiques de la transformation donne un maximum d'assurance pour une réponse correcte en ce qui concerne l'activité transformante des agents chimiques. Il y a plus de risques d'avoir de faux-négatifs que de faux-positifs. Les faux-négatifs peuvent être soit la conséquence d'une faible activité, la transformation induite se rapprochant trop près de la transformation spontanée, soit l'absence d'une activation métabolique nécessaire à leur action. Les risques pour les premiers faux-négatifs sont diminués si les contrôles de transformation sont bien faits. Pour les seconds si le cas se présente on peut avoir recours au traitement par voie transplacentaire (19).

SUMMARY

We believe that there are four criteria for the evaluation of cell transformation in culture : development of transformed colonies, appearance of altered foci when cells sensitive to contact inhibition are used, formation of colonies in agar, and the capacity to induce tumors in animals (tumorigenic potentiality).

The formation of transformed colonies provides a short term test (t. of BERWALD-SACHS) to determine the transformating activity of chemical carcinogens. Judging by the published results there is a good correlation between this in vitro test and in vivo results. Nevertheless this test does not seem to me appropriate to studies on the activity of complex mixtures. But the use of a short term test to monitor transformation in the course of long term tests does see justified and useful.

The formation of altered foci is the basis of two intermediate term tests (t. of CHEN-HEIDELBERGER, and t. of HUEBNER'S group). The value of these two tests have been experimentally proven, but their only disadvantage is the necessity of using cells that are not readily available.

Formation of colonies in agar is a criterion useful only for the control of cell transformation during long term tests.

Tumorigenicity constitutes a long term test. It is the absolute proof of cell transformation. We propose a long term experimental system combined with periodic checks of transformation by cell cloning in liquid medium (criteria of the transformed colonies or altered foci), and by cloning in agar.

The reported results show that it is possible to determine cellular transformation accurately without recourse to the animal. This combined long term test permits the classification of treatment based on their transformating capacity.

BIBLIOGRAPHIE

(1) EARLE, W.R., : Production of malignancy "in vitro". IV. The mouse fibroblast cultures and changes seen in the living cells. J. Nat. Cancer Inst., 1943, 4 : 165-212.

(2) EARLE, W.R., and NETTLESHIP , A., : Production of malignancy "in vitro". V. Results of injections of cultures into mice. J. Nat. Cancer Inst., 1943, 4 : 213-227.

(3) MACPHERSON I. The characteristics of animal cells transformed "in vitro". Adv. Cancer Res., 1970, 13 : 169-215, Ac. Press, New-York.

(4) BARKER, B.E. and SANFORD,K.K. : Cytologic manifestations of neoplastic transformation "in vitro". J. Nat. Ca. Inst., 1970, 44 : 39-63.

(5) PUCK, T.T., and MARCUS, P.I. : A rapide method for viable cell titration and clone production with HeLa cells in tissue culture : The use of X-irradiated cells to supply conditioning factors, Proc. Nat. Aca. Sci. U.S.A., 1955, 41 : 432-437.

(6) BERWALD, Y., and SACHS, L. : "In vitro" cell transformation with chemical carcinogens, Nature (London), 1963, 200 : 1182-1184.

(7) BERWALD, Y., and SACHS, L. : "In vitro" transformation of normal cells to tumor cells by carcinogenic hydrocarbons . J. Nat. Cancer Inst., 1965, 35 : 641-661.

(8) ABERCROMBIE, M.,. Contact inhibition : The phenomenon and its biological implication. Nat. Cancer Inst. Monogr., 1967, 26 : 249-277.

(9) CHEN, T.T., and HEIDELBERGER, C., : Quantitative studies on the malignant transformation of mouse prostate cells by carcinogenic hydrocarbons "in vitro". Int. J. Cancer, 1969, 4 : 166-178.

(10) FREEMAN, A.E., and IGEL, H.J. : Carcinogenesis "in vitro". I. "In vitro" transformation of rat embryo cells : Correlation with the known tumorigenic activities of chemicals in rodents. In vitro, 1975, 11 : 107-116.

(11) HUBERMAN, E., and SACHS, L., Cell susceptibility to transformation and cytotoxicity by the carcinogenic hydrocarbon benzo(a)pyrène, Proc. Nat. Acad. Sci., 1966, 56 : 1123-1129.

(12) DI-PAOLO, J.A., DONOVAN, P., and NELSON, R. Quantitative studies of "in vitro" transformation by chemical carcinogens. 1969. J. Nat. Cancer Inst.. 42 : 867-876.

(13) DI-PAOLO, J.A., NELSON, R.L., and DONOVAN, P.J. : Sarcoma-producing cell lines derived from clones transformed "in vitro" by benzo(a)pyrene. Science. 1969, 165 : 917-918.

(14) DI-PAOLO, J.A., DONOVAN, P.J., and NELSON, R.L. : Transformation of hamster cells "in vitro" by polycyclic hydrocarbons without cytotoxicity. Proc. Nat. Acad. Sci. U.S.A., 1971, 68 : 2958-2961.

(15) DI-PAOLO, J.A., DONOVAN, P.J., and NELSON, R.L. : X-irradiation enhancement of transformation by benzo(a)pyrene in hamster embryo cells. Proc. Nat. Acad. Sci. U.S.A., 1971, 68 : 1734-1737.

(16) DI-PAOLO, J.A., TAKANO, K., and POPESCU, N.C. : Quantitation of chemically induced neoplastic transformation of Balb/3T3 cloned cell lines. Cancer Res., 1972, 32 : 2686-2696.

(17) HUBERMAN, E., DONOVAN, P.J., and DI-PAOLO, J.I.: Mutation and transformation of cultured mammalian cells by N-acetoxy-N-2-fluorenylacetamide. J. Nat. Cancer Inst., 1972, 48 : 837-840.

(18) HUBERMAN, E., KUROKI, T., MARQUARDT, H., SELKIRK, J.K., HEIDELBERGER, C., GROVER, P., and SIMS, P. : Transformation of hamster embryo cells by epoxide and other derivatives of polycyclic hydrocarbons. Cancer Res., 1972, 32 : 1391-1396.

(19) DI PAOLO, J.A., NELSON, R.L., DONOVAN, P.J. and EVANS, C.H. Host-mediated "in vivo-in vitro" assay for chemical carcinogenesis. Arch. Pathol., 1973, 95 : 380-385.

(20) MARQUARDT, H., KUROKI, T., HUBERMAN, E., SELKIRK, J.K., HEIDELBERGER, C., GROVER, P., and SIMS, P. : Malignant transformation of cells derived from mouse prostate by epoxides and other derivatives of polycyclic hydrocarbons. Cancer Res., 1972, 32 : 716-720.

(21) REZNIKOFF, C.A., BRANKOW, D.W. and HEIDELBERGER, C. : Establishment and characterization of a cloned line at C_3H mouse embryo cells sensitive to postconfluence inhibition of division. Cancer Res. 1973, 33 : 3231-3228.

(22) REZNIKOFF, C.A., BERTRAM, J.S., BRANKOW, D.W. and HEIDELBERGER C. : Quantitative and qualitative studies of chemical transformation cloned C_3H mouse embryo cells sensitive to postconfluence inhibition of cell division. Cancer Res., 1973, 33 : 3239-3249.

(23) GROVER, P.L., SIMS, P., HUBERMAN, E., MARQUARDT, H., KUROKI T., and HEIDELBERGER, C. : "In vitro" transformation of rodent cells by K-region derivatives of polycyclic hydrocarbons. Proc. Nat. Acad. Sci. U.S.A., 1971, 68 : 1098-1101.

(24) MARQUARDT, H., SODERGREN, J.E., SIMS, P., and GROVER, P. : Malignant transformation "in vitro" of mouse fibroblasts by 7,12 dimethylbenz(a)anthracene and 7-hydroxymethylbenz(a)anthracene and by their K-region derivatives. Int. J. Cancer, 1974, 13 : 304-310.

(25) RHIM, J.S., CREASY, P., and HUEBNER, R.J. : Production of altered cell Foci by 3-<u>Methylcholanthrene</u> in mouse cells infected with AKR leukemia virus. Proc. Nat. Acad. Sci. U.S.A., 1971, 68 : 2212-221

(26) FREEMAN, A.E., WEISBURGER, E.K., WEISBURGER, J.H., WOLFORD, R.G., MARYAK, J.ill and HUEBNER, R.J., : Transformation of cell cultures as an indication of the carcinogenic potential of chemicals. J. Nat. Cancer Inst., 1973, 51 : 799-808.

(27) RHIM, J.S., VASS, W., CHO, N.Y. and HUEBNER, R.J. : Malignant transformation induced by 7,12-dimethylbenz(a)anthracene in rat embryo cells infected with Rauscher leukemia virus. Int. J. Cancer, 1971, 7 65-74.

(28) RHIM, J.S., and HUEBNER, R.J. : In vitro transformation assay of major fractions of cigarette smoke condensate (CSC) in mammalian cell lines. Proc. Soc. Exp. Biol. and Med., 1973, 142 : 1003-1007.

(29) MACPHERSON, I., and MONTAGNIER, L. : Agar suspension culture for the selective assay of cells transformed by polyoma virus. Virology, 1964, 23 : 291-294.

(30) MONTAGNIER, L. : La croissance sélective en gélose de cellules transformées par un virus cancérogène. Path. Biol., 1966, 14 : 244-251.

(31) LASNE, C., GENTIL, A., and CHOUROULINKOV, I. : Two stage malignant transformation of rat fibroblasts in tissue culture. Nature, 1974, 247 : 490-491.

(32) FREI, J.V. and OLIVER, J. : Influence of methylnitrosourea on malignant transformation of mouse embryo cells in tissue culture. J. Nat. Cancer Inst., 1971, 47 : 857-863.

(33) ELIAS, P.M., YUSPA, S.H., GULLINO, M., MORGAN, D.L., BATES; R.R., and LUTZNER, M.A. : "In vitro" neoplastic transformation of mouse skin cells : Morphology and ultrastructure of cells and tumors. J. Invest. Dermatol., 1974, 62 : 569-581.

(34) RHIM, J.S. and HUEBNER, R.J. : Transformation of rat embryo cells "in vitro" by chemical carcinogens. Cancer Res., 1973, 33 : 695-700.

(35) KIRKLAND, D.J. and PICK, C.R. : The histological appearance of tumours derived from rat embryo cells transformed "in vitro" spontaneously and after treatment with nitrosomethylurea. Br. J. Cancer, 1973, 28 : 440-452.

(36) NAMBA, M., MASUJI, H., and SATO. Carcinogenesis in tissue culture. IX. Malignant transformation of cultured rat cells treated with 4-nitroquinoline-1-oxide. Japan J. Exp. Med., 1969, 31 : 253-265.

(37) TAKII, M., TAKAKI, R., and OKADA, N. Carcinogenesis in tissue culture, XVI. Malignant transformation of rat cells by N-méthyl-N-nitro-N-nitrosoguanidine. Jap. J. Exp. Med., 1971, 41 : 563-579.

(38) HASHIMOTO, Y., and KITAGAWA, H.S. : "In vitro" neoplastic transformation of epithelial cells of rat urinary bladder by nitrosamines. Nature, 1974, 252 : 497-499.

(39) MARKOVITS, P., COPPEY, J., PAPADOPOULO, D., MAZABRAUD, A. et HUBERT-HABART, M. : Transformation maligne de cellules d'embryon de hamster en culture par les dimethyl-7,10 benzo(c)acridine. Int. J. Cancer, 1974, 14 : 215-225.

(40) EVANS, Ch. H., and DI-PAOLO, J.A. : Neoplastic transformation of guinea Pig fetal cells in culture induced by chemical carcinogens. Cancer Res., 1975, 35 : 1935-1044.

(41) FREEMAN, A.E., PRICE, P.J., BRYAN, R.J., GORDON, R.J., : Transformation of rat and hamster embryo cells by extracts of city smog. Proc. Nat. Acad. Sci. U.S.A., 1971, 68 : 445-449.

MUTAGENICITY TESTS IN CHEMICAL CARCINOGENESIS

H. BARTSCH

Unit of Chemical Carcinogenesis
International Agency for Research on Cancer
150 cours Albert Thomas, 69008 Lyon, France

During the past decades, it has become increasingly evident that chemically inert foreign compounds can be converted in the body to chemically reactive metabolites that combine with tissue macromolecules and thereby cause various adverse biological effects including carcinogenesis and mutagenesis[49]. The recent development of rapid and simple mutagenicity assays, in which mammalian metabolism is taken into account[39,40], has raised the question of the validity of data from such tests for the toxicological evaluation of compounds in terms of human risk[16,43,48,49] as they offer a promising tool to detect, remove and thus prevent carcinogenic hazards to man by chemicals since it is recognised that environmental factors play an essential role in the aetiology of human cancer.

The somatic-cell mutation theory of carcinogenesis has recently gained more attention, since it is becoming more and more evident that the majority of chemical carcinogens, which had been considered non-mutagenic, are mutagens[3,34,57,45]; and many mutagens, the carcinogenicity of which had not yet been investigated, have now been shown to be carcinogens. This convergence of mutagens and carcinogens, which has strengthened dramatically since 1960, has resulted from the discovery that many such chemicals require metabolic activation in order to show a biological activity. This metabolic activation is in many cases associated with binding to nucleophilic sites in nucleic acids and proteins in the organism[10]. The growing experimental evidence linking the carcinogenic activity of numerous chemicals to their capacity to be converted into electrophilic derivatives that may also exert a mutagenic effect, had led to the suggestion that an empirical relationship between chemical carcinogenesis and mutagenesis exists[5,35,44].

However, such a correlation has so far been limited to those changes of the genotype which appear as a consequence of structural or functional alterations of nucleic acids and thus are genetically transmissible.

Despite the diversity in chemical structures of known carcinogens and mutagens, such alkylating agents, N-nitrosamines and N-nitrosamides, naturally occurring compounds, aromatic hydrocarbons, etc., recognition of a common element in chemical carcinogens and mutagens has rapidly progressed since it was understood that the majority of carcinogens and many mutagens need metabolic activation in the host for transformation to their so-called ultimate reactive forms[36]. Comparison of these ultimate reactive metabolites of carcinogens has revealed that the common denominator of these substances is their electrophilicity, i.e., they are compounds whose atoms have an electron-deficiency which enables them to react with electron-rich sites in cellular nucleic acids and proteins causing mutagenic effects paralleled in many cases by the onset of DNA repair processes. The specificity of certain chemical compounds to produce tumours only in certain organs or in certain animal species, is primarily determined by the available concentration of ultimate reactive metabolites which is itself a consequence of a critical balance between activation and detoxication processes.

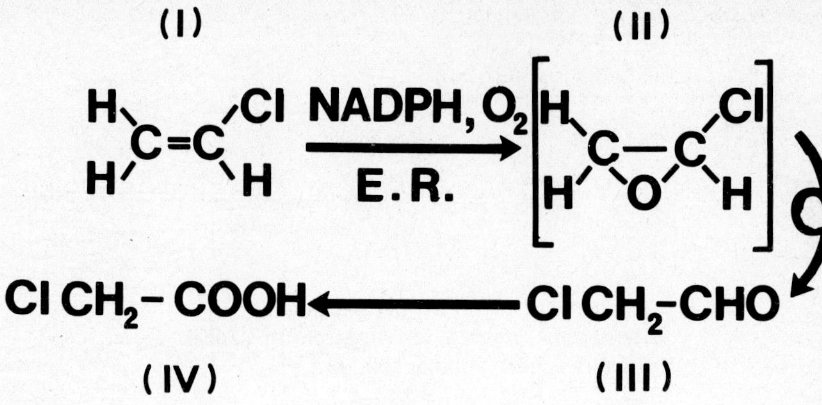

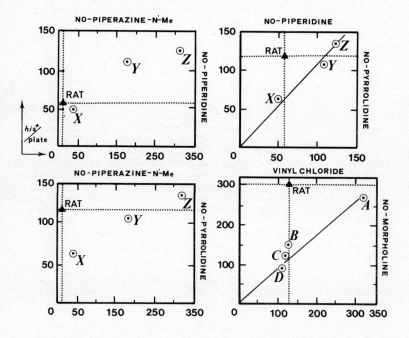

Fig. 1

Fig. 2. Liver-enzyme-mediated mutagenicity (his^+ revertant colonies of *S. typhimurium* 1530/plate) of BD-VI rat tissue and individual human biopsy samples, represented by different characters. N-nitrosamines were assayed according to Ames et al[3] and vinyl chloride as described by Bartsch et al[7]. The concentration of the carcinogens used was within a linear range of the dose-response curves obtained in an assay system containing 9,000 x g supernatant equivalent to 39 mg liver/plate. Number of spontaneous his^+ revertant colonies from appropriate controls (no co-factors or no substrate) was subtracted. Data from Bartsch et al[9] and unpublished work.

In view of the properties of ultimate reactive forms of chemical carcinogens, and, in particular, the growing evidence for an empirical relationship between the carcinogenic and the mutagenic activity of chemicals, it is now possible to use this relationship in short-term tests for the detection of potential carcinogens. There are many systems available at present which involve different genetic indicators and metabolic activation systems for detecting mutagenic activity; they all, however, have individual advantages and limitations. Ideally, an appropriate mutagenicity test system would include the full metabolic competency of the intact organism, that is, humans. Thus, the conclusion has been reached that a battery of test systems is needed in order to detect the genetic hazards caused by chemicals[16,41,44].

In evaluating these methods, the major question is which screening tests, singly or collectively, can serve as reliable indicators or predictors of the potential carcinogenic hazard of a chemical. This can only be achieved by testing a representative number of different classes of carcinogens and non-carcinogens. A valid test should be positive for compounds with known carcinogenic activities and negative for negative compounds, within the limits of the test. Extensive studies are being carried out in Japan and in the United States to evaluate short-term tests[40].

Results accumulated up to the present time[1,34,45] using a rat-liver microsome test *in vitro* with *S. typhimurium* strains developed by Bruce Ames have shown that about 70-80% of the carcinogens tested were also mutagens, while the number of false positives and false negatives was much lower, ranging from 5 to 10%. In tests used for preliminary screening, a small proportion of false positive and false negative results may be acceptable; but, for a final test, no false negative results can be accepted.

At present, mutagenicity assays, especially with sub-mammalian indicators, can be utilized in the following ways: a) to trace carcinogens and/or mutagens in the complex environment of man; b) to prescreen compounds for their mutagenic action *per se* or after metabolic activation, in order to select high priority compounds for long-term carcinogenicity tests. Special attention should be given to chemicals for which a substantial human exposure is known to occur and to newly-developed drugs used over a period of years in children, young adults or pregnant women; c) the *in vitro* systems allow investigation of the capability of human tissues to generate electrophilic intermediates from the compound under test. The usual type of metabolic studies, e.g., studies on distribution or excretion of the drug, are of limited value if the chemical acts by a partial conversion to a mutagenic or carcinogenic intermediate which interacts with DNA and is not excreted as such. This technique may also aid in the selection for long-term carcinogenicity testing of those animal species most similar to man; d) urinary metabolites can be tested for their mutagenic action following deconjugation by enzymes and reactivation by liver fractions. These assays, which have been recently developed by Ames[2,11], permit the detection of mutagens in rats treated with carcinogens and can be used for monitoring the exposure of human populations. As recently shown by Legator[29], in patients treated with the anti-schistosomal agent Niridazole, mutagenic activity for *S. typhimurium* strain was demonstrable in the urine.

The validity and/or the practical application of such microbial mutagenicity tests has also been demonstrated in the IARC laboratory with vinyl chloride, which has recently been recognised as a carcinogen in animals and in man[19].

As previously demonstrated, in rat and mouse the major enzyme activity for conversion of vinyl chloride monomer into mutagens is located in liver and kidney, sites at which angiosarcoma or nephroblastoma are induced in animals which have been exposed to vinyl chloride[7,31].

Compound	Organ fraction	Activity % human[a]	mouse
Vinyl[b] chloride	liver	A 170 B 70 C 64 D 46	(100)
Vinyl[b] bromide	liver	Z 32 Y 36 X 23	(100)
Vinylidene[b] chloride	liver	K 38 Z 17 Y 16 X 11	(100)
2-Chloro[b] 1,3-butadiene	liver	K 22 Z 0 Y 0 X 0	(100)
	lung	L,M 0 N,O 0 P 0	(0)
1,4-Di[c] chloro-butene-2	liver	Z 83 Y 38 X 17	(100)

TABLE 1

MUTAGENICITY MEDIATED BY HUMAN AND BY ANIMAL TISSUES

[a]Samples from human individuals are represented by different characters. Data collated from Bartsch et al[9] and unpublished work

[b]Mutagenicity assay (Bartsch et al[7,8]) with S. typhimurium TA100 or TA1530

[c]Plate incorporation assay (Ames et al[3])

Compound	Organ fraction	Activity % human[a]	rat
N-nitroso-[b] morpholine	liver	A 84 B 47 C 37 D 28	(100)
	lung	E,F 0 G,H	(0)
N-Nitroso-pyrrolidine	liver	Z 117 Y 91 X 54	(100)
N-Nitroso-piperidine	liver	Z 216 Y 180 X 86	(100)
N-Nitroso-N'-methyl-piperazine	liver	Z 3,180 Y 1,800 X 370	(100)

TABLE 2

MUTAGENICITY MEDIATED BY HUMAN AND BY ANIMAL TISSUES

[a]Samples from human individuals are represented by different characters. Data collated from Bartsch et al[9] and unpublished work

[b]Mutagenicity assay (Ames et al[3]) with S. typhimurium TA1530

Results obtained by Barbin[4], Bartsch and Montesano[6], Bartsch[7], Göthe[14] and Kappus[25], lend strong support to the theory that hepatic mixed-function oxidase of rat, mouse and human liver converts vinyl chloride (I) into chloroethylene oxide (II), which is known to rearrange spontaneously to 2-chloroacetaldehyde (III)[50] (Fig. 1). A further oxidation product, monochloroacetic acid (IV) is an identifiable urinary metabolite in rats and man[15,17].

The mutagenic action of these compounds was studied in *Salmonella typhimurium* strains[7,31,33,42] or in Chinese hamster V79 cells, using 8-azaguanine or oubain resistance as genetic markers[18]. The results indicated that chloroethylene oxide and 2-chloroacetaldehyde were the most mutagenic compounds of this series in both systems.

Chloroethylene oxide is a primary reactive metabolite formed from vinyl chloride by mouse liver microsomes, oxygen and a NADPH generating system and is a rapidly reactive alkylating agent[4]. This evidence was obtained by comparison of the absorption spectra of the adducts obtained after reaction of 4-(4-nitrobenzyl)pyridine with chloroethylene oxide with the spectrum of a volatile vinyl chloride metabolite, which turned out to be identical to the one obtained with chloroethylene oxide. In the absence of the NADPH generating system, no such product was formed. Since chloroethylene oxide rearranges spontaneously to chloroacetaldehyde, those compounds are strong mutagens in several systems, and both now appear to be carcinogenic metabolites of vinyl chloride[6].

The validity of the microbial test system using *S. typhimurium* strains to predict the possible carcinogenic action of structurally related chemicals such as vinylidene chloride (1,1-dichloroethylene) and 2-chloro-1,3-butadiene was verified[8]. 2-Chloro-1,3-butadiene has been used for the production of rubber, and vinylidene chloride is used as a co-polymer with vinyl chloride in the manufacture of plastics. 2-Chloro-1,3-butadiene has been reported to induce chromosomal aberrations in lymphocytes of peripheral blood from exposed workers[26] and in the USSR it has been associated with the induction of skin and lung tumours in humans[27,28]. In the presence of a mouse liver microsomal fraction, the mutagenic effect as expressed as the number of histidine revertant colonies per μmole of substrate per hour of exposure per plate measured with two different *S. typhimurium* strains, TA1530 and TA100, was highest with 2-chloro-1,3-butadiene followed by vinylidene chloride and vinyl chloride. These findings indicated that the epidemiological monitoring of workers exposed to vinylidene chloride and 2-chloro-1,3-butadiene may be necessary. No mutagenic effect was detectable with vinyl acetate. This compound was used by Maltoni and Lefemine[32] as the control substance in long-term bioassays for vinyl chloride and was not found to be tumorigenic.

Table 1 summarizes the relative capacity of human biopsy samples to convert the halogenated hydrocarbons into mutagenic intermediates. The results are expressed as a percentage of an appropriate animal control. Samples from human individuals are represented by different characters.

The enzymic activities of four different human liver biopsy samples are expressed in relation to that of mouse liver, which is given as 100%. Sample A was twice as active in converting vinyl chloride into mutagenic metabolites, sample B was equally active as mouse liver, and the two other samples, C and D, were about 50% as active. From this data, it becomes clear that rat, mouse and human liver, the target organ in which vinyl chloride causes cancer, can convert this carcinogen into electrophilic and mutagenic intermediates. These results, at least in our opinion, would have been sufficient to classify this compound as biologically hazardous to man. Similarly, human liver specimens were active in converting vinyl bromide, vinylidene chloride, 2-chloro-1,3-butadiene and 1,4-dichlorobutene-2 into mutagens, the activity being, in general, lower than that in mouse liver. For the latter compound, Van Duuren et al[46] reported a local sarcomogenic action in mice.

A number of N-nitrosamines, some examples of which are listed in Table 2, are known to be potent carcinogens to various animal species, and man is exposed to some of them (for a review, see Magee[30]). Although no epidemiological investigations or case reports have so far given evidence that they have actually induced cancer in man, there is a strong possibility that they are hazardous to man[38].

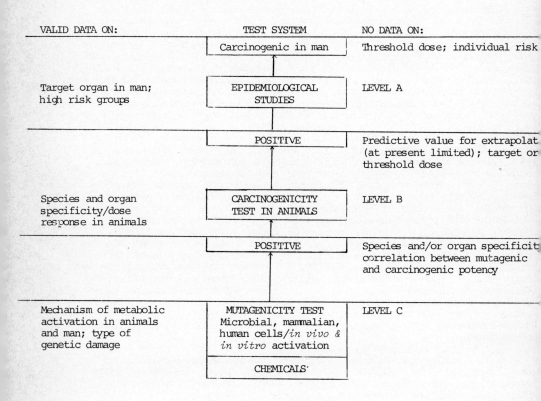

TABLE 3

FRAMEWORK OF CARCINOGENICITY TEST PROCEDURES

N-nitroso-morpholine, for example, was the most active of a series of cyclic nitrosamines in inducing point mutations in *S. typhimurium* TA1530 strain after metabolic conversion by rat liver enzymes[9]. The possible biological hazard to man of this compound becomes quite evident since liver samples A, B, C and D from four humans efficiently converted nitrosomorpholine into mutagenic intermediates (Table 2); the average enzymic activity was closest to that of rat liver, in which N-nitroso-morpholine is a potent carcinogen[13].

The differences in the enzyme activities of samples A, B, C and D indicate a promising means of obtaining valuable information on the variations within human populations, if a large number of human individuals is analysed. From the genetic background or the induced state of certain enzymes in those individuals one might obtain an idea of their susceptibility towards certain carcinogens[12]. We are pursuing such a study in the IARC laboratory; some results are shown in Figure 2.

The relative enzymic activities of the liver samples described in Table 3 were the same when, instead of N-nitroso-morpholine, vinyl chloride was utilised as substrate. Sample A converted vinyl chloride and N-nitroso-morpholine into mutagenic intermediates the most efficiently while the other samples, B, C and D, were less active. These data, although preliminary, already allow one important conclusion to be drawn: individuals A and Z appear to be liable to a higher risk due to their higher enzymic activity to convert not only one carcinogen into reactive metabolites, but, as measured for individual Z, several carcinogens, which even belong to different chemical classes.

The identification of people who may be at an exceptionally high risk for cancer because of a genetically determined enzyme profile may in future be of some help in establishing regulatory measures to admit only certain levels of carcinogens in the environment. Such levels should be established by taking into consideration the possible risk for the most susceptible individuals.

The last part of this discussion concerns the usefulness of these mutagenicity tests in predicting the possible carcinogenic effects of chemicals in man.

Despite the converging tendency of chemicals to be both carcinogenic and mutagenic, it is not known at present whether in future all carcinogens will be found to be mutagens and all mutagens, carcinogenic. For the strong mutagens, such as base analogues, nitrous acids and hydroxylamine, no carcinogenic effect in animals has been reported so far[35]. They do not act via electrophilic intermediates. On the other hand, synthetic steroidal sex hormones, which are carcinogenic in animals[19] have not yet been reported to be mutagenic, and we have also been unable in our own laboratory to detect a mutagenic effect in *in vitro* tissue-mediated assays using the human and animal carcinogen α,α'-diethylstilboestrol as substrate with the TA100 strain of *S. typhimurium* (Bartsch, unpublished). These findings would suggest that for this class of compounds, a different cancer-inducing mechanism may be implied.

There are various limitations of the mutagenicity test systems, since some of the factors which determine the processes of cancer development *in vivo* cannot be duplicated in, for example, *in vitro* mutagenicity systems. Some of the most important of these determining factors are the concentration of ultimate reactive metabolites available for reaction in organs and animal species with cellular macromolecules, which is a consequence of a critical balance between metabolic activation and detoxication processes; an organ-specific release of proximate or ultimate carcinogens by enzymic deconjugation; the biological half-life of metabolites; and organ-specific DNA repair or frequency of DNA replication in target or non-target organs and immuno-surveillance.

The use of mutagenicity tests in the assessment of the carcinogenic risk due to chemicals justifies their availability, since, although an enormous number of new chemicals are entering the environment, the world capacity for testing for carcinogenicity is only about 400 compounds per year, and it is not possible to test all of them in long-term carcinogenicity tests[20,21,22,23,24]. It is therefore necessary to prescreen in order to establish priorities for the long-term testing. In evaluating these tests, four points should be emphasized: firstly, positive results from mutagenicity tests given at level C (Table 3) cannot automatically be taken to imply a definite carcinogenic effect in man (level A); secondly, positive or negative results from these tests cannot substitute for long-term carcinogenicity testing in animals (level B); thirdly, the present methodology, in particular, tissue-mediated mutagenicity assays, is effective in predicting the carcinogenic potential of chemicals with a high per cent probability, ranging at present between 60 and 80%, but gives no indication of target organs or species specificity of the carcinogenic action of the chemical; and, fourthly, the relative potency of a chemical as a mutagen cannot at present be correlated with its potency as a carcinogen in experimental animals or man.

Taking into account present reproducibility, cost, number of chemicals that can be examined in a short time and the scientific basis of the test, tissue-mediated mutagenicity procedures using well-characterized genetic indicators and a metabolically defined *in vitro* activation system (apart from the *Drosophila* test [Vogel[47]]), appear at present to be the most promising short-term tests for the detection of adverse biological effects produced by chemicals. The present methodology still incurs the chance of false negative results, depending on the systems utilized, either due to mutagen specificity, lack of appropriate cofactors for activation or to extreme reactivity and volatility of the compound or its metabolites[9]. However, the number of false negative results obtained in the *in vitro* tissue-mediated assay is very small in comparison to that obtained with other mutagenicity test procedures. It has often been mentioned, theoretical limitations are implied in the short-term test currently used since the development of cancer *in vivo* is determined by factors which cannot yet be duplicated in many of the mutagenicity test systems.

Although there are still many problems involved in the interpretation of results of mutagenicity tests in terms of evaluating the carcinogenicity of chemicals, short-term tests for carcinogens can already be used in detecting environmental, aetiological cancer-causing agents with a sensitivity which did not exist ten years ago. They could thus be a powerful tool, when used in combination with epidemiological studies in environmental control. This possibility raises, therefore, two important issues: firstly, the validity of the short-term bioassay must be based on reliable corroboration by data from long-term tests in experimental animals; so, first of all, the long-term data must be critically analysed and accepted to provide solid and definite background information; and, secondly, the scientific community, as well as health authorities, must reach a consensus on the usefulness of results from these test systems in predicting the possible carcinogenic effects of compounds in man. Otherwise, and if no public health measures are taken, for instance, to remove the agent from the environment or greatly to reduce exposure to it, cancer in man can only be treated and not prevented.

ACKNOWLEDGEMENTS

The author's experimental work was partially supported by the National Cancer Institute of the USA, Contract No. 1CP-55630. For valuable discussion and/or experimental assistance, the author is indebted to L. Tomatis, R. Montesano, C. Malaveille, A. Barbin, A. Camus, G. Planche, G. Brun and H. Brésil. The help of Miss P. Stafford Smith and Mrs E. Ward in preparing the manuscript is gratefully acknowledged.

RESUME

Si l'on n'a pas montré que tous les mutagènes chimiques sont cancérogènes, on a maintenant constaté que la plupart des cancérogènes chimiques, dont plusieurs provoquent un cancer chez l'homme, sont mutagènes lorsqu'on les soumet à une des techniques d'épreuve de la mutagénicité qui associent des systèmes microbiens, mammaliens ou d'autres cellules animales, en tant que cibles génétiques, à un système d'activation métabolique *in vitro* ou *in vivo*.

Cette activation métabolique est, dans bien des cas, associée avec la fixation à des sites nucléophiles dans les acides nucléiques et les protéines. L'accumulation des données expérimentales qui lient l'activité cancérogène de nombreuses substances chimiques à leur capacité à être converties en dérivés électrophiles pouvant aussi exercer un effet mutagène, conduit à penser qu'il existe une relation empirique entre la cancérogenèse et la mutagenèse chimiques. Cette corrélation s'est jusqu'ici limitée aux modifications du génotype qui résultent des altérations structurales ou fonctionnelles des acides nucléiques et qui sont donc génétiquement transmissibles.

Les résultats de telles épreuves de mutagénicité sont à l'heure actuelle considérés comme insuffisants pour permettre de prévoir l'action cancérogène organo-spécifique de cancérogènes agissant sur l'organisme de l'animal d'expérience et de l'homme en particulier. Toutefois, les épreuves de mutagénicité constituent un moyen excellent pour a) détecter les cancérogènes et/ou les mutagènes dans l'environnement complexe de l'homme; b) présélectionner les substances chimiques environnementales à soumettre à une expérimentation animale de longue durée; c) étudier les mécanismes de l'activation métabolique des substances chimiques et d) comparer la capacité d'une série de tissus humains à convertir les substances exogènes en produits réagissants électrophiles.

A l'aide d'une épreuve de mutagénicité *in vitro*, avec des souches de *S. typhimurium*, sont présentés des exemples précis d'utilisation de ce test aux fins mentionnées sous a) à d) ci-dessus. Sur la base des résultats d'épreuves de mutagénicité à l'aide d'échantillons biopsiques humaines, il est procédé à une comparaison du métabolisme de certains cancérogènes chez l'animal d'expérience et chez l'homme. Certains aspects théoriques des relations entre cancérogenèse et mutagenèse sont examinés, ainsi que les applications et les inconvénients pratiques des tests de mutagénicité pour déterminer les risques cancérogènes auxquels l'homme est exposé du fait des substances chimiques.

References

(1) Ames, B.N. and McCann, J.: Carcinogens are mutagens: A simple test system In Screening Tests in Chemical Carcinogenesis, Montesano, R., Bartsch, H. and Tomatis, L. eds., 1975, IARC Scientific Publications No. 12, Lyon.

(2) Ames, B.N.: A simple method for the detection of mutagens in urine: studies with the carcinogen 2-acetylaminofluorene. Proc. Nat. Acad. Sci. (Wash.) 1974, 71, 737-741.

(3) Ames, B.N., Durston, W.E., Yamasaki, E. and Lee, D.F.: Carcinogens are mutagens: a simple test combining liver homogenates for activation and bacteria for detection. Proc. Nat. Acad. Sci. (Wash.) 1973, 70, 2281-2285.

(4) Barbin, A., Brésil, H., Croisy, A., Jacquignon, P., Malaveille, C., Montesano, R. and Bartsch, H.: Liver-microsome-mediated formation of alkylating agents from vinyl bromide and vinyl chloride. Biochem. Biophys. Res. Commun., 1975, 67, 596-603.

(5) Bartsch, H. and Grover, P.L.: Chemical carcinogenesis and mutagenesis In Scientific Foundations of Oncology, Symington, T. and Carter, R.L., eds., 1975 (in press).

(6) Bartsch, H. and Montesano, R.: Mutagenic and carcinogenic effects of vinyl chloride. Mutation Res., 1975 (in press).

(7) Bartsch, H., Malaveille, C. and Montesano, R.: Human, rat and mouse liver-mediated mutagenicity of vinyl chloride in *S. typhimurium* strains. Int. J. Cancer, 1975, 15, 429-437.

(8) Bartsch, H., Malaveille, C., Montesano, R. and Tomatis, L.: Tissue-mediated mutagenicity of vinylidene chloride and 2-chlorobutadiene in *Salmonella typhimurium*, Nature, 1975, 255, 641-643.

(9) Bartsch, H., Malaveille, C. and Montesano, R.: The predictive value of tissue-mediated mutagenicity assays to assess the carcinogenic risk of chemicals In Screening Tests in Chemical Carcinogenesis, Montesano, R., Bartsch, H. and Tomatis, L. eds., 1975, IARC Scientific Publications No. 12, Lyon.

(10) Brookes, P. and Lawley, P.D.: Effects on DNA: Chemical methods In Chemical Mutagens, Hollaender, A., ed., 1971, Vol. 1, pp. 121-144, Plenum, New York.

(11) Commoner, B., Vithayathil, A.J. and Henry, J.I.: Detection of metabolic carcinogen intermediates in urine of carcinogen-fed rats by means of bacterial mutagenesis. Nature (Lond.), 19741, 249, 850-852.

(12) Conney, A.H., Kapitulnik, J., Levin, W., Dansette, P. and Jerina, D.: Use of drugs in the evaluation of carcinogen metabolism in man In Screening Tests in Chemical Carcinogenesis, Montesano, R., Bartsch, H. and Tomatis, L. eds., 1975, IARC Scientific Publications No, 12, Lyon.

(13) Druckrey, H., Preussmann, R., Ivankovic, S. and Schmahl, D: Organotrope carcinogene Wirkungen bei 65 verschiedenen *N*-Nitroso-Verbindungen an BD-Ratten, Z. Krebsforsch., 1967, 69, 103-201.

(14) Gothe, R., Callemna, C.J., Ehrenberg, L. and Wachmeister, C.A.: Trapping with 3,4-dichlorobenzenethiol of reactive metabolites formed *in vitro* from the carcinogen vinyl chloride, AMBIO, 1974, 3, 234-236.

(15) Grigorescu, I. and Toba, G.: Clorura di vinil. Aspecte de toxicologie industriala. Rev. Chim. Rom., 1966, 17, 499-501.

(16) Group 17, Environmental Mutagenic Hazards, Mutagenicity screening is now both feasible and necessary for chemicals entering the environment. Science, 1975, 187, 503-514.

(17) Hefner, R.E., Watanabe, P.G. and Gehring, P.G.: Preliminary studies of the fate of inhaled vinyl chloride monomer in rats. Ann. N.Y. Acad. Sci., 1975, 246, 135-148.

(18) Huberman, E., Bartsch, H. and Sachs, L.: Mutation induction in Chinese hamster V79 cells by two vinyl chloride metabolites chloroethylene oxide and 2-chloroacetaldehyde. Int. J. Cancer, 1975, 16, 639-644.

(19) IARC, IARC Monographs on the Evaluation of the Carcinogenic Risk of Chemicals to Man, 1974, Vol. 7, pp. 291-318, Lyon.

(20) IARC, Information Bulletin on the Survey of Chemicals being tested for Carcinogenicity, August 1973, No. 1.

(21) IARC, Information Bulletin on the Survey of Chemicals being tested for Carcinogenicity, December 1973, No. 2.

(22) IARC, Information Bulletin on the Survey of Chemicals being tested for Carcinogenicity, May 1974, No. 3.

(23) IARC, Information Bulletin on the Survey of Chemicals being tested for Carcinogenicity, November 1974, No. 4.

(24) IARC, Information Bulletin on the Survey of Chemicals being tested for Carcinogenicity, July 1975, No. 5.

(25) Kappus, H., Bolt., H.M., Buchter, A and Bolt, W.: Rat liver microsomes catalyse covalent binding of ^{14}C-vinyl chloride to macromolecules, Nature, 1975, 257, 134-135.

(26) Katosova, L.D.: Cytogenic analysis of peripheral blood of workers engaged in the production of chloroprene. Gigiena truda i professional'nye zabol., 1973, 17, 30-33.

(27) Khachatrian, E.A.: The role of chloroprene in the process of skin neoplasm formation. Gigiena truda i professional'nye zabol., 1972, 18, 54-55.

(28) Khachatrian, E.A.: The occurrence of lung cancer among people working with chloroprene. Problems in Oncology, 1972, 18, 85-86.

(29) Legator, M.S., Connor, T.H. and Stoeckel, M.: Detection of mutagenic activity of metronidazole and niridazole in body fluids of humans and mice, Science, 1975, 188, 118-119.

(30) Magee, P.N., Montesano, R. and Preussmann, R.: N-nitroso compounds and related carcinogens In Chemical Carcinogens, Searle, C. ed., 1975 (in press).

(31) Malaveille, C., Bartsch, H., Montesano, R., Barbin, A., Camus, A.M., Croisy, A. and Jacquignon, P.: Mutagenicity of vinyl chloride, chloroethylene oxide, chloroacetaldehyde and chloroethanol. Biochem. Biophys. Res. Commun., 1975, 63, 363-370.

(32) Maltoni, C. and Lefemine, G.: Carcinogenicity bioassays of vinyl chloride, Environm. Res., 1974, 7, 387-405.

(33) McCann, J., Simmon, V., Streitwieser, D. and Ames, B.N.: Mutagenicity of chloroacetaldehyde, a possible metabolic product of 1,2-dichloroethane (ethylene dichloride), chloroethanol (ethylene chlorohydrin), vinyl chloride and cyclophosphamide (environmental carcinogens/alkylhalides). Proc. Nat. Acad. Sci. (Wash.), 1975, 72, 3190-3193.

(34) McCann, J., Choi, E., Yamasaki, E. and Ames, B.N.: Detection of carcinogens as mutagens in the *Salmonella*/microsome test: Part 1, Assay of 300 chemicals. Proc. Nat. Acad. Sci. (Wash.), 1975, 72 (in press).

(35) Miller, E.C. and Miller, J.A.: The mutagenicity of chemicals carcinogens: correlations, problems and interpretations In Chemical Mutagens, Hollaender, A. ed., 1971, Vol. 1, pp. 83-119, Plenum New York.

(36) Miller, J.A.: Carcinogenesis by Chemicals: an overview. Cancer Res., 1970, 30, 559-576.

(37) Miller, J.A. and Miller, E.C.: Chemical carcinogenesis, mechanisms and approaches to its control. J. Natl Cancer Inst., 1971, 47, v-vix.

(38) Montesano, R. and Bartsch, H.: Mutagenesis and carcinogenesis of N-nitroso compounds: possible environmental hazards. Mutation Res., 1976 (in press).

(39) Montesano, R. and Tomatis, L. eds.: Chemical Carcinogenesis Essays, IARC Scientific Publications No. 10, 1974, Lyon.

(40) Montesano, R., Bartsch, H. and Tomatis, L. eds.: Screening Tests in Chemical Carcinogenesis, IARC Scientific Publications No. 12, 1975, Lyon.

(41) Ramel, C. ed.: Report of a symposium held at Skokloster, Sweden, 1972, AMBIO Special Report, 1973, No. 3.

(42) Rannug, U., Johansson, A., Ramel, C. and Wachmeister, C.A.: The mutagenicity of vinyl chloride after metabolic activation. AMBIO, 1974, 3, 194-197.

(43) de Serres, F.J. and Sheridan, W. eds.: The Evaluation of Chemical Mutagenicity Data in Relation to Population Risk, Environmental Health Perspectives, 1973, No. 6.

(44) Stoltz, D.R., Poirier, L.A., Irving, C.C., Stich, H.F., Weisburger, J.H. and Grice, H.C.: Evaluation of short-term tests for carcinogenicity, Toxicol. Appl. Pharmacol., 1974, 29, 157-180.

(45) Sugimura, T., Yahagai, T., Nagao, M., Takeuchi, M., Kawachi, T., Hara, K., Yamasaki, E., Matsushima, T., Hashimoto, Y. and Okada, M.: Validity of mutagenicity tests using microbes as a rapid screening method for environmental chemicals In Screening Tests in Chemical Carcinogenesis, Montesano, R., Bartsch, H. and Tomatis, L. eds., IARC Scientific Publications No. 12, 1975, Lyon.

(46) van Duuren, B.L., Goldschmidt, B.M. and Seidman, I.: Carcinogenic activity of di- and trifunctional α-chloro ethers and of 1,4-dichlorobutene-2 in ICR/HA Swiss mice. Cancer Res., 1975, 35, 2553-2557.

(47) Vogel, E.: The relation between mutational pattern and concentration by chemical mutagens in *drosophila* In Screening Tests in Chemical Carcinogenesis Montesano, R., Bartsch, H. and Tomatis, L. eds., IARC Scientific Publications No. 12, 1975, Lyon.

(48) WHO, Evaluation and Testing of Drugs for Mutagenicity: Principles and Problems. WHO Techn. Rep. Ser., 1971, No. 482.

(49) WHO, Assessment of the Carcinogenicity and Mutagenicity of Chemicals. WHO Techn. Rep. Ser., 1974, No. 546.

(50) Zief, M. and Schramm, C.H.: Chloroethylene oxide. Chem. Ind., 1964, 660-661.

APPLICATION OF THE RESULTS
OF CARCINOGEN BIOASSAYS TO MAN

P. SHUBIK and D.B. CLAYSON

Eppley Institute for Research in Cancer
University of Nebraska Medical Center
42nd & Dewey Avenue, Omaha,
Nebraska 68105 U.S.A.

As a result of epidemiological studies, between 20 and 50 chemicals, or mixtures of substances, are known to induce cancer in man. The precise number depends on the level of evidence deemed necessary to incriminate a chemical of carcinogenicity to man. This information has been gathered as a result of studies of the consequences of occupational exposure to specific substances, the use of drugs or natural food contaminants, and personal habits (Table 1).

Table I: Examples of Carcinogens in Man

Occupational Exposure	Tumor Site
Soot, tar, pitch and oil	Skin, lung
Benzidine, 2-naphthylamine, 4-aminobiphenyl	Bladder
Asbestos	Lung, pleura
Vinyl chloride	Angiosarcoma
Drugs	
Chlornaphazine	Bladder
Diethylstilbestrol	Vagina (offspring)
Analgesic (abuse)	Renal pelvis
Natural Food Contaminant	
Aflatoxin	Liver
Habits	
Cigarette smoking	Lung, bladder, pancreas, etc.

The epidemiological discovery of a carcinogen in man often does not follow until the substance has been in the environment for as long as 20-40 years, during which time hundreds, thousands or even millions of people may have been exposed and are at risk from its effects. Clearly, therefore, a more rapid method of discovering potential human carcinogens is required. Practically every substance known to induce cancer in man also produces tumors when administered experimentally to animals. Inorganic arsenicals appear to be the major exception (Clayson, 1962). The basis of carcinogen bioassay is the assumption that chemicals known to induce tumors in experimental animals will, in all probability, do so in man. It is dangerous to ignore this precept.

In this account, the problems of the application of animal tests to man will be considered from three standpoints: (1) the design of the animal experiment; (2) our present knowledge of the effect of species differences on the induction of tumors and (3) quantitative aspects of carcinogenesis. In conclusion, a number of problems in experimental design and interpretation will be considered.

The Design of Carcinogenicity Bioassays

The design of carcinogenicity bioassays involves several compromises. Nowadays, the usual objective in testing a chemical is to establish that it is safe for use, that is to say, is not a carcinogen. Proving an absolute negative is scientifically impossible and, therefore, a decision has to be reached on the scale of the investigation necessary to give a reasonable certainty of safety in use. The scale of the investigation is determined by balancing the desire for assured safety against the cost of maintaining large numbers of animals for the greater part of their lifespan. Smaller, short-lived animals, such as rats, mice or hamsters, are used. The numbers of animals must be high enough to detect tumor yields of 10% or more, yet not so high as to make the test prohibitively expensive. The scheme recommended by the National Cancer Institute (Sontag et al., 1975) is shown (Table II).

Table II: National Cancer Institute Bioassay Program

Number of Animals		Male	Female
Species I	Level 1	50	50
	Level 2	50	50
Species II	Level 1	50	50
	Level 2	50	50

(200 untreated animals serve as controls to 4 tests in each species)

There are two dose levels: near the maximum tolerated, a somewhat lower dose, and a group of untreated animals, both sexes and two species. Current knowledge suggests this and similar protocols will pick out the majority of carcinogenic substances which may affect man.

This small number of animals serves to screen compounds which may be widespread in the human environment and which, if they only induced tumors in one percent of the exposed population, would cause a major tragedy. Therefore in carcinogenesis, it is usual to administer far larger amounts of the substance than would normally be found in the environment. This is justified because in many experiments, it has been shown that there is a direct doseresponde relationship between the amount of test substance and the yield of tumors. Nevertheless, it is important to ensure the absence of possibly carcinogenic impurities in the test substance and to avoid levels which lead to "chemical intoxication."

Carcinogenic impurities in substances to be examined for carcinogenicity are of special concern when these substances have a low level of toxicity and can therefore be fed at very high levels. Such impurities may arise because impure reactants are used in the synthesis of the substance. Economics militate against using pure chemicals in industry except for special purposes, such as the synthesis of drugs or food additives. If a carcinogenic intermediate is used in the manufacture of a substance, some unreacted intermediate may find its way into the final product and lead to an environmental hazard. This may be illustrated in the example of the dyestuffs, Yellow OB and Yellow AB, which have contained unreacted 2-naphthylamine. The question of carcinogenic impurities is made more difficult because their concentration may vary appreciably with different batches of material. Careful analytical monitoring of industrial chemicals for impurities is the only solution to this problem.

Administration of high levels of test chemicals to animals can also lead to difficulties in the biological interpretation of the results. The effects of high levels of a test substance, chemical intoxication, may take many forms from debilitation of the test animals to more specific biological effects. Doses of chemical which lead to changes in body weight should be avoided as the incidence of tumors in experimental animals may be affected by the nutritional state of the test animals (reviewed by Clayson, 1975a). Paired-feeding studies may be needed in extreme cases, but should be avoided if possible as they increase the complexity and the cost of the bioassay. Another effect is seen in the feeding of 4% diethyleneglycol which led to urinary bladder stones and tumors in male, but not female, rats (Fitzhugh and Nelson, 1946; Weil, et al., 1965). The bladder stone rather than the chemical, in this example, was probably responsible for the tumor (Clayson, 1974). Overloading the subcutaneous tissues with iron dextran or food colorants has led to the induction of local sarcomas in rats and mice which appear to be dependent on the physical properties of the substance and therefore, in all probability, have little relevance to the human situation.

Other areas in the design of bioassays which are important in judging their applicability to man include the route of administration, and the part of life over which the test substance is to be administered. In general, it is best if the route of administration is identical to that by which man is thought to be exposed. This most certainly does not mean that if a test substance is shown to induce tumors by another conventional route of administration, the carcinogenicity of this substance can be safely ignored.

Conventional methods of administering carcinogens (feeding, painting, parenteral, inhalation) should be used in bioassay studies of substances of environmental importance which are to be related to man. Other methods such as bladder implantation, the use of threads soaked in carcinogen, and so on, are research methods, but are not necessarily applicable to the environmental problem.

In many experiments, carcinogens are administered only to adult animals, although these substances may affect man at any time during life. For example, diethylstilbestrol (DES) has been used in the United States for the maintenance of pregnancy in women with a history of spontaneous abortion. The female children of these mothers have developed clear cell adenocarcinoma of the vagina and other proliferative lesions in the reproductive tract which appear capable of progressing to malignancy (Herbst and Scully, 1970; Herbst et al., 1971). A similar effect of transplacentally-administered DES has been observed in hamsters (Rustia, personal communication) and shown to induce more severe, proliferative lesions in female than in male offspring. There are many other examples of the increased sensitivity of the fetus and the neonatal experimental animal to carcinogens, and it should be asked whether these periods of life should be included in standard test protocols. It appears logical, at this time, to consider two-generation tests, i.e., tests in which one generation is treated from conception to death, for substances which are added to foodstuffs or used as drugs for man, or are general environmental contaminants, and thus may affect man at any time during life. The two-generation test has not, however, been particularly well researched and its general application at this stage may lead to further problems in interpretation, as in the experiments with high levels of saccharin which have led to incidences of bladder tumors in rats in the two-generation test, but not when rats were treated as adults (Munro et al., 1975; Taylor & Friedman, 1974).

The Choice of Species

Carcinogens act differently in different species, as may be demonstrated by a consideration of the occupational bladder carcinogens, 4-aminobiphenyl, 2-naphthylamine and benzidine (Table III).

Table III: Response of Different Species to Three Industrial Bladder Carcinogens

	2-Naphthylamine	Benzidine	4-Aminobiphenyl
Man	Bladder Tumors	Bladder Tumors	Bladder Tumors
Dog	Bladder Tumors	? Bladder Tumors	Bladder Tumors
Monkey	Bladder Tumors	Not Known	Not Known
Hamster	Bladder Tumors	Liver	Not Known
Mouse	Hepatomata	Hepatomata	Hepatomata (Bladder Tumors)
Rat	None	Acoustic Intestinal Tumors Hepatomata	Mammary Tumors Breast Tumors Mammary Tumors

The reason for the different response is complex, but the most important factor is undoubtedly that each species has its own complement of enzymes to activate and detoxify carcinogens (Miller, 1970).

With the aromatic amines, the first step in the activation reaction is N-hydroxylation and the second, esterification of the N-hydroxy compounds. Other reactions such as aromatic ring hydroxylations and conjugation appear only to be detoxification reactions (Figure 1).

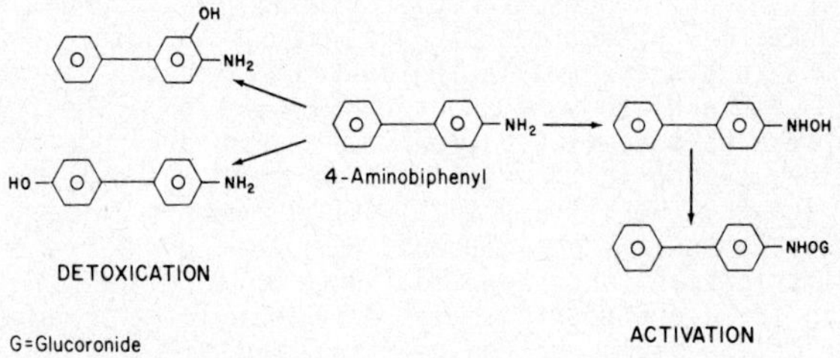

Figure 1. Metabolism of 4-aminobiphenyl (Dog) (Radomski et al., 1973)

The conjugated N-hydroxy compound dissociates to a positively-charged fragment or electrophile (Figure 2)

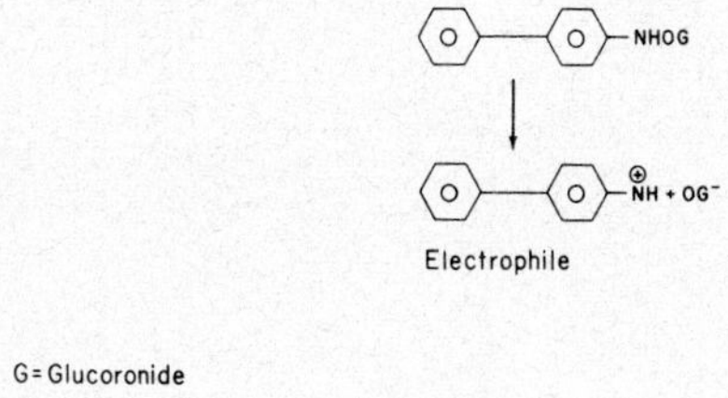

⬡—⬡—NHOG
↓
⬡—⬡—$\overset{\oplus}{N}H + OG^-$
Electrophile

G = Glucoronide

Figure 2. Production of Electrophile from Activated Aromatic Amine

which is the effective, or ultimate, carcinogen. If a species is unable to N-hydroxylate aromatic amines, as for example, the guinea pig, it develops tumors when the N-hydroxy derivative is administered, but not when the original amine is given (Miller, 1970). The guinea pig has not yet been shown to develop aromatic amine-induced tumors. Although the original metabolic approach to the activation of the aromatic amines appeared to be comparatively straigtforward (Miller, 1970), subsequent investigation of the second, conjugation, stage demonstrated that it is complex and a variety of conjugation reactions now appear to be implicated (reviewed by, Clayson and Garner, in press).

Other carcinogens (Figure 3)

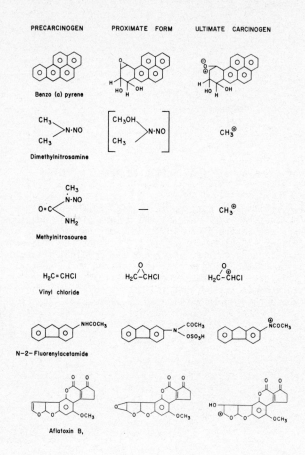

Figure 3. Activation of Chemical Carcinogens

are activated to similar highly reactive electrophiles which then interact with the tissues. This mechanism has proved of considerable use in understanding species and tissue differences in response to individual carcinogens. Unfortunately, at this time, not enough is known, either in man or animals, about the metabolism of carcinogens to predict safely that chemicals carcinogenic in animals cannot be activated by man. In our present stage of knowledge, we must assume that man possesses the full capability to activate carcinogens demonstrated in animals.

A further difficulty which arises in the interpretation of animal tests is that the most used experimental animals, rats, mice or hamster, live for 2-3 years, man for 70 years. It has been customary to assume that the latency of tumors is more or less directly related to lifespan. The basis for this assumption is hard to define and probably rests on the observation that most naturally-occurring cancers arise in the second half of life. Nevertheless, powerful carcinogens can induce tumors with a much shorter latency. Possibly, species-dependent variation in DNA repair gives a clue to the apparent relation between tumor latency and lifespan. This mechanism has been shown to be more effective in human cells than in those of the shorter-lived rodents (Strauss, 1974).

Quantitative Aspects

There is presently a strong, but not necessarily well informed, opinion that any level of a carcinogen, however small, cannot be tolerated in the environment unless it is impracticable to avoid it. Yet nobody seems willing to calculate the cancer risk inherent in exposure to a single hour's sunlight, or the smoke from just one cigarette in a lifetime.

Animal experiments clearly demonstrate a dose-response relationship exists for carcinogens as for all other biological responses (Bryan and Shimkin, 1941, 1943). The difficulty is that few published experiments are capable of detecting tumor yields of less than 10%, or possibly 5%. Unless there are very high occupational exposures of limited numbers of men, there are few instances of carcinogen exposure which affect such a high proportion of the human population. We have assumed that the dose-response curve can be extrapolated downwards to zero from the 5-10% level, and using statistical devices such as those proposed by Mantel et al. (1975), have tried to calculate an acceptable risk. The nature of the extrapolation, arising from the necessity to extrapolate downwards, would be sufficient to fail an elementary student in a first year university science examination. The 95% confidence limits shown on the hypothetical extrapolation curve (Figure 4)

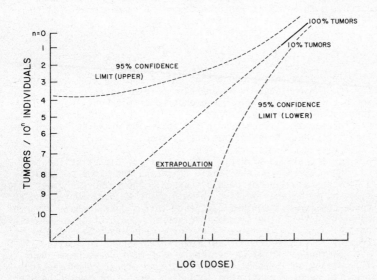

Figure 4. Extrapolation of Results of Carcinogenicity Data Towards Zero Dose

clearly indicates the weakness of the present position. Nevertheless, we are forced to accept this position because of the great financial cost and the lack of facilities to demonstrate the shape of the lower reaches of this dose-response curve in experimental animals. The study, if it were to be commissioned, would utilize the facilities needed for assaying many hundreds of hitherto untested substances.

The justification for the concept that no dose of a carcinogen is safe appears to rest on the work of Druckery (1959) showing that the effect of carcinogens persist in an irreversible fashion, and the support for this concept gained from the two-stage initiation-promotion model of mouse skin carcinogenesis (Berenblum and Shubik, 1947, 1949). It must be remembered, however, that these observations were made with large experimental doses of carcinogens and do little to answer the question of what happens at really low exposures, for example, ppb or ppt levels of carcinogens that possess weak carcinogenicity in animals. The theoretical, and partly political, concept of zero tolerance to carcinogens was initiated before DNA repair, the ability of the cell to repair lesions induced in its own genetic material, was discovered. We should be looking into the practicability of determining whether pre-replication DNA repair mechanisms are capable only of removing a certain proportion of induced lesions in the genome no matter how few are present, or whether DNA repair becomes progressively more efficient as the number of induced lesions diminishes.

As more and more carcinogens are discovered, it is becoming increasingly difficult to ensure that they do not contaminate man's environment to some extent and there is a desperate need to have some rational guidance as to levels which can be tolerated that is more satisfactory than the one reached politically after often ill-judged, conflicting statements by both environmentalists and industry representatives. Alternates for many carcinogens exist (Clayson, 1975b) and the disappearance of such carcinogens cannot hurt mankind one iota. In some cases, such as aflatoxin or asbestos, a complete refusal to use such substances might produce more harm than good.

Specific Examples

It should be clear from this account that animal testing is a quantitatively inexact model for the human disease. Nevertheless, intelligently used, it provides a reasonable (though costly) screen and should pick out the majority of potent carcinogens. Our lack of ability to utilize this method rationally in specific instances may be illustrated in the following examples, which demonstrate the difficulties that are sometimes encountered.

The evidence for the effect of very large amounts of diethylstilbestrol (5-125 mg/day) during pregnancy on the female children, and its experimental confirmation in hamsters has already been discussed. There are two other major uses of this estrogen, as a feed additive (growth promoter) in cattle, and as a post-coital contraceptive. The first use leads to residues in beef liver of the order of 3 ppb (3 µg/kg), the second involves the consumption of about 200 mg diethylstilbestrol as a single dose. Many governments have banned the use of diethylstilbestrol in cattle feed as an unwarranted danger to health. The contraceptive use continues. Animal experiments over many years have shown that estrogens, usually diethylstilbestrol, have a profound effect in increasing the development of tumors of many tissues including those such as the hamster kidney, which are not generally regarded as hormone responsive. The first question which has to be asked is whether diethylstilbestrol induces cancer because it is estrogenic or because it reacts directly with cellular targets in the random manner associated with conventional carcinogens, such as the aromatic amines. It must, at this stage, be suspected that diethylstilbestrol exerts its tumorigenic action as an estrogen, because it can be substituted in tumorigenesis experiments in animals by other estrogens. Evidence of its lack of significant direct carcinogenic interaction is still lacking. Until this information is generated, it would seem rational to find substitutes for its clinical uses rather than to worry about low levels in beef liver. If evidence can be produced to demonstrate that its action is as an estrogen, rather than as a directly acting carcinogen, it may be possible to develop permissible limits based on the normal levels of circulating estrogens in the blood of the general population. The rationale utilized in the determination of a permissible level would be similar to that used in radiation protection in which natural radiation presents an unavoidable minimum.

The oral contraceptive presently available consist of estrogenic and/or progestogenic components. These medicaments are used by a frighteningly large number of women throughout the world (10 million is the current estimate in the U.S.A.). For this reason, we should expect that these substances would have been subject to the most careful and rigorous animal testing before use. It appears that the regulatory agencies have ignored the induction of liver tumors in rodents by some of these agents on the questionable grounds that they were benign or occurred only in one sex (Committee on Safety of Medicines, 1972). Now there are reports of findings of similar lesions in a small proportion of women using these drugs (Lancet, 1974). Again, we need to know whether some of these agents are direct-acting carcinogens or we are observing only a hormonal effect. The possibility of removing oral contraceptives from the market is not politically feasible, even if it is desirable from the viewpoint of cancer prevention. It would be necessary to weigh the threatened population increase against the tumor burden that might be induced, in justifying their withdrawal on the present evidence.

The third example to be discussed is DDT, one of the most useful pesticides the world has ever known. Its major disadvantage is its persistence at low residue levels in the biosphere after use. This alone has not been judged sufficient to compel complete cessation of its use. DDT gives liver tumors in mice, but not in other mammalian species tested (IARC Monographs, 1974). This has been sufficient reason for its banning in the United States. As has been explained above, it is not yet possible to decide whether man reacts to DDT in the same way as the mouse, or is like the hamster which tolerates much higher doses for long periods. It is possible to find out. DDT has been in use for many years, and given the will and the financial support, it should not be difficult to identify groups of men who have received high exposures to this substance either during its manufacture or application. If this is done, we should be able to determine whether the action taken in completely banning an important and useful chemical is, or is not, justified. The epidemiological studies suggested for DDT in man are possible because DDT has had an extensive use in the environment before its tumorigenic properties in the mouse were discovered. With newer substances which are properly tested in animals before their introduction into the environment, we shall not have the capability to verify the significance of the animal experiments. It will be scientifically tragic if this opportunity is missed.

Because of the complexities of carcinogenesis and the differences between man and animals, it is difficult to make quantitative comparisons from animal experiments that have validity for man. The potential dangers in trying to do this are well illustrated by the search for a safer cigarette. The demonstration that cigarette smoke is carcinogenic rests on epidemiological observations in man (e.g., Doll and Hill, 1964) and on the induction of skin tumors by the smoke condensate in mice (Wynder et al., 1953). Should we expect the mouse skin to react to carcinogens in the presence of many other chemicals in the same way as the human lung? What allowances should be made for the facts that a different material (condensate rather than smoke), a different tissue (skin versus lung) and different species (mouse versus man) are affected? It is distressing to read that a high level government committee in the United Kingdom, rather than discussing these differences which make the interpretation of results so hazardous, prefer to describe, cookbook fashion, mouse skin painting tests for the evaluation of the safety of new smoking materials. They admit that the final outcome can only be resolved by epidemiological investigations of an experiment in man (Hunter, 1974), a sad comment for those who believe in cancer prevention.

Conclusions

Animal experiments to test for the possible carcinogenic activity of chemicals provide the best and only deeply researched method for the detection of environmental carcinogens for man. The fact that most known human carcinogens give tumors in animals encourages the belief that these tests have validity. However, there are significant differences in the numbers of the exposed populations of men and animals, in the part of the lifespan during which each is exposed, in the metabolic activation of the carcinogens, and in the longevity of men and experimental rodents. For regulatory purposes, we must assume that the results of bioassays in rodents will closely parallel tumor induction in man, although we cannot be sure of this. In some cases, such as rodent bladder tumors associated with bladder stone, or subcutaneous sarcomas arising locally to massive injection of food dyes, there may be reason to reject an association. It is only by continued research into the way in which both man and laboratory animals react to carcinogens that we may hope to refine our methodologies and obtain an accurate, well-defined net to trap potential environmental carcinogens without depriving the community of chemicals, through false associations or false positive results that may be of great value, sociologically or economically.

RESUME

Les épreuves sur l'animal de l'activité cancérogène des substances chimiques constituent la meilleure et la seule méthode approfondie de détection des cancérogènes environnementaux chez l'homme. Le fait que la plupart des cancérogènes humains connus provoquent des tumeurs chez l'animal renforce la croyance en la valeur de ces épreuves. Toutefois, il existe d'importantes différences dans les chiffres des populations exposées d'hommes ou d'animaux, la période de la vie au cours de laquelle chacune est exposée, l'activation métabolique des cancérogènes et la longévité de l'homme et des rongeurs expérimentaux. Aux fins de réglementation, nous devons présumer que les résultats des épreuves biologiques chez les rongeurs reflètent étroitement l'induction tumorale chez l'homme, encore que nous ne puissions en être sûrs. Dans certains cas, comme les tumeurs vésicales des rongeurs associées au calcul vésical, ou les sarcomes sous-cutanés naissent localement d'une injection massive de colorants alimentaires, on peut avoir des raisons de rejeter une association. Ce n'est que par des recherches continues sur la manière dont l'homme et les animaux de laboratoire réagissent aux cancérogènes que nous pouvons espérer affiner nos méthodes et tisser le filet qui permettra de piéger les éventuels cancérogènes environnementaux, sans priver la collectivité, par de fausses associations ou des résultats faussement positifs, de substances qui peuvent avoir une grande utilité, sur le plan sociologique ou économique.

References

(1) Berenblum I. and Shubik, P.: New quantitative approach to the study of stages of chemical carcinogenesis in the mouse's skin. Brit. J. Cancer, 1947, 1, 383-391.

(2) Berenblum I. and Shubik P.: The persistence of latent tumour cells induced in the mouse's skin by a single application of 9,10-dimethyl-1,2-benzanthracene. Brit. J. Cancer, 1949, 3, 384-386.

(3) Bryan W.R. and Shimkin M.B.: Quantitative analysis of dose-response data obtained with carcinogenic hydrocarbons. J. Natl. Cancer Inst., 1975, 1, 807-833.

(4) Bryan W.R. and Shimkin M.B.: Quantitative analysis of dose-response data obtained with 3 carcinogenic hydrocarbons in strain C3H mice. J. Natl. Cancer Inst., 1943, 3, 503-531.

(5) Clayson D.B.: Chemical Carcinogenesis. Churchill, London.

(6) Clayson D.B.: Guest Editorial. Bladder carcinogenesis in rats and mice: possibility of artifacts. J. Natl. Cancer Inst., 1974, 52, 1685-1689.

(7) Clayson D.B.: Nutrition and experimental carcinogenesis: a review. Cancer Res., 1975a, 35, in press.

(8) Clayson D.B.: Benzidine and 2-naphthylamine: voluntary substitution or technological alternatives. New York Acad. Sci., 1975b, in press.

(9) Clayson D.B. and Garner R.C.: Aromatic amines and related carcinogens. In: Chemical Carcinogenesis edited by C.S. Searle. American Chemical Society Monographs Series, Washington, D.C., in press.

(10) Committee on Safety of Medicines: Carcinogenicity tests of oral contraceptives. A report by the Committee on Safety of Medicines. H.M. Stationery Office, London, 1972.

(11) Doll R. and Hill A.B.: Mortality in relation to smoking: ten years' observations of British doctors. Brit. Med. J., 1964, 5395 & 5396, 1399-1410 & 1460-1467.

(12) Druckrey H.: Pharmacological approach to carcinogenesis. In: Ciba Foundation Symposium on Carcinogenesis: Mechanisms of Action. Ed. Wolstenholme and O'Connor, Churchill, London, 1959, p. 110.

(13) Fitzhugh O.G. and Nelson A.A.: Comparison of the chronic toxicity of triethylene glycol with that of diethylene glycol. J. Indus. Hyg. Toxicol., 1946, 28, 40-43.

(14) Herbst A.L. and Scully R.E.: Adenocarcinoma of the vagina in adolescence: A report of 7 cases including 6 clear-cell carcinomas (so-called mesonephromas). Cancer, 1970, 25, 745-757.

(15) Herbst A.L., Ulfedler H. and Poskanzer D.C.: Adenocarcinoma of the vagina. Association of maternal stilbestrol therapy with tumor appearance in young women. New Engl. J. Med., 1971, 284, 878-881.

(16) Hunter R.B.: First report of the independent scientific committee on smoking and health - tobacco substitutes and additives in tobacco products, their testing and marketing in the United Kingdom. H.M. Stationery Office, London, 1974.

(17) IARC Monographs on the Evaluation of Carcinogenic Risk of Chemicals to Man, Some organochlorine pesticides - DDT and associated substances, IARC, Lyon, 1974, 5, 83-124.

(18) Lancet: Liver tumours and steroid hormones. Lancet, 1973, ii, 7844, 1481.

(19) Mantel N., Bohidar N.R., Brown C.C., Ciminera J.L. and Tukey J.W.: An improved Mantel-Bryan procedure for "safety" testing of carcinogens. Cancer Res., 1975, 25, 865-872.

(20) Miller J.A.: Carcinogens by chemicals: an overview. G.H.A. Clowes Memorial Lecture. Cancer Res., 1970, 30, 559-576.

(21) Munro I.C., Moodie C.A., Krewski D. and Grice H.C.: A carcinogenicity study of commercial saccharin in the rat. Toxicol. Appl. Pharmacol., 1975, 32, 513-526.

(22) Radomski J.L., Rey A.A. and Brill E.: Evidence for a glucuronic acid conjugate of N-hydroxy-4-aminobiphenyl in the urine of dogs given 4-aminobiphenyl. Cancer Res., 1973, 33, 1284-1289.

(23) Sontag J.M., Page N.P. and Saffiotti U.: Guidelines for carcinogen bioassay in small rodents. U.S. Dept. of Health, Education and Welfare, Washington, D.C., 1974.

(24) Strauss B.S.: Minireview: Repair of DNA in mammalian cells. Life Sciences, 1974, 15, 1685-1693.

(25) Taylor J.M. and Friedman L.: Combined chronic feeding and three-generation reproduction study in the rat. Annual Meeting - Society of Toxicology, Washington, D.C., 1974.

(26) Weil C.S., Carpenter C.P. and Symth H.F.: Urinary bladder response to diethylene glycol. Calculi and tumors following repeated feeding and implants. Arch. Environ. Health, 1965, 11, 569-581.

(27) Wynder E.L., Graham E.A. and Croninger A.B.: Experimental production of carcinoma with cigarette tar. Cancer Res., 1953, 13, 855-864.

Environmental Pollution and Carcinogenic Risks
Pollution de l'environnement et risques cancérogènes

CAPACITE DE CELLULES EN CULTURE D'ACCUMULER DES HYDROCARBURES POLYCYCLIQUES A UN NIVEAU DECELABLE MICROSPECTROFLUORIMETRIQUEMENT
RESULTATS PRELIMINAIRES

J.M. SALMON (1), P. VIALLET (1),
E. KOHEN (2) et F. ZAJDELA (3)

(1) Laboratoire de Chimie Physique — Centre Universitaire Avenue de Villeneuve, 66025 Perpignan, France.

(2) Clin. Fac., Dept. Pathology and Adj. Fac., Dept. Physiology School of Med., Univ. of Miami, Papanicolaou Cancer Research Institute — P.O. BOX 23-6188, Miami (Flo.) U.S.A.

(3) Unité de Physiologie Cellulaire, INSERM U 22 — Fondation Curie — Institut du Radium, Bât. 110, 91405 Orsay, France.

Au cours d'études plus générales des métabolismes cellulaires par microspectrofluorimétrie, nous avons été amenés à étudier le comportement de cellules vivantes dans un milieu de culture contenant des traces de benzo(a)pyrène (BP). Nous avons constaté un passage rapide du BP dans les cellules avec une accumulation suffisante pour permettre un enregistrement aisé du spectre de l'hydrocarbure à l'échelon d'une cellule isolée ; ceci est difficilement réalisable à partir du spectre de fluorescence du milieu de culture initial (1).

Cette observation nous a conduit à étudier les conditions expérimentales nécessaires à la détermination du BP intracellulaire par la microspectrofluorimétrie.

I - MATERIEL ET METHODES

L'appareil utilisé est un microspectrofluorimètre multicanal permettant l'enregistrement d'un spectre complet de la fluorescence émise par une cellule isolée vivante dans son milieu de culture (2).

Cet enregistrement peut être effectué en 32 m sec, vitesse apparemment suffisante pour pouvoir suivre des phénomènes biologiques.

Le microspectrofluorimètre (Fig. 1) se compose :
- d'une source monochromatique de radiations électromagnétiques "1",
- d'un microscope inversé muni d'un dispositif ultropak "2",
- d'un dispositif d'analyse de la fluorescence émise par la cellule isolée "3", constitué par :
   (a) un système dispersif "4" (prisme d'Amicci ; association prisme réseau

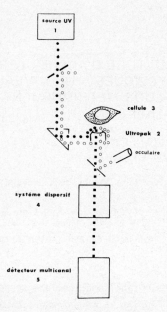

Fig. 1 : Schéma du microspectrofluorimètre
.... radiation excitatrice
▬▬▬ fluorescence émise par la cellule
ooo radiation de visualisation.

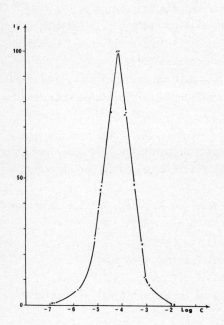

Fig. 2 : Variation de l'intensité de fluorescence en fonction de Log C (pour couvrir un plus grand domaine de concentrations) : x solution mère 1,3 10^{-2} M/l
solution mère 7,5 10^{-4} M/l.

type Carpenter),

(b) un détecteur multicanal "5" relié à une centrale de mesure et d'exploitation des données.

Pour la commodité des manipulations (positionnement de cellules etc...), on peut superposer à la radiation d'excitation U.V., une radiation dans le visible (visualisation en champ noir) ; un miroir dichroïque renvoie cette radiation vers l'oculaire du microscope.

Les cellules utilisées étaient des cellules d'ascites E12 (3). Ces cellules, comme l'a montré leur étude au microscope électronique, ne possèdent qu'un réticulum endoplasmique rudimentaire. Ceci suggère qu'elles sont peu aptes à métaboliser des substances du type benzopyrène. C'est pour cette raison qu'elles ont été choisies pour la présente étude.

Le protocole de culture ayant été décrit précédemment (3), nous ne signalons ici que les traitements particuliers qu'elles ont subis. Les cellules traitées en dose "aiguë" sont mises en présence d'un milieu contenant du BP (10^{-6} M) ; au bout de six minutes, la microchambre qui les contient est lavée trois fois avec un milieu de culture exempt de BP. D'autres cellules sont cultivées (exposition chronique) dans un milieu de culture contenant du BP (10^{-7} M) et 0,01 % d'éthanol. Le milieu de culture est renouvelé tous les deux jours. Avant l'analyse spectrofluorimétrique, les cellules sont placées dans un milieu de culture sans BP.

Le BP utilisé a été spécialement purifié dans le laboratoire de P. JACQUIGNON, Institut de Chimie des Substances Naturelles, Gif-sur-Yvette (France).

II - AVANTAGES DE LA TECHNIQUE DE MICROSPECTROFLUORIMETRIE

La méthode permet l'enregistrement rapide (soit toutes les 32 m sec) du spectre de la fluorescence émise par la cellule.

1) Effet de la concentration :

Lorsque l'on observe la variation de l'intensité de fluorescence du BP à une longueur d'onde donnée (427 nm) en fonction de la concentration, nous obtenons la courbe de la figure 2 qui ne permet pas de doser le BP. En revanche, si l'on prend en considération la totalité du spectre de fluorescence du BP et sa variation en fonction de la concentration et si l'on règle l'amplification du signal de façon à maintenir constante l'intensité du pic à 427 nm, on peut représenter la variation du rapport   Intensité de fluorescence à 408 nm/Intensité de fluorescence à 427 nm (soit I_{408}/I_{427}) en fonction de la concentration (Fig. 4). L'utilisation conjointe des courbes des figures 2 et 4 permet alors de doser le BP sans ambiguïté.

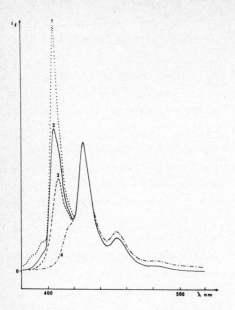

Fig. 3 : Variation du spectre de fluorescence du BP dans l'éthanol en fonction de la concentration.
Courbes : 1) $1,3 \ 10^{-6}$ M/l ; 2) $1,3 \ 10^{-4}$ M/l ;
3) $6,5 \ 10^{-4}$ M/l ; 4) $1,3 \ 10^{-2}$ M/l .

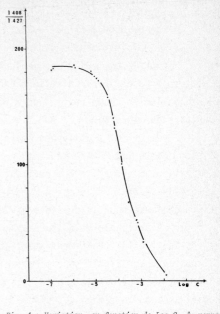

Fig. 4 : Variation, en fonction de Log C, du rapport $I \ 408 / I \ 427$.
Les diverses concentrations sont obtenues à partir de
x sol. mère $1,3 \ 10^{-2}$ M/l.
. sol. mère $7,5 \ 10^{-4}$ M/l.

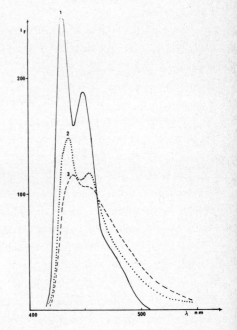

Fig. 5 : Spectres de fluorescence de cellules traitées en dose aigüe avec BP.
Courbes : 1) BP 10^{-4} M/l dans l'éthanol
2) et 3) deux exemples de spectres observés pour les cellules traitées au BP.

Fig. 6 : Spectres de fluorescence des cellules cultivé trois mois dans un milieu contenant BP
Courbes : 1) BP 10^{-4} M/l dans l'éthanol
2) cellules traitées en dose aigüe avec BP
3) cellules non traitées.

2) Présence de plusieurs fluorochromes :

La présence de plusieurs fluorochromes ou l'existence d'un seul fluorochrome dans des environnements différents (p. ex. le spectre de fluorescence du BP subissant un glissement de 6 nm vers les plus courtes λ lorsque BP est dissout en présence de caféine (4)) entraine toujours une modification plus ou moins profonde du spectre d'émission.

Le fait de disposer de la totalité du document spectral permet soit l'identification de l'un ou de l'autre des fluorochromes, soit même, après identification, la détermination des quantités relatives des divers composés fluorescents. C'est ainsi qu'il nous a été possible d'évaluer pour des cellules de lignées différentes, le rapport   Pyridines nucléotides réduites libres/Pyridines nucléotides réduites liées.

III - APPLICATION A L'ETUDE DES SPECTRES DE FLUORESCENCE DU BP ABSORBE PAR LES CELLULES

L'appareil utilisé dans cette étude préliminaire, n'ayant pas été conçu pour cet usage, le premier pic du BP (408 nm) ne figure pas dans les spectres. Cela est sans importance pour la suite de notre exposé. Cet inconvénient n'existe plus avec l'appareillage dont nous disposons désormais.

(a) Présentation des résultats :

Lorsque nous étudions la fluorescence des cellules ayant absorbé BP, nous devons nous attendre à observer simultanément la fluorescence propre aux cellules, c'est-à-dire celle des pyridines nucléotides réduites NAD(P)H et la fluorescence du BP absorbé. Le rendement quantique de fluorescence du BP étant nettement plus grand que celui de NAD(P)H, il est possible de diminuer suffisamment l'intensité de la radiation excitatrice de façon à rendre négligeable la contribution de NAD(P)H au spectre de fluorescence.

Deux types d'expériences ont été menés à bien :
- étude de la fluorescence de cellules mises en contact pendant un temps bref (6 mn) avec des doses élevées de BP appelées doses aiguës,
- étude de la fluorescence de cellules cultivées pendant de longues périodes (1 à 3 mois) dans un milieu à faible concentration en BP.

Dans les deux cas, des différences importantes peuvent apparaître entre le spectre de fluorescence enregistré dans les cellules et celui d'une solution de BP dans l'éthanol.

En ce qui concerne les cellules traitées en doses aiguës (BP, 10^{-6} M), nous observons une certaine diversité dans la position du pic situé vers les plus courtes longueurs d'onde (Fig. 5). Pour certaines cellules, les spectres sont très voisins de celui du BP en solution dans l'éthanol (maxima 430-431 nm et 448-450 nm).

Pour d'autres cellules les maxima ne sont pas déplacés, mais le spectre subit une déformation (variation des intensités relatives du premier et du deuxième pic). Enfin certaines cellules présentent un spectre représenté par la courbe 3 de la Fig. 5 avec un glissement important du maximum à 434-436 nm et la transformation deuxième pic en un épaulement ; l'ensemble du spectre semble avoir subi un glissement vers les plus courtes longueurs d'onde.

Pour les cellules cultivées pendant trois mois dans un milieu contenant BP, 10^{-7} M, les spectres de fluorescence des cellules (Fig. 6) présentent une plu grande homogénéité (courbe 3) et sont très différents de celui de BP en solution dans l'éthanol (maxima 440 nm, aspect très peu structuré, et glissement vers les courtes longueurs de l'ensemble du spectre).

Ces mêmes cellules traitées en dose aiguë avec du BP, 10^{-6} M donnent la courbe 2 de la Fig. 6 : spectre plus structuré dont la position du premier maximu (435-437 nm) est intermédiaire entre celui du BP dans l'éthanol et celui du BP da les cellules cultivées dans un milieu contenant BP.

(b) <u>Interprétation</u> :

A priori nous pouvons nous attendre à retrouver le BP sous deux formes au moins :

- soit stocké dans les structures lipidiques de la cellule,
- soit engagé dans des interactions avec des constituants chimiques du milieu intracellulaire.

Le spectre de cellules ayant reçu BP en dose aiguë présente un caractèr plus typique du BP que les cellules cultivées longtemps en présence de BP. Il est assez logique d'attribuer le spectre représenté par la courbe 3 de la Fig. 6 à la fraction du BP en interaction avec des constituants du milieu intracellulaire. Lo de l'addition de BP en dose aiguë, on augmente la quantité de BP libre (BP en solution) et nous obtenons alors un spectre composite plus proche de celui du BP libre (courbe 2 de la Fig. 6).

Au cours de la pénétration du BP dans les cellules (doses aiguës) celui ci doit se répartir en BP libre et BP en interaction et le spectre de fluorescenc obtenu dépend de l'importance relative de l'un ou de l'autre. Il n'est donc pas étonnant d'obtenir des spectres différents selon le mode opératoire et le type de cellules utilisées.

IV - <u>CONCLUSIONS ET DISCUSSION</u>

La complexité de la fluorescence observée rend difficile la déterminati quantitative du BP intracellulaire.

Plusieurs approches sont pourtant possibles :

- On peut envisager de rechercher des cellules chez lesquelles les interactions entre le BP et les constituants cellulaires sont fortement réduites ; ainsi on devrait pouvoir observer la seule fluorescence du BP libre. Nous n'avons pas fait des essais dans ce sens pour l'instant.

- Une élaboration préalable de modèles d'interactions les plus probables devrait permettre la décomposition des spectres de fluorescence des cellules ayant absorbé du BP en spectre du BP libre et celui attribuable au BP en interaction.

Deux autres points sont à considérer si l'on envisage la détermination quantitative du BP absorbé par les cellules:

-Il faut éviter des cellules qui présentent un métabolisme intense des hydrocarbures polycycliques, sinon les lectures seront rendues encore plus complexes par la présence de métabolites fluorescents.

-Il est évident que le glissement vers les courtes longueurs d'ondes et la déformation du spectre que nous avons observé dans les cellules ayant été cultivées pendant longtemps en présence de BP peuvent être interprétés par des perturbations plus complexes qu'une simple interaction. Nous n'avons pas pour l'instant abordé cet aspect.

En conclusion, on peut dire que l'utilisation de la méthode microspectrofluorimétrique pour le dosage du benzopyrène intracellulaire nécessite la définition d'un protocole expérimental très précis tant pour la préparation et le choix des cellules que pour l'exploitation des documents spectroscopiques ; documents qui ne peuvent en aucun cas être réduits à des observations effectuées à une seule longueur d'onde.

BIBLIOGRAPHIE

(1) SALMON J.M., KOHEN C. and GUNNAR B. : Microspectrofluorometric study of benzo(a)pyrene metabolisation in benzo(a)pyrene grown single living cells. Histochemistry, 1974, 42, 85-98.

(2) KOHEN E., KOHEN C., THORELL B. and SALMON J.M. : Studies on metabolic events in localized compartments of the living cell by rapid microspectrofluorometry, in Biological and Medical Physics. LAWRENCE J.M. and GOFMAN J.W. Eds 1974, vol. 15, pp 271-305, Academic Press, N.Y.

(3) SALMON J.M., KOHEN E., KOHEN C. and BENGTSSON G. : A microspectrofluorometric approach for the study of Benzo(a)Pyrene and Dibenzo(a,h) Anthracene Metabolization in single living cells. Histochemistry 1974, 42, 61-74.

(4) BOYLAND E. and GREEN B. : The interaction of polycyclic hydrocarbons and purines. Brit. J. Cancer 1962, 16, 2, 347-360.

SUMMARY : CAPACITY OF CELLS IN CULTURE TO ACCUMULATE POLYCYCLIC AROMATIC HYDROCARBONS AT A LEVEL SUFFICIENT TO BE DETECTABLE BY SPECTROFLUORIMETRIC METHODS - PRELIMINARY RESULTS.

In the course of more general microspectrofluorometric studies on intracellular metabolic control mechanisms, the behaviour of living cells was investigated in culture media containing traces of Benzo(a)Pyrene (BP) difficultly recordable by spectrofluorimetry. A rapid transfer of BP to the intracellular phase with accumulation was observed. In this manner the recording of the hydrocarbon fluorescence emission spectrum from one single cell became very easy.

This fact has led to the investigation of the experimental conditions required for a quantitative detection of trace amounts of aromatic hydrocarbons, and the results presented here are the initial finding of such analysis.

In a first part, the requirement for a recording of the complete fluorescence emission spectrum will be emphasized as a condition for a quantitative result

The second part of the presentation is devoted to observations effectivel made on fluorescence emission spectra recorded from cells grown in the presence of traces of BP. Due to the aim pursued it is evident that the cells selected for this type of studies should exhibit a relatively slow metabolization of the hydrocarbon.

Two significant facts emerge :

(a) in most cases, the fluorescence spectrum recorded from BP-medium grown cells is not identical to that of BP in solution (less defined structure, displaced maxima, and modified relative intensities of maxima). The spectrum observed in the BP medium grown cells is apparently the result of two spectra, one corresponding to a free BP fraction, the other to a BP fraction interacting with cellular constituents.

(b) When the cells were maintained for periods up to one to three months in presence of low amounts of BP, it was noticed that the BP fractions "interacting" seemed relatively more significant than in cells non "adapted" to BP.

Both results are evidently subject to biological interpretations outside the scope of this communication. However, they permit to draw the attention on the fact, that the quantitative determination of the hydrocarbon in the living cell can be achieved only after definition of a sufficiently precise experimental protocol.

EXTRAPOLATION OF TEST RESULTS TO MAN

L. TOMATIS

International Agency for Research on Cancer
150, cours Albert Thomas, 69008 Lyon, France

The task of evaluating the possible hazard for man of environmental chemicals, and in general of quantitatively estimating the total environmental carcinogenic load, would be greatly facilitated if we had proof that all cancers arose following the same mechanism and if we knew more about the possible role of modifying factors in carcinogenesis.

The origin of cancer, either from one single cell or from a number of cells which transform independently, is compatible with its development following a similar or same mechanism. If we accept the hypothesis that cancers of clonal (single) cell origin are mainly spontaneous and that cancers originating from multiple cells are in many cases induced, we should then accept that spontaneous and induced cancers share the same or similar mechanistic steps. This implies that exposure to environmental carcinogens can add to and speed up a process already spontaneously initiated, as well as initiate it *de novo*.

This hypothesis does not make prevention any easier. Precisely because it is very difficult to calculate a minimal- or an undetectable-effect dose of a carcinogen, especially when it is acting in combination with another carcinogen which may modify its activity; it is not likely to be possible to establish "acceptable levels" for many carcinogens that will, in fact, be acceptable.

There is good evidence in experimental carcinogenesis that, following chronic exposure, tumour incidence and induction time are dose-related - the lower the dose, the longer the period required for the tumours to appear and the lower their incidence. At very low dose levels it might be expected that the induction time becomes so long as to exceed life expectancy. There are good grounds, therefore, to argue in favour of reducing exposure to carcinogens.

Scientists and health officials argue at considerable length on risk versus benefit evaluations, as well as cost versus benefit ratios, and reinvigorate their argument either with scientific statements or with statements which just indicate how imprecise or inadequate biological science is. It seems as if it is always taken for granted that a technological innovation or, for instance, a marginal increase of productivity of the soil, are absolute untouchable priorities to which the health of the people, or at least some people, could be sacrificed, since the innovation or productivity increases are benefits which far exceed any risk or any "limited" health hazards.

Has animal experimentation ever contributed to prevent human cancer? The example most often quoted is that of acetylaminofluorene (AAF), which, ready to be marketed on a wide scale as an insecticide, was withdrawn because one experiment had shown its carcinogenicity. Was this experiment so well planned and were the results so convincing that it would have been impossible for any industry or any economic interests to exert sufficient pressure to put this chemical on the market? Apart from the fact that AAF later became one of the better known experimental carcinogens, it was withdrawn basically for two reasons: (1) because substitutes were available and (2) because it had not been widely marketed. The second reason was probably the decisive one, because even when substitutes are available, it would be unrealistic to believe that a chemical produced on a large scale would be withdrawn before all the investments made on it had produced an expected or near-to-expected profit. In fact, very rarely, if ever (with the possible exception of some medical drugs) are there chemicals for which a substitute would not be available.

Examples to be added to that of AAF may exist, since in the past fifteen years a large number of food additives, pesticides and herbicides have been reviewed with regard to their toxicity, and for a number of them the experimental evidence of carcinogenicity following the present regulations was considered sufficient to recommend a zero tolerance in food. In so far as food additives or chemical residues present in food are concerned, extrapolation from experimental data to man is considered as valid. When, however, the withdrawal of chemicals present in the industrial environment is considered, extrapolation from experimental data to man is considered unscientific and not valid.

It would seem obvious that once an aetiological factor for cancer has been identified or a causal relationship between a given exposure and cancer occurrence has been established, the aetiological factor should be removed or, at least, its presence reduced to the minimum achievable with present technology, and exposure to it avoided. This, however, is not happening and there are some well-known examples of this.

There seems to be little merit in trying to develop or improve short-term testing procedures if their efficiency is measured solely on the capacity of matching the results obtained in long-term bioassays, which are often not accepted as final evidence of carcinogenicity. It is hard to understand why the diffuse discontent and distrust of long-term bioassays has been replaced by an euphoric enthusiasm for short-term tests, as if the latter had found a key to making unfailing predictions. In terms of cancer prevention, the validity and limitations of long-term experimentation depend on the use made of the results. It would be hard to justify the present emphasis which is being laid on much-needed short-term tests if their usefulness is *a priori* restricted and thus undermined by equating their validity to that of long-term bioassays.

RESUME

Notre ignorance du mécanisme réel de la cancérogenèse rend plus complexe l'évaluation des risques cancérogènes, mais il y a de bonnes raisons de croire à l'existence d'une relation dose/réponse entre l'exposition et l'induction tumorale.

Des arguments qui tendent à comparer le risque ou le coût aux avantages viennent encore compliquer le problème.

Lorsqu'on a montré que l'acétylaminofluorène, qui devait être commercialisé comme insecticide, était cancérogène chez l'animal d'expérience, on a arrêté la production de cette substance. Les cas de ce genre sont rares, car on tend à n'accepter les indices expérimentaux d'une cancérogénicité que lorsqu'il s'agit d'additifs alimentaires et non de produits chimiques industriels.

Pour que la mise au point d'épreuves de brève durée soit utile, nous devons être certains qu'elles permettent de prévoir avec exactitude une cancérogénicité, mais aussi que leurs résultats seront admis comme indices d'une cancérogénicité chez l'homme Elles seront d'une faible utilité si on assimile simplement leur valeur à celle des épreuves de longue durée.

Environmental Pollution and Carcinogenic Risks
Pollution de l'environnement et risques cancérogènes

INSERM, 1976, Vol. 52, pp. 263-272

MUTAGENICITY AND POSSIBLE CARCINOGENICITY OF HAIR COLOURANTS AND CONSTITUENTS

S. VENITT and C.E. SEARLE (*)

Pollards Wood Research Station,
Nightingales Lane, Chalfont St. Giles,
Bucks HP8 4SP, United Kingdom

(*) The Medical School, Birmingham B15 2TJ, United Kingdom.

INTRODUCTION

Preparations for colouring the hair are now used by many millions of women and a considerable number of men. The most widely used (1) are the "permanent" colourants in which dye intermediates of the phenylenediamine type are oxidized immediately before use, forming quinone-imines which then react rapidly with various couplers to produce coloured products.

"Semipermanent" colourants many of which contain nitrophenylenediamines penetrate and dye the hair directly without the use of an oxidant. A third class of "temporary" colourants contain acid dyes of the type used in wool dyeing, while some heavy-metal salt preparations are also sold for producing gradual darkening of greying hair.

A middle-aged woman who presented with anaemia and neutropenia in 1972 was found to have made heavy use of two semipermanent hair colourants (2). When she later developed acute myeloid leukemia it was decided to carry out some tests on the two colourants to investigate the possibility of a connection between the use of these colourants and her disease.

Carcinogenicity tests of two semipermanent colourants in mice

The colourants tested were "GS", which contains 2-nitro-p-phenylenediamine (2NPPD) and 4-nitro-o-phenylenediamine (4NOPD), and "RB" containing Cl Acid Black 107 (an azo-dye metal complex) and 4-amino-2-nitrophenol. These dye constituents are incorporated in a detergent base containing perfume and other additives.

Carcinogenicity tests were conducted using young adult male and female mice of the DBAf and A strains. Each colourant was diluted tenfold in 50% aqueous acetone shortly before use, and applied twice-weekly to the clipped dorsal skin. Strain A mice received 0.4 ml per application but DBAf mice received only 0.2 ml per application after some toxic effects were observed during treatment at the higher dose level.

Results to date (November 1975) are shown in table 1; figure 1 is a plot of percentage of animals with tumours vs duration of treatment. Tumours, mostly of lymphoid origin, occurred in all groups of mice but it is apparent from figure 1 that in both dye-treated groups tumours arose consistently earlier than in control groups. This is most clearly seen in the DBAf groups, where only one tumour has so far occurred (1/32 at 72 weeks) in the controls, whereas there have been 5/48 in

TABLE 1   Tumours in 2 strains of mice
                       treated topically with diluted
                           hair-dye preparations

Strain	Group	Mice with tumours	Tumours at weeks				
A	Control	7/32 (22%)	61 80	75* 80	75	80*	80*
	Dye GS	9/52 (17%)	48* 79	51* 80*	55* 80	57* 80	57*
	Dye RB	8/32 (25%)	38* 80*	38* 80*	72* 80	74*	80*
DBAf	Control	1/32 (3%)	72*				
	Dye GS	5/48 (10%)	26*	37*	41*	66+	69+
	Dye RB	3/32 (9%)	41*	47*	71*		

* Lymphoid tumours.
+ Genital tract sarcomas.

FIGURE 1. Cumulative tumour incidence in two strains of mice treated topically with tenfold diluted hair colourants.

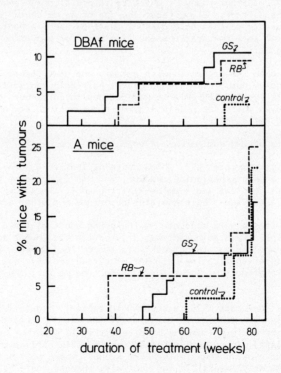

the GS-treated group, of which 3 tumours arose before or at 41 weeks of treatment. Two female GS-treated mice have developed genital tract sarcomas. Several tumours also arose early in the dye-treated A strain mice, although late in the experiment there was no difference in the incidence of tumours between treated and control groups. Clearly the results obtained so far are suggestive of a weak tumorigenic effect, and the earlier appearance of tumours in some of the dye-treated mice is thought to be of more significance than the final incidences. It should also be pointed out that some dye-constituents of colourants GS and RB were present only at approximately 0.01% in the applied solutions, and in experiments now in progress higher doses of individual constituents are under test.

In view of the suggestive evidence of carcinogenicity outlined above, it was decided to test both colourants, together with a range of similar products, for mutagenicity in bacteria, in order to get some idea of the number and kinds of potentially hazardous chemicals used in hair colourants. A preliminary account of these experiments has been published (3). There is now a considerable body of evidence that most carcinogens in their active form are mutagenic, and that bacterial mutagenicity tests are the best validated assays currently available for the rapid prescreening of environmental and industrial chemicals (4).

Bacterial mutation studies with hair colourants

The bacterial strains used in these studies were *Salmonella typhimurium* TA 1535 and TA 1538. These are two of a series of tester strains developed by Ames and co-workers (5) for the rapid screening of chemicals for mutagenicity. Both strains are unable to synthesize histidine *(his⁻)* due to a base-pair substitution (TA 1535) or a frameshift mutation (TA 1538). Both strains are defective in DNA-repair (*uvrB*) and also have a defective lipopolysaccharide cell wall (*rfa*) which renders them more permeable to large organic molecules. Mutations are scored by plating treated bacteria on minimal medium containing a trace of histidine which allows only a few generations of growth. Reverse mutants to prototrophy will, however, continue to grow and form visible colonies against a background lawn of auxotrophic bacteria.

Many carcinogens exert their biological effects only when converted to reactive chemical species by metabolic activation by microsomal enzymes. As bacteria do not possess such enzyme systems, a supernatant fraction from rat liver combined with an NADPH generating system ("S-9 mix") is incorporated in the assay system. Indicator bacteria, a trace of histidine, test compound, and the S-9 mix are combined in dilute molten agar. This is then poured on to a suitable selective agar plate and allowed to solidify. Under these conditions the microsomal enzymes present in the S-9 mix remain active for several hours and the transfer of reactive chemical species to the indicator bacteria is facilitated. Assays are routinely performed with or without the addition of the S-9 mix.

Mutagenicity of semipermanent hair colourants and constituents

The colouration imparted to the hair by semipermanent colourants is achieved by using low-molecular weight dyes which penetrate the hair cortex (1). These are predominantly nitrophenylenediamines, aminophenols and aminoanthroquinones. This class of colourant includes the two products GS and RB which were tested for carcinogenicity in mice, as described above.

A series of colourants from different manufacturers, including GS and RB mentioned above were diluted tenfold in deionized water, filter-sterilized, and 0.1 ml samples were assayed for mutagenicity in TA 1535 and 1538. Positive results were obtained only in TA 1538, indicating that the mutagenic constituents are frameshift mutagens. Results obtained using TA 1538 are shown in Table 2. Colourant RB, which contains

TABLE 2. Mutagenicity of some semipermanent hair colourants in S. typhimurium

Colourant	Induced his^+ revertant colonies per plate (mean of three plates, after subtraction of spontaneous revertants, about 15 - 30 per plate)	
	- S-9 mix	+ S-9 mix
GS	1071	208
RB	27	388
B	975	28
D	807	614
G	144	11
L	956	876
X	1073	739
1	222	35

4-amino-2-nitrophenol, and a metal-azo-dye complex, was mutagenic only after metabolic activation, whereas the rest of the series tested were directly mutagenic. 4-amino-2-nitrophenol was inactive when tested alone. Colourants GS, B, G and 1 were markedly less mutagenic when assayed in the presence of the S-9 mix.

Most of the colourants in the series contain nitrophenylenediamines (2NPPD) and/or 4NOPD. Both compounds (purity >97%) were tested and found to be mutagenic in TA 1538, both in plate-incorporation assays as described above and also in experiments where bacteria were treated with dye in liquid suspension, thoroughly washed by filtration, and then assayed for both mutation and survival of colony-forming ability. The results of such an experiment are shown in figure 2. It is

FIGURE 2. Mutagenicity of nitrophenylenediamines in S. typhimurium TA 1538

Log-phase bacteria were suspended in buffer and treated with graded concentrations of 2NPPD or 4NOPD for 1 hour at 37°C. The bacteria were then washed free of dye by filtration. Samples were assayed for survival of colony-forming ability and for mutation (reversion to his^+).

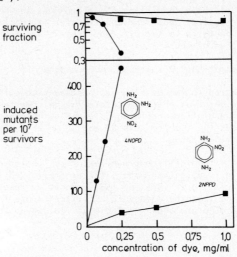

clear that both compounds are direct-acting mutagens which exert their mutagenic effects at relatively low levels of cell killing. 4NOPD is about ten times more mutagenic than 2NPPD at equivalent doses.

Mutagenicity of permanent hair colourants and constituents

Permanent hair colourants are the most widely used type of hair colouring preparation, accounting for three-quarters of the USA market (1). Colouration is produced within the hair shaft by oxidation of colourless *para*phenylenediamine-type intermediates in the presence of couplers such as resorcinol and *meta*phenylenediamines, using hydrogen peroxide as the oxidant. The compounds in general use include phenylenediamines, diaminotoluenes, diaminoanisoles, and aminophenols (6).

A series of proprietary permanent hair colourants was tested for mutagenicity in TA 1538: each colourant was mixed with its appropriate oxidant according to the manufacturer's instructions and the mixture was then diluted tenfold in deionized water. 0.1 ml samples were assayed for mutagenicity in the presence or absence of S-9 mix. The results are shown in Table 3. These results were obtained using an

TABLE 3   Mutagenicity of some permanent hair colourants in *S. typhimurium* TA 1538, assayed with rat-liver S-9 fraction

Each colourant was mixed with the oxidant supplied by the manufacturer, diluted 10-fold in deionized water, and 0.1 ml assayed by the plate-incorporation method.

	Brand	Induced his^+ revertant colonies per plate, (mean of three plates, after subtraction of spontaneous revertants, about 30 per plate)
supplied for domestic use	C	340
	M	67
	N	65
	O	45
	U	459
	2	224
	3	154
	Y	223
	Z	74
supplied to hairdressers	11	90
	33	95
	44	62
	88	380

effective 100-fold dilution of the original colourant mixtures, and much higher mutational yields were obtained with less diluted samples. None of the colourants was mutagenic in the absence of metabolic activation, and they were much less active when tested without prior oxidation. The oxidants supplied with each colourant were not detectably mutagenic, although they were toxic, as shown by their ability to inhibit the growth of the background lawn of auxotrophic bacteria. None of the colourants tested were mutagenic in TA 1535.

A number of individual constituents were also tested, using the plate-incorporation method and S-9 mix. Results are shown for the isomeric phenylenediamines (*ortho* (OPD), *meta* (MPD), and *para* (PPD) in figure 3. All three phenylenediamines were mutagenic, in the order of potency MPD > OPD > PPD. No activity was observed in the absence of metabolic activation. Addition of an equal volume of 6% H_2O_2 to each compound prior to assay caused a marked increase in mutagenic yield, especially in the case of MPD (figure 3, right hand panel). Similar results have been obtained with diaminotoluenes and diaminoanisoles and in both cases the *meta*-substituted diamines (2,4-diaminotoluene and 2,4-diaminoanisole) were markedly more mutagenic than the other isomers, and addition of H_2O_2 again dramatically increased the rate of mutation. No mutagenic activity was detected in TA 1535. Our mutagenicity results are in good agreement with those reported by Ames *et al* (7) for a similar range of hair colourant constituents.

FIGURE 3. Mutagenicity of phenylenediamines in S. typhimurium TA 1538

Graded concentrations of *ortho-*, *meta-*, and *para*-phenylenediamine were assayed for mutagenicity (reversion to his^+) in the presence of rat-liver S-9 mix. The left-hand panel shows results obtained without the addition of H_2O_2, and the right-hand panel shows the effects of mixing each phenylenediamine (at twice the initial concentration) with an equal volume of 6% H_2O_2, prior to assay.

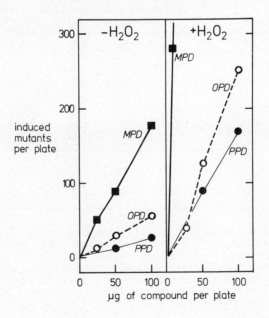

Mutagenicity of 'Colour-set' hair colourants

These products are intended to provide temporary coloured highlights to the hair and at the same time act as setting agents. They are said to contain mixture of acid dyes of the type used in wool dyeing (1).

Each colour set lotion was filter sterilized and 0.1 ml samples were assayed for mutagenicity in TA 1535 and TA 1538 in the presence or absence of S-9 mix. Mutagenic activity was observed in samples of all six colourants tested, and this activity was dependent on the presence of the S-9 liver fraction (Table 4).

TABLE 4   Mutagenicity of 6 colour-set hair colourants in
S. typhimurium TA 1535 and 1538 assayed with rat-liver S-9 fraction

Manufacturer	Product	Induced his^+ revertant colonies per plate, (mean of three plates, after subtraction of spontaneous revertants, about 30 per plate)	
		TA 1535	TA 1538
A	P	200	74
	Q	162	0
	R	185	0
	4	48	134
B	S	5	214
	0	0	86

The identity of the mutagenic constituents of these preparations is not known to us, but it appears that there are at least two different mutagens involved, since both brand B lotions are mutagenic only in TA 1538, whereas all four brand A lotions are mutagenic in TA 1535, and 2 are also active in TA 1538. Since none of the six products were detectably mutagenic in the absence of S-9 mix it is clear that the mutagenic constituents of these colourants require metabolic activation to exert their mutagenic effects.

DISCUSSION

The skin painting experiments showed that two semipermanent hair colourants applied at low concentrations caused an early appearance of tumours in two strains of mice. Both colourants were mutagenic in *S. typhimurium,* and further investigation revealed that three categories of hair colourant contained mutagenic constituents.

It has already been shown that 2NPPD causes chromosome damage in cultured human lymphocytes (3) and Kirkland and Venitt have recently shown that both 2NPPD and 4NOPD cause chromosome damage in cultured Chinese hamster cells (8). In addition, both compounds induce elevated incidences of sister-chromatid exchanges in a line of Chinese hamster cells (9). The evidence so far suggests, therefore, that many hair colourants contain constituents which possess biological effects characteristic of carcinogens, namely mutagenicity in bacteria and the ability to cause chromosome damage and allied effects in mammalian cells. One hair colourant constituent, 2,4-diaminotoluene, which is mutagenic in bacteria, has been shown to be carcinogenic when fed to rats, causing a high incidence of hepatocarcinoma (10).

In view of the very widespread use of hair colourants, both in the home and by professional hairdressers, and bearing in mind that a small percentage of applied colourant constituents can be absorbed through the skin (11), it seems essential that the suspect individual constituents be rigorously tested in lifetime animal studies. Animal testing of permanent colourants raises special difficulties, however, since the main user-contact is with the complex products derived from the oxidation and coupling reactions of the primary intermediates and couplers. Carcinogenicity studies of model colourant formulations have been performed by the USA and German Cosmetics manufacturers, and the results have been negative (12,13).

However, these experiments are open to criticism on the grounds that the doses applied and the numbers of animals used were too low to detect a weak carcinogenic effect, and that adequate positive controls were lacking.

The difficulties surrounding relevant animal testing of the wide range of hair colourant constituents underlines the importance of epidemiological studies of exposed populations. There is already some fragmentary evidence suggesting a possible link between the occupation of hairdresser and excess cancer incidence. Three epidemiological studies of bladder cancer (14,15,16) noted an excess of hairdressers and barbers with tumours of the urinary tract. In each study taken separately the excess was based on very small numbers of cases and statistical evidence was therefore lacking. However, the fact that three studies, separated in place and time, have mentioned this excess strengthens the argument for the existence of a hazard. Moreover, in the Leeds study (16), it was noted that the mean age at diagnosis was consistently low in hairdressers, a characteristic of occupationally induced tumours. A fourth study (17), however, found less cases of bladder cancer than expected in barbers, while Menkart (18) has denied that any such excess could be attributable to hair dyes.

Another population potentially at risk would be people engaged in the manufacture of hair colourants. A recent analysis of cancer mortality in the USA noted elevated rates for both bladder and liver cancer (in men) in counties associated with chemical industries including the manufacture of cosmetics (19).

In such cases it is obviously impossible to ascribe any excess cancer incidence to any one chemical or group of chemicals. However, having regard to the evidence presented here and by Ames *et al.* (7), the well-known susceptibility of the human bladder epithelium to carcinogenesis by some aromatic amines, and the great increase in the use of synthetic hair colourants in recent years, it is clear that more comprehensive epidemiological studies are now urgently required.

REFERENCES

(1) Corbett, J.F. and Menkart, J. : Hair coloring. Cutis, 1973, 12, 190-197.

(2) Gyde, O.H.B. : personal communication. 1973.

(3) Searle, C.E., Harnden, D.G., Venitt, S., and Gyde, O.H.B., : Carcinogenicity and mutagenicity tests of some hair colourants and constituents. Nature, 1975 225, 506-507.

(4) McCann, J., Choi, E., Yamasaki, E., and Ames, B.N. : Detection of carcinogens as mutagens in the *Salmonella*/microsome test: assay of 300 chemicals. Proc. Nat. Acad. Sci. USA, 1975, 72, in press.

(5) Ames, B.N., Durston, W.E., Yamasaki, E. and Lee, F.D. : Carcinogens are mutagen a simple test system combining liver homogenates for activation and bacteria for detection. Proc. Nat. Acad. Sci. USA, 1973, 70, 2281-2285.

(6) Corbett, J.F. : Hair dyes, in 'The chemistry of synthetic dyes', Venkataraman, K., ed., 1971, Vol. 5, pp. 475-534, Academic Press, New York and London.

(7) Ames, B.N., Kammen, H.O., and Yamasaki, E. : Hair dyes are mutagenic: identification of a variety of mutagenic ingredients. Proc. Nat. Acad. Sci. USA, 1975 72, 2423-2427.

(8) Kirkland, D.J. and Venitt, S. : Cytotoxicity of hair colourant constituents: chromosome damage induced by two nitrophenylenediamines in cultured Chinese hamster cells. Mutation Res., 1976, in press.

(9) Perry, P. : personal communication, 1975.

(10) Ito, N., Hiasa, Y., Konishi, Y., and Marugami, M. : The development of carcinoma in liver of rats treated with m-toluylenediamine and the synergistic and antagonistic effects with other chemicals. Cancer Res., 1969, 29, 1137-1145.

(11) Kiese, M., and Rauscher, E. : The absorption of p-toluylenediamine through human skin in hair dyeing. Toxicol. Appl. Pharmacol., 1968, 13, 325-331.

(12) Kinkel, H.J., and Holzman, S. : Study of long-term percutaneous toxicity and carcinogenicity of hair dyes (oxidizing dyes) in rats. Fd. Cosmet. Toxicol., 1973, 11, 641-648.

(13) Burnett, C., Lanman, B., Giovacchini, R., Wolcott, G., Scala, R. and Keplinger, M. : Long term toxicity studies on oxidation hair dyes. Fd. Cosmet. Toxicol., 1975, 13, 353-357.

(14) Wynder, E.L., Onderdonk, J. and Mantel, N. : An epidemiological investigation of cancer of the bladder. Cancer, 1963, 16, 1388-1407.

(15) Dunham, L.J., Rabson, A.S., Stewart, H.L., Frank, A.S. and Young, J.L. : Rates, interview, and pathology study of cancer of the urinary bladder in New Orleans, Louisiana., J. Nat. Cancer Inst., 1968, 41, 683-709.

(16) Anthony, H.M., and Thomas, G.M. : Tumors of the urinary bladder: an analysis of the occupations of 1,030 patients in Leeds, England. J. Nat. Cancer Inst., 1970, 45, 879-896.

(17) Cole, P., Hoover, R. and Friedell, G.H. : Occupation and cancer of lower urinary tract. Cancer, 1972, 29, 1250-1260.

(18) Menkart, J. : Excess bladder cancer in beauticians? Science, 1975, 190, 96-98.

(19) Hoover, R. and Fraumeni, J.F. : Cancer mortality in US counties with chemical industries. Environmental Res., 1975, 9, 196-207.

ACKNOWLEDGEMENTS

We thank Dr. E.L. Jones for reporting on the animal histology and Carole Bushell for technical assistance. This work was supported by a grant to C.E. Searle from the Cancer Research Campaign and by grants to the Institute of Cancer Research from the Cancer Research Campaign and the Medical Research Council.

SUMMARY

Skin painting experiments showed that two semi-permanent hair colorants applied at low concentrations caused an early appearance of tumours in two strains of mice. Both colorants were mutagenic in *S. typhimurium* and further investigation revealed that three categories of hair colorant contained mutagenic constituents.

RESUME

Des expériences de badigeonnage cutané ont montré que deux colorants capillaires semi-permanents, utilisés à de faibles concentrations, provoquaient l'apparition précoce de tumeurs dans deux souches de souris. Les deux colorants avaient une action mutagène sur *S. typhimurium* et d'autres recherches ont révélé la présence de constituents mutagènes dans trois catégories de colorants capillaires.

VI

ASSESSMENT OF CARCINOGENIC RISK
ÉVALUATION DU RISQUE CANCÉROGÈNE

Chairman / *Président :* Dr. C. ROSENFELD

LES ASPECTS EPIDEMIOLOGIQUES DANS L'EVALUATION DES RISQUES CANCERIGENES DUS A LA POLLUTION

R. FLAMANT

Unité de Recherches Statistiques INSERM U.21 et Institut Gustave-Roussy, 94800 Villejuif, France

Nous allons passer en revue les différents méthodes classiques en épidémiologie (1), en étudiant leur utilisation dans le cas particulier des risques cancérigènes dus aux divers facteurs polluants.

Deux difficultés doivent être soulignés d'emblée. La première est due au caractère propre du développement du cancer qui n'apparaît que après une longue période par rapport à l'exposition au risque. La seconde est due à la mesure des facteurs polluants qui ne peuvent pas toujours se faire à l'échelon individuel mais seulement au niveau d'une collectivité ou d'une population.

Epidémiologie déscriptive

Nous étudierons successivement trois types d'études :

l'utilisation des statistiques de mortalité et de morbidité
la pathologie géographique
l'étude des migrants

Utilisation des statistiques de mortalité et de morbidité

On peut par exemple établir des corrélations entre l'évolution du taux de mortalité des cancers et l'évolution de différents facteurs polluants mais on connaît le caractère souvent grossier des données de mortalité et seuls peuvent être réellement etudiés les cancers à forte mortalité, tel le cancer bronchique.

Les données de morbidité, lorsqu'elles existent, sont sur ce plan de bien meilleure qualité. Elles seules permettent d'étudier des cancers à faible mortalité, tels les épithéliomas cutanés. La constitution de registres du cancer de plus en plus nombreux doit permettre dans l'avenir d'attirer l'attention sur une augmentation de telle ou telle localisation cancéreuse et de la corréler avec une variation d'un phénomène extérieur.

Pathologie géographique

Il s'agit de comparer la fréquence des cancers entre des régions dont on mesure par ailleurs le degré de pollution.

Un exemple peut en être tiré d'une étude de l'INSERM sur la morbidité par cancer en relation avec la pollution atmosphérique (2). Quatre départements français avaient été retenus, deux non pollués (Hérault et Ille-et-Vilaine) et deux pollués (Bouches du Rhône et Rhône). L'index retenu pour définir le degré de pollution était l'importance relative des zones urbaines et rurales et la dimension du parc automobile. Bien entendu, il était prévu de mesurer la pollution atmosphérique tout au long de l'étude au niveau de postes urbains et ruraux, les trois indices retenus à l'époque étant la concentration en microgrammes par m^2 des particules contenues dans l'atmosphère, la concentration en anhydride sulfureux (SO_2) et la concentration en oxyde de carbone (CO).

Il était également apparu essentiel aux auteurs de tenir compte de la consommation de tabac et d'alcool, dont le rôle est important dans la survenue des cancers pour lesquels on peut suspecter un rôle cancérigène de la pollution et dans la mesure où ces deux habitudes diffèrent entre milieu urbain et milieu rural. En effet si on trouve une différence de fréquence d'un cancer donné entre ville et campagne, ce n'est pas seulement les différences de pollution qui peuvent être invoquées mais d'autres facteurs qui différencient le citadin du rural, notamment la consommation de tabac et d'alcool.

On peut alors dans l'analyse statistique des données étudier les liaisons entre l'incidence d'un cancer et le degré de pollution, à égalité des autres facteurs différenciant ville et campagne.

L'étude des migrants

On a trouvé par exemple chez les immigrants anglais en Afrique du Sud que leur taux de mortalité par cancer bronchique était plus élevé que celui des hommes blancs nés sur place et celui des autres immigrants. Ce n'était pas une question de tabac car les hommes d'Afrique du Sud sont d'aussi gros fumeurs que les Anglais. Par ailleurs, les plus âgés des immigrants (soixante-cinq ans et plus) avaient une mortalité analogue aux autres. La conclusion de l'auteur était que ces sujets, ayant quitté l'Angleterre avant 1910, avaient échappé à l'intense pollution atmosphérique, apparue plus tard.

Quelque soit l'intérêt de ces types d'enquête, il faut souligner leur limitation dans la portée des conclusions. En effet celles-ci se font au niveau de populations et même si l'on trouve une forte corrélation entre la survenue d'un cancer et l'existence d'un facteur de population au niveau de groupes d'individus, rien ne prouve que cette même relation existe au niveau de l'individu même. On peut concevoir qu'il existe à la fois dans une région des individus exposés à un haut risque de pollution et des individus différents ayant pour d'autres raisons un taux élevé de cancer.

Ces types d'études peuvent seulement fournir des hypothèses de travail qui devront être confirmés par des études épidémiologiques plus fines.

Epidémiologie analytique

Elle consiste à faire des enquêtes en prenant comme cible non plus un groupe mais les individus, en évaluant pour chacun d'eux son risque cancer et son exposition à un facteur de risque.

Les enquêtes analytiques peuvent être classées :

a) sur une base chronologique, selon que le facteur est mesuré avant l'apparition du cancer (enquêtes prospectives) ou après (enquêtes rétrospectives).

b) selon le mose d'échantillonnage en trois types : dans le type I, on prend un échantillon représentatif de la population, dans le type II on choisit le nombre de sujets exposés et non exposés au risque de pollution, dans le type III on choisit le nombre de cancers et de témoins.

Ces deux critères permettent de classer les enquêtes en cinq catégories
(une des combinaisons n'existant pas).

CLASSIFICATION DES ENQUETES ANALYTIQUES

Type d'échantillonnage / Chronologie	I échantillon représentatif de la population	II un groupe exposés, une groupe non exposés	III un groupe de cancers, un groupe de témoins
prospective			////////
rétrospective			

Chaque catégorie d'enquêtes a ses avantages et inconvénients :

- la sécurité est plus grande avec le mode prospectif qu'avec
le mode rétrospectif, en raison de son caractère "aveugle"
(quand le facteur exposition au risque est mesuré, on ignore
si le sujet sera malade ou non).

- le nombre de sujets, très élevé dans le type I, peut être réduit en
utilisant le type II (si le facteur est rare) ou III (si la
maladie est rare).

Les enquêtes les plus couramment utilisées dans le domaine de la pollution sont de type II ou III. Les enquêtes de type II consistent par exemple à selectionner dans une usine des ouvriers exposés à un facteur de pollution et des ouvriers, de même sexe et de même âge, travaillant à des endroits non exposés et d'étudier, de façon rétrospective ou prospective, l'incidence des cancers dans les deux groupes. Dans l'enquête de type III, toujours rétrospective, on prend des sujets atteints de cancers et des sujets témoins, non cancéreux, là encore de même sexe et de même âge et on les interroge sur leurs expositions à des facteurs de pollution, notamment professionels.

Les conclusions de ces enquêtes sont schématiquement de deux genres différents selon l'esprit dans lequel elles ont été entreprises.

Ou bien on étudie un facteur précis, pour lequel on soupçonne un effet causal dominant dans un type de cancer, par exemple les polychlorures de vinyl dans les angiosarcomes du foie. Toutes les modalités de l'enquête vont être adaptés au but précis d'isoler le rôle de ce facteur parmi d'autres possibles qui peuvent le masquer plus ou moins complètement. Finalement on peut aboutir, non pas à une démonstration absolue de causalité que seule une expérience pourrait obtenir, impensable chez l'homme, mais à une forte présomption de causalité appuyée souvent par des recherches expérimentales ou laboratoires.

Dès lors, la conviction devient suffisamment forte pour entraîner la suppression du facteur de risque et la disparition de la maladie apporte a posteriori la preuve du rôle causal de cet agent.

Ou bien, on étudie un certain nombre de facteurs dont on ne sait pas exactement s'ils sont directement la cause, ou seulement le reflet d'une cause encore inaccessible. Il s'agit de facteurs de risque qui, s'ils n'ont pas le même intérêt scientifique que les facteurs causaux, ont néanmoins un intérêt pragmatique évident. Ils permettent de reconnaître des populations dites à haut risque chez lesquelles une surveillance particulière pourra être exercée en vue de dépistage et des mesures préventives pourront être adoptées. Dans certains cas, on a même pu entreprendre de véritables essais de prévention, selon une méthodologie expérimentale voisine de celles des essais thérapeutiques comparatifs (3).

Résumé

L'épidémiologie descriptive qui étudie des groupes de population est d'un grand intérêt en permettant d'attirer l'attention sur tel ou tel facteur susceptible d'être carcinogène.

Elle doit être complétée par des enquêtes analytiques dont certaines parmettent d'aboutir à une quasi certitude concernant la responsabilité du facteur mis en cause et d'autres permettent néanmoins d'isoler des populations à haut risque en vue de dépistage et de prévention.

A tous les moments, ces recherches épidémiologiques doivent être confrontées avec les résultats des recherches expérimentales.

Summary

Descriptive epidemiology in which population groups are studied can be of considerable value in drawing attention to a potential carcinogen.

This must be followed by analytical studies which might either lead with a fair degree of certainty to the identification of the responsible factor or at least help to identify the high-risk group who could then be screened for the suspected cancer.

Such epidemiological research must always take into account the results of experimental studies.

BIBLIOGRAPHIE

(1) ROUQUETTE C., SCHWARTZ D.
   Méthodes en épidémiologie
   1970, Flammarion, Paris.

(2) LASSERRE O., GARBE E., FLAMANT R.
   L'enquête française sur la morbidité par cancer en relation avec certains facteurs étiologiques : l'alcoolisme et la pollution atmosphérique
   1968, Bulletin de l'INSERM, 23, 5, 1301 - 1312.

(3) SCHWARTZ D., FLAMANT R., LELLOUCH S.
   L'essai thérapeutique chez l'homme
   1970, Flammarion, Paris.

FEASIBILITY OF MONITORING POPULATIONS TO DETECT ENVIRONMENTAL CARCINOGENS

C.S. MUIR (1), R. MACLENNAN (1),
J.A.H. WATERHOUSE (2) et K. MAGNUS (3)

(1) Unit of Epidemiology and Biostatistics, IARC, 150 cours Albert Thomas, 69008 Lyon, France.

(2) Regional Cancer Registry, Queen Elizabeth Medical Centre, Birmingham B15 2TJ, England.

(3) Cancer Registry of Norway, Montebello, Oslo 3, Norway.

This paper considers how populations can be monitored to detect the possible entry of a carcinogen into the environment and some of the implications of such monitoring programmes. The word "detect" is used advisedly, as not all monitoring systems will permit identification of the agent responsible.

In relation to cancer, monitoring can be defined as "continuing observation in order to decide when changes in incidence or mortality may have occurred".

Monitoring can be undertaken in two ways:

(a) passive or routine monitoring

Study of the cancer experience of populations in general over time to ascertain whether the level of cancer is significantly changing. For this purpose, routinely collected data are generally used.

(b) active or exposure-oriented monitoring

Study of the cancer experience of populations which have been defined on the basis of specific exposures and comparison with the experience of persons not so exposed. For this type of monitoring it is usually necessary to identify exposed groups and follow their subsequent fate.

Teppo (1) describes these two concepts of monitoring as "monitoring without a hypothesis" and "monitoring with a hypothesis" respectively.

PASSIVE (ROUTINE) MONITORING

In passive monitoring time-series of either incidence or mortality data are examined at suitable intervals and by appropriate geographical or other class to detect temporal variation. The value of morbidity and mortality data for this type of monitoring varies: each has its advantages and disadvantages (see Table 1).

The interpretation of time-series, whether morbidity or mortality, is difficult. Two questions have to be answered - is the rise significant in the statistical sense and, if significant, is it artefactual. Several statistical techniques can be used, e.g. linear regression (2). As a rule of thumb for infrequent cancers, say, 50 cases a year, it would require an annual increase (or decrease) of 26% to attain

TABLE 1. COMPARISON OF USE OF MORTALITY AND MORBIDITY DATA FOR PASSIVE MONITORING

	MORTALITY	MORBIDITY
AVAILABILITY SPACE TIME	Available for most of the world and for some countries, long periods of time (70 years or more), but degree of detail and accuracy varies over time (ICD Revisions) and place.	Reasonably accurate incidence data available for some 80 populations in 30 countries, but majority of series begin around 1960.
EFFECT OF TREATMENT	For some sites with poor survival, e.g. oesophagus, lung and stomach, mortality data give good reflection of impact of disease on population, the more so as survival has not materially improved with time. For other sites trends are influenced by improving therapy and falling or stationary mortality may mask true increase.	Unaffected by treatment.
QUALITY OF INFORMATION	Not subject to distortion due to duplication. Detailed information rarely available and even if necropsy performed, this information may not be on death certificate. Information on histological type infrequent.	Subject to distortion due to failure to eliminate duplicate registration of same individual. Quality of information usually good with more detail than in mortality data. For deceased person, often includes results of post mortem examination. Information on histological type usually available.
INFORMATION ON POSSIBLE EXPOSURE:		
RESIDENCE	Usually available	Usually available
OCCUPATION	Usually available but often poor for retired persons. Last occupation prior to retirement may not be the "usual" one due to transfer to lighter work. Statement of occupation by widow often subject to "promotion" distortion.	Usually available but often poor quality. Occupation at time of diagnosis may not be "usual" one due to ill-health.
CENSUS INFORMATION	Linkage rarely possible for individuals but may be possible for groups permitting comparison of proportion of persons with cancer within a given occupation with proportion of that occupation in general population in the region	Linkage rarely possible for individuals but may be possible for groups permitting comparison of proportion of persons with cancer within a given occupation with proportion of that occupation in general population in the region.

significance at the 1% level if a 5 year period be considered. If figures are available for 15 years then the change needed is in the order of 2½% per annum. In terms of, say, cancer of the eye, a population of 7-10 million would be needed to yield 50 such neoplasms a year. In other words sizeable populations would have to be studied.

While the longer time period increases statistical sensitivity, it also increases the chance of artefact. Diagnostic criteria and methods may change during this time and such changes were long held by many to account for the rise in lung cancer incidence. Today the same arguments are advanced for pancreatic and other deep-sited neoplasms. Yet in England and Wales there was a statistically significant increase in mortality from malignant neoplasms of the respiratory tract between 1900 and 1905 (3). There are further confounding variables, notably the changes in the International Classification of Disease and coding rules. Thus a significant 2% increase in lung cancer mortality in the US noted in 1968 was due to change in coding rules (4). Many of the changes found by Burbank (2) in US mortality data from 1950-1967 remain to be explained.

If the incidence of a given type of cancer is already high and a new carcinogen with the same target organ appears in the general environment it would require to be very widespread and powerful for the increase to be detected, and, if detected, the cause would still of course have to be determined. This statement can be exemplified by a total population of one million persons with a male lung cancer rate of 80 per 100,000 per annum, i.e. 400 cases a year. An increase of 15% to 460 cases, corresponding to a rate of 92 per 100,000 per annum would be highly significant but it would remain to be shown that this increase was not due to previously existing causes such as smoking.

Routine passive monitoring normally deals with the experience of whole populations and thus a very significant rise in a common cancer in a small group would not normally be detected. This is an inherent limitation of the method.

It has been suggested that passive monitoring is most likely to be successful when the change affects a rare or unusual form of cancer such as the angiosarcoma of liver (associated with heavy exposure to vinyl chloride monomer) or vaginal adenocarcinoma in young women and girls (the daughters of women given diethylstilbestrol for threatened abortion). Even for such neoplasms it is problematical whether passive monitoring of mortality data would have been efficacious. It is unlikely that the histological diagnoses would have been coded, even if entered on the death certificate. Such discoveries are more likely to be made by cancer registries. Based on the results of the 3rd US National Cancer Survey it has been estimated that the average annual incidence of hepatic angiosarcoma is around 0.014 per 100,000 per annum (5) or one case per 7 million population. Under such circumstances even one case becomes worth investigating. To have established that the first case observed worked in a vinyl chloride plant would have aroused suspicion. A second case in the same plant or industry would constitute a strong presumption of a causal association and the need for thorough investigation. To be able to assess the significance of the occurrence of rare cancers it is necessary to have a notion of incidence by histological type, a requirement which implies good registration.

Consideration of the above suggests that passive monitoring would be most effective when it is possible to look not only at site but also histological type and age-group. The recent decision of WHO to include in the 9th Revision of the International Classification of Diseases, Injuries, and Causes of Death (ICD) a code for the histology of neoplasms is a step of considerable potential importance. For each cancer registry or vital statistics office, to publish tabulations by site, age, sex, age-group and histological type would be very expensive, although the Swedish

Cancer Registry has shown what can be done (6). It would thus seem most appropriate to have magnetic tapes of such information available for analysis by a designated agency. Separate analyses for urban and rural areas should normally be undertaken as chemical agents entering them may be quite different.

Yet the more finely the cake is sliced, the more likely it is that differences will emerge and the less likely it becomes that there will be sufficient numbers in the cells in the resulting tables to attain statistical significance. A balance will have to be struck.

Teppo (1) has pointed out that computers can never take all possibilities into account and careful examination of cancer registry and mortality data by experts with a medical background is of great importance. Artefact is most likely to be detected by those who know where the figures come from.

Detection of the agent

If a rise in incidence or mortality can be accepted as real, the cause remains to be determined. The potential of cancer registry data is well illustrated by malignant melanoma of skin, the incidence of which has nearly trebled in Norway over 20 years (Fig. I). Detailed analysis of survival information showed that this was not due to a growing tendency to remove suspicious moles or to increasing recognition by pathologists of amelanotic forms of the disease. Distribution by age, sex, anatomical site, time and geographic region give strong support to the belief that increased exposure to solar radiation was responsible (7). Such a progression of events is all too rare. Monitoring of gastric cancer (falling) and colon cancer (rising) has not yet been followed by identification of the agents responsible.

The long and variable latent period of cancer poses special problems. Thus, paradoxically, a carcinogen may be disappearing from the environment at a time when the yield of the related cancer is increasing. Such an effect may be occurring in Japan where although the number of persons dying with gastric cancer is still increasing, the rates in the younger age-groups are falling suggesting that recent generations are no longer exposed to the causal agents or, perhaps less likely, that the latent period has become much longer. Such cohort studies are of great value but require data to be available over long periods (Table 2, Fig. II).

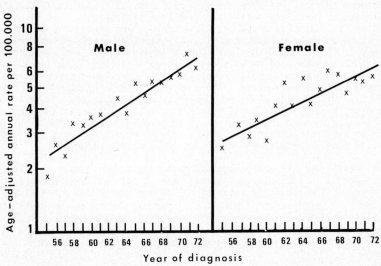

FIG. I   Incidence of malignant melanoma of the skin in Norway, 1955-1972 (7).

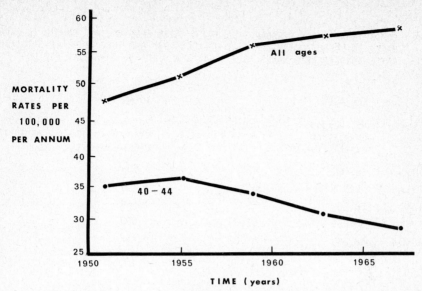

FIG. II Age-adjusted mortality from gastric cancer for all ages is rising in Japan while the age-specific rates in the age-group 40-44 are slowly falling.

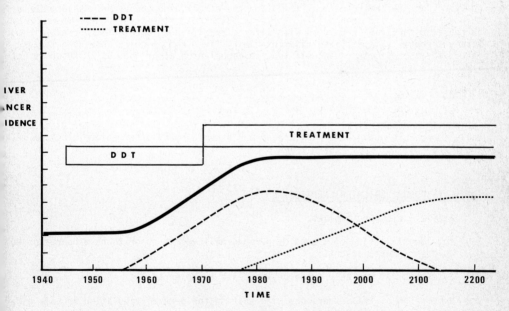

FIG. III The incidence of liver cancer is shown to rise some 10-15 years after the introduction of, say, DDT. Due to the latent period incidence associated with this exposure continues to rise for 10 years after withdrawal before gradually falling. If another hepatic carcinogen: "treatment X" enters the environment about the same time as DDT is withdrawn then there would be no change in the overall liver cancer incidence.

TABLE 2. COMPARISON OF AGE-STANDARDIZED DEATH RATES FROM GASTRIC CANCER IN JAPANESE MALES IN SELECTED TIME PERIODS BETWEEN 1950 AND 1967 WITH AGE-SPECIFIC RATES FOR JAPANESE MALES AGED 40-44 YEARS IN THE SAME TIME PERIODS

Time	ALL AGES		40-44 YEARS	
	No. persons dying from gastric cancer	Age-standardized mortality rate	No. persons dying from gastric cancer	Age-specific mortality ra[te]
1950-51	38,888	47.3	1,569	35.2
1954-55	45,015	51.6	1,686	36.3
1958-59	50,908	56.0	1,564	34.5
1962-63	53,869	57.3	1,500	31.8
1966-67	58,083	59.6	1,752	28.5

Different carcinogens may have the same target organ and the result of the disappearance of one agent from the environment may be masked by an increase in the leve[l] of another. Thus if DDT were a human hepatic carcinogen (as it is for the mouse) with a 20 year latent period, its effect would begin to appear in, say, 1970. If removed from the environment at this time the effect might nevertheless last for a f[ur]ther 40 years. If, around 1970, there is introduced "treatment X" which results in increase in survival in persons with liver cirrhosis and, due to the increase in survival, there is an associated rise in primary liver cancer which appears a few years later, the two effects could well tend to cancel each other out, as is shown in Fig. III.

Comment

Passive monitoring can rarely do more than indicate an area for further study. There will undoubtedly be false alarms. Many of these will be rejected as biologically improbable but others will appear worth pursuing by other epidemiological meth[ods.] The method has not really been tried.

ACTIVE (EXPOSURE ORIENTED) MONITORING

Active monitoring implies:

(a) the identification of groups of persons believed to be or to have been exposed to environmental agents;

(b) ascertainment of their subsequent disease experience;

(c) comparison of this experience with that in the general population or some other suitable comparison group.

To illustrate the width of applicability, several groups are considered: occup[a]tional and industrial, persons taking therapeutic drugs over long time periods, and migrants.

TABLE 3. SUMMARY COMPARISONS OF PASSIVE AND ACTIVE MONITORING (see also text)

	PASSIVE	ACTIVE
BASE MATERIAL	Data collected routinely for other purposes (see Table 1).	Cohorts of persons with specified exposures need to be defined and assembled.
COST OF BASE MATERIAL	Relatively cheap as already collected.	Usually fairly expensive unless census defined cohorts used.
COST OF MONITORING	Relatively cheap.	Follow-up may be very expensive unless cohort readily linked to death records or cancer registry.
EVALUATION OF FINDINGS	(a) Difficult unless for very rare unusual cancers. (b) Artefact may be difficult to exclude (c) May be difficult to decide whether result warrants further study.	Usually sufficiently specific to warrant further studies.
FAILURES	(a) Will not detect change in risk in small groups. (b) Difficult to allow for effect of confounding variables.	(a) Inability to link records or effect follow-up. (b) Impracticable to monitor all groups likely to be at risk. (c) Unless cohorts properly defined will not detect changes in risk in small groups. (d) Difficult to allow for effects of confounding variables.
VALUE	Systematically undertaken could give earlier warning of changes which need to be investigated.	(a) Results usually sufficiently specific to warrant further studies. (b) May point to need to protect worker and public from product in question until risk fully explored.
PRIORITY AREAS	Regions with good quality cancer registration.	(a) Industries producing significant amounts of chemicals shown to be carcinogenic in animals and other chemicals in general use or in contact with foodstuffs. (b) Drugs for chronic disease. (c) Migrants.

There are of course formidable logistic problems associated with this type of study. Drugs prescribed are not always taken and many drugs such as aspirin are freely available[1]. Unless the drug is highly carcinogenic large numbers will have to be followed for long periods of time. It can be argued that most drugs are introduced slowly and it may be many years before usage is sufficiently widespread to merit the expense of study. While as few as 5,000 person-years of observation would yield one case of cancer for a site with a general population rate of 20 per 100,000 per annum and to observe 4 cases would be significant (23), this is an over-simplification as with the latent period of cancer length of exposure must be taken into account. To follow 5,000 persons for the year after a short course of treatment would be meaningless in terms of cancer; to follow 500 for 10 to 30 years could yield information of value. The disease for which the drug was given might be such that few would survive the 10 to 30 years needed. Nevertheless, we believe that the time has come to tackle this problem systematically and the existence of schemes whereby drugs prescribed are paid for by the state or an insurance agency provides the administrative framework for cohort assembly. The conflict of opinion as to what constitutes a control group when this problem is approached by the case-control method is well illustrated by the reserpine/breast cancer controversy (24).

Transplacental effects

Medication in pregnancy poses problems of record linkage as effects may appear in both mother and child. The transplacental carcinogenic action of diethylstil-bestrol (25) was detected only because the significance of a cluster of a very rare cancer occurring in an unusual age-group was realized by an astute clinician. As there is general believe that foetal tissues are more susceptible to carcinogens than adult tissue, there is a case to be made for the issue, as in South Australia, of prescription books containing details of drugs prescribed and taken during pregnancy. The reported association between influenza during pregnancy and increased cancer risk in the offspring, notably leukaemia (26), could perhaps be due to the use of medication for the viral disease.

If the effect is slight it will be difficult to detect. Thus, although Neute and Buck (27) found a slight excess of cancer in the offspring of 9,000 smoking mothers followed for 7-10 years from birth, the difference was not statistically significant. The later such cancers appear, unless they are distinctive, like adenocarcinoma of the vagina mentioned above, the more difficult it will be to separate them from those due to post-natal exposures.

Migrants

Migrants form a unique resource. Often coming from countries with completely different cancer patterns, their subsequent cancer experience in the host country may provide indicators to the existence and possibly the nature of environmental carcinogens. Thus the pioneer studies of Haenszel and Kurihara (28) showed that the low level of large bowel cancer in Japanese was not genetically ordained since it rose very quickly (in terms of cancer) on residence in the United States of America.

The existence of Japanese-born Japanese and US-born Japanese has permitted study of the factors involved - for colon cancer beef and string beans have been suggested as possible agents (29).

Every effort should be made to identify migrants and their offspring and compare their cancer experience with that of the indigenous population. It is to be regretted that in many countries there are barriers to the identification of such cohorts (30).

---

[1]It is not only prescription drugs which may be dangerous. Prolonged abuse of phenetecin - a mild analgesic - has been associated with cancer of the renal pelvis (22)

TABLE 3. SUMMARY COMPARISONS OF PASSIVE AND ACTIVE MONITORING (see also text)

	PASSIVE	ACTIVE
BASE MATERIAL	Data collected routinely for other purposes (see Table 1).	Cohorts of persons with specified exposures need to be defined and assembled.
COST OF BASE MATERIAL	Relatively cheap as already collected.	Usually fairly expensive unless census defined cohorts used.
COST OF MONITORING	Relatively cheap.	Follow-up may be very expensive unless cohort readily linked to death records or cancer registry.
EVALUATION OF FINDINGS	(a) Difficult unless for very rare unusual cancers. (b) Artefact may be difficult to exclude (c) May be difficult to decide whether result warrants further study.	Usually sufficiently specific to warrant further studies.
FAILURES	(a) Will not detect change in risk in small groups. (b) Difficult to allow for effect of confounding variables.	(a) Inability to link records or effect follow-up. (b) Impracticable to monitor all groups likely to be at risk. (c) Unless cohorts properly defined will not detect changes in risk in small groups. (d) Difficult to allow for effects of confounding variables.
VALUE	Systematically undertaken could give earlier warning of changes which need to be investigated.	(a) Results usually sufficiently specific to warrant further studies. (b) May point to need to protect worker and public from product in question until risk fully explored.
PRIORITY AREAS	Regions with good quality cancer registration.	(a) Industries producing significant amounts of chemicals shown to be carcinogenic in animals and other chemicals in general use or in contact with foodstuffs. (b) Drugs for chronic disease. (c) Migrants.

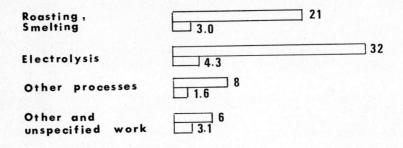

FIG. IV   Comparison of the number of cancers observed (upper bar) in Norwegian nickel workers with number expected (lower bar) on basis of general population experience (13).

Industrial and occupational exposure

There have been few attempts to systematically examine risk for all occupations. The best known of these are the decennial publications of the Registrar General of England and Wales (8), which link occupation as stated on the death certificate with the census distribution of occupations, thus permitting computation of standardized mortality ratios. Guralnick (9, 10) has undertaken similar studies in the US. In Sweden, as in other Scandinavian countries, the potential for such studies is very great and the wide variety of information which can be linked has been described by Bolander (11).

Milham (12) in the State of Washington (USA) has developed systems (matched case control, proportionate mortality ratio, standardized mortality ratio) for the estimation of occupational risk; the method used (9, 10) is particularly flexible.

Stated occupation does not always indicate possible exposures and two workers at the same plant could well have completely different risks. Some large industrial concerns have begun to keep chronological records of the place of work and work station for each employee, usually in relation to the rate of pay for the job. If such records followed the worker from employer to employer, assessment of hazard would be much easier. Such records are, however, expensive to maintain.

Numerous studies of the risk attached to certain occupations have been completed or are underway. Thus Pedersen et al. (13) reviewed cancer experience of workers in the nickel industry in Norway, showing differential risks by process in the plant (Fig. IV); Fox et al. (14) have described the cancer mortality in 40,867 persons employed in the rubber and cable making industries on 1 February 1967; Matanoski et al (15) examined mortality from cancers and other causes in radiologists, comparing their experience with that of other physicians; Doll and his colleagues continue to follow the fate of British doctors.

Each of these studies was in a sense tailor-made and did not form part of a systematic effort to assess industrial risk. Today a 1% census sample of workers in selected occupations in England and Wales are being followed for possible cancer risk through the National Cancer Registry. Individuals are "tagged" on the basis of the stated employment at the time of the 1971 census and their appearance in the Register watched for.

In addition to these screening studies, which may be rather non-specific as the stated occupation may not indicate industry or likely exposure, it would seem reasonable to pay particular attention to workers in plants producing chemicals which have been shown to be carcinogenic in animals and which are produced in large quantities and for which there is known human exposure. Among 164 chemicals reviewed in the IARC monograph series (16) significant human exposure is known to occur in 98 which have at least some carcinogenic activity in experimental animals. Exposure is predominantly or partly occupational in 60 (17). The number of workers exposed to any one chemical in any one country is likely to be fairly small, hence the need for international collaboration to accumulate sufficient person/years to assess risk.

In the design of such monitoring studies the dilution effect needs to be borne in mind. Take, for example[1], a town with a male population of 100,000 persons where the lung cancer incidence is 70 per 100,000 per annum. A factory located in this town has a work force of 10,000 men. If the factory work force had the same cancer experience as the town one could expect 7 cases of lung cancer to arise each year. If, within that factory, there is a process employing 1,000 men who have a 3 fold lung cancer risk, 2.1 cases of lung cancer would be expected among them in one year compared to 0.7 in 1,000 men working elsewhere. It would require five years observation to be sure that this difference was statistically significant at the 99% level. However, if it were not possible to follow the 1,000 men at threefold risk separately their excess would be pooled with that of the other 9,000 workers and 6.3 + 2.1 or 8.4 cases would be observed instead of the 7 expected. Under these circumstances, it would require about 12 years to show that the overall factory incidence was likely to be greater than that of the town at the 95% level. Knowledge of the process(es) at which an individual was employed and duration of that employment is of paramount importance for such work.

Ideally, monitoring of any group should continue to death. The survivors of the Hiroshima/Nagasaki atomic bomb explosion showed an excess of leukemia and reticuloendothelial neoplasms beginning some 8 years after the exposure. This excess risk fell but some 15 years after irradiation an excess of cancers of other sites such as stomach or breast is now appearing. The same phenomenon could well occur with other exposures.

Iatrogenic neoplasia

The long-term effects of medical treatment are unlikely to be detected by the treating physician. The studies of Court-Brown and Doll (18) on the fate of persons treated for ankylosing spondylitis by radiation began long after the course of therapy and were not planned at the time of treatment. Interest in the fate of recipients of transplanted kidneys has uncovered the carcinogenic consequences of continued use of the immunosuppressives (19), but there are few long-term studies of this nature. The subsequent cancer experience of children surviving modern chemotherapy for leukaemia would be of considerable interest.

In monitoring activities, drugs taken for long periods (arbitrarily three months or more) should have priority. Normally these would be used for chronic disease such as hypertension, chronic arthritis, tuberculosis and mental disorder. To cite but one example, Clemmesen et al. (20), noting that phenobarbitone had produced liver cancer in mice, examined the cancer risk of 9,136 patients treated for epilepsy at the Filadelfia Hospital, Denmark, from 1933 to 1963 and failed to detect any oncogenic effect. Similar studies of persons taking the so-called tranquillizers are indicated. Oral contraceptives should be added to the list (21).

---

[1] This example is very simplified in that it does not take into account age-structure, etc.

There are of course formidable logistic problems associated with this type of study. Drugs prescribed are not always taken and many drugs such as aspirin are freely available[1]. Unless the drug is highly carcinogenic large numbers will have to be followed for long periods of time. It can be argued that most drugs are introduced slowly and it may be many years before usage is sufficiently widespread to merit the expense of study. While as few as 5,000 person-years of observation would yield one case of cancer for a site with a general population rate of 20 per 100,000 per annum and to observe 4 cases would be significant (23), this is an over-simplification as with the latent period of cancer length of exposure must be taken into account. To follow 5,000 persons for the year after a short course of treatment would be meaningless in terms of cancer; to follow 500 for 10 to 30 years could yield information of value. The disease for which the drug was given might be such that few would survive the 10 to 30 years needed. Nevertheless, we believe that the time has come to tackle this problem systematically and the existence of schemes whereby drugs prescribed are paid for by the state or an insurance agency provides the administrative framework for cohort assembly. The conflict of opinion as to what constitutes a control group when this problem is approached by the case-control method is well illustrated by the reserpine/breast cancer controversy (24).

Transplacental effects

Medication in pregnancy poses problems of record linkage as effects may appear in both mother and child. The transplacental carcinogenic action of diethylstilbestrol (25) was detected only because the significance of a cluster of a very rare cancer occurring in an unusual age-group was realized by an astute clinician. As there is general believe that foetal tissues are more susceptible to carcinogens than adult tissue, there is a case to be made for the issue, as in South Australia, of prescription books containing details of drugs prescribed and taken during pregnancy. The reported association between influenza during pregnancy and increased cancer risk in the offspring, notably leukaemia (26), could perhaps be due to the use of medication for the viral disease.

If the effect is slight it will be difficult to detect. Thus, although Neute and Buck (27) found a slight excess of cancer in the offspring of 9,000 smoking mothers followed for 7-10 years from birth, the difference was not statistically significant. The later such cancers appear, unless they are distinctive, like adenocarcinoma of the vagina mentioned above, the more difficult it will be to separate them from those due to post-natal exposures.

Migrants

Migrants form a unique resource. Often coming from countries with completely different cancer patterns, their subsequent cancer experience in the host country may provide indicators to the existence and possibly the nature of environmental carcinogens. Thus the pioneer studies of Haenszel and Kurihara (28) showed that the low level of large bowel cancer in Japanese was not genetically ordained since it rose very quickly (in terms of cancer) on residence in the United States of America.

The existence of Japanese-born Japanese and US-born Japanese has permitted study of the factors involved - for colon cancer beef and string beans have been suggested as possible agents (29).

Every effort should be made to identify migrants and their offspring and compare their cancer experience with that of the indigenous population. It is to be regretted that in many countries there are barriers to the identification of such cohorts (30

---

[1]It is not only prescription drugs which may be dangerous. Prolonged abuse of phenacetin - a mild analgesic - has been associated with cancer of the renal pelvis (22)

Comment

The reader will realize that the groups and cohorts described above and considered to be worth active monitoring are essentially univariate. It is possible to allow for the effect of age, sex, and perhaps duration of exposure or employment, but rarely is it possible to examine the effect of other factors such as smoking, dietary habits (including consumption of alcoholic drinks) etc. These will need to be allowed for in subsequent, more detailed studies. Active monitoring of such cohorts can thus only indicate the possibility of an environmental risk and generally other confirmatory studies including the experience of similar cohorts elsewhere (case-control and ad hoc prospective studies) will be needed.

DISCUSSION

Routine passive monitoring, while relatively inexpensive in that the data have already been collected, may not yield many clues to the causes of cancer (Table 3). At best it may point to situations needing to be investigated. When looked at internationally, a rise in a certain cancer believed to be due to a particular exposure becomes less likely if no increase is seen in another country where this exposure has also occurred for a similar period of time.

Active monitoring, dealing in general with more precisely defined groups and exposures, could be expected to yield more specific information (Table 3). The major expense lies not in the creation of such cohorts but in their follow-up and such studies on a routine basis in our opinion make ready linkage with mortality files and preferably a cancer registry system covering the same population virtually mandatory. A cancer registry is like an old master - the longer it has been in existence the more valuable it becomes.

Priority for cohort assembly should, we believe, currently be given to occupational exposures, particularly those involving chemicals shown by animal studies to have carcinogenic activity, as an occupational exposure often precedes and exceeds a more general spread of an agent into the population. *Monitoring everybody for everything is likely to result in nothing.*

Social responsibility

The implementation of monitoring systems discussed above implies acceptance of a series of responsibilities by various groups.

(a) the monitors

Keepers of monitoring systems, active or passive, have the responsibility to ensure that if increased risk is considered to exist in a population as a whole or in some defined segment of that population, then this finding must be brought to the attention of the authorities and, if relevant, both sides of industry in a detached manner. The possibilities of artefact, that the results may be due to statistical chance, to confounding variables or to sampling error must be considered and included in any report. Further investigations necessary should be mentioned and a timetable for the review of further evidence established.

(b) the authorities

The authorities are responsible for establishing mechanisms for independent evaluation of reports emanating from monitoring systems and ensuring that the interests of all sections of the community are taken into consideration. It is the government which has the final responsibility for initiating and policing the necessary ensuing regulations.

(c) industry

The social responsibility of manufacturers lies in their willingness to maintai[n] and <u>keep</u> records of employees, past as well as present, their work stations and levels of exposure and to make these available to investigators. Industry should not use the entirely proper caveats and reservations which investigators, particular[ly] when a risk is demonstrated for the first time, include in their publications as an excuse for inactivity or concealment. Concealment of identified hazard will undoubtedly become rarer if only because the courts are likely to award increasingly punitive damages. Several manufacturers' associations have shown the way by commissioning independent investigation of health in their industries.

Labour has social responsibility for ensuring that workers are adequately protected, recognizing that on occasion protection may require closing of a plant or process. Protection would seem to be a better way of spending funds than payment o[f] "danger money" or compensation to the man who develops an industrial cancer.

(d) the public

The active monitoring of specified groups of the population implies that sets o[f] records will at some point be linked and for such linkage to be effective, each must bear sufficient identifying information to ensure that they pertain to the same ind[i]vidual. Such identification may be made by means of name, date of birth, place of brith, etc., or by numbers, whether allocated by social security or other agencies, or the unique personal numbering system used in several Scandinavian countries. Such links are regarded by some as intolerable invasion of personal privacy, yet published scientific reports based on such links never identify individuals, only t[he] experience of groups. The monitoring of the kind discussed in this paper has noth[ing] to do with the activities of credit-rating agencies and the like. To be protected from cancer risks the public must accept the need for such linkage.

<u>CONCLUSION</u>

Monitoring, active or passive, does not represent a quick road to the identifi[-] cation of environmental carcinogens. If systematically and wisely applied it is likely to point to products or processes which are likely to carry a risk of human cancer - a risk which would have to be confirmed or otherwise by other methods.

SUMMARY

In relation to cancer, monitoring, i.e. "continuing observation in order to decide when changes in incidence or mortality may have occurred" can be of two types:

(a) passive (or routine) monitoring using routinely collected cancer statistics for populations in general

(b) active (or exposure-oriented) monitoring in which the cancer experience of groups with specific exposures is compared with that of the non-exposed.

The advantages and disadvantages of mortality and morbidity data for passive monitoring are listed. It is concluded that passive monitoring can rarely do more than indicate areas for further study.

The potential of active monitoring is examined by consideration of industrial and occupational exposures, the follow-up of persons on prolonged drug therapy (including transplacental effects) and changes in cancer risk following migration. As such studies are often univariate, it may not be possible to allow for the effect of other factors, and confirmatory studies will often be required. It is believed that priority should currently be given to monitoring occupational exposures, particularly those involving chemicals shown by animal studies to have carcinogenic activity. To monitor everybody for everything will result in nothing.

The social responsibilities which devolve on monitors, the authorities, both sides of industry and the general public as a result of the establishment of monitoring systems are discussed. It is concluded that monitoring, active or passive, does not represent a quick road to the identification of environmental carcinogens but can point to products or processes likely to have a cancer risk - a risk which would have to be confirmed or otherwise by other methods.

RESUME

En matière de cancer, la surveillance, c'est-à-dire "l'observation continue afin de déterminer quand des changements d'incidence ou de mortalité peuvent s'être produits" peut être de deux types :

a) surveillance passive (ou systématique) à l'aide des statistiques couramment recueillies sur les populations en général ;

b) surveillance active (ou axée sur l'exposition) où les antécédents cancérologiques de groupes soumis à des expositions particulières sont comparés à ceux de groupes non exposés.

Les avantages et les inconvénients des données de mortalité et de morbidité pour la surveillance passive sont énumérés. Il en est déduit que la surveillance passive peut rarement faire plus qu'indiquer les domaines de futures études.

Les possibilités de la surveillance active sont étudiées, moyennant l'examen d'expositions industrielles et professionnelles, l'étude suivie de sujets soumis à une chimiothérapie prolongée (effets transplacentaires compris) et celle de l'évolution du risque de cancer consécutivement aux migrations. Comme ces études sont souvent à une seule variable, il se peut qu'on ne puisse pas tenir compte des effets d'autres facteurs, et des études confirmatives sont fréquemment nécessaires. On estime qu'il convient actuellement de donner la priorité à la surveillance des expositions professionnelles, particulièrement celles dues à des substances chimiques qui sont apparues cancérogènes dans les études sur l'animal. Surveiller tout le monde pour tout détecter n'aurait pas d'utilité.

Il est procédé à un examen des responsabilités sociales qui incombent aux contrôleurs, aux autorités, aux partenaires dans l'industrie et à la population générale après la création des systèmes de surveillance. La conclusion est que la surveillance, active ou passive, n'offre pas un plus court chemin pour identifier des cancérogènes environnementaux mais peut désigner les produits ou les processus susceptibles de comporter un risque cancérogène - risque qui devra être confirmé ou non par d'autres méthodes.

References

(1) Teppo L. : Monitoring of Cancer. Personal communication, 1975

(2) Burbank F. : Patterns in cancer mortality in the United States: 1950-1967. Nat. Cancer Inst. Monogr. 33, 1971, Government Printing Office, Washington DC

(3) Day N.E. : Personal communication

(4) Percy C., Garfinkel L., Krueger D.E. and Dolman A.B. : Apparent Changes in Cancer Mortality: A study of the effects of the introduction of the Eighth Revision, International Classification of Disease, 1968. Pulb. Hlth Rep., Sept Oct., 1974, Vol. 89, 418-428

(5) International Agency for Research on Cancer. IARC Internal Technical Report 75/001, 1975. Report of a Working Group on Epidemiological Studies on Vinyl Chloride Exposed People, 1975, Lyon

(6) National Board of Health and Welfare : Cancer Incidence in Sweden, 1959-1965. Stockholm, 1971-81

(7) Magnus K. : Incidence of malignant melanoma of the skin in Norway, 1955-70 - Variations in time, space and solar radiation. Cancer, 1973, 32, 1275-86

(8) Office of Population Censuses and Surveys : Registrar General's Decennial Supplement. England and Wales. Occupational Mortality Tables. HMSO London, 1

(9) Guralnick L. : Mortality by Occupation and Industry. Among Men 20-64 Years of Age. United States, 1950 : Vital Statistics - Special Reports, 1962, Vol. 53, No. 2

(10) Guralnick L. : Mortality by Occupation and Cause of Death. Among Men 20-64 Yea of Age. United States, 1950 : Vital Statistics - Special Reports, 1963, Vo. 33 No. 2

(11) Bolander A.M. : Linkage of census and death records to obtain mortality registe for epidemiological studies in Sweden. Proceedings of XIth International Cance Congress, Florence, 1974. Vol. 3 : Cancer Epidemiology, Environmental Factors. P. Bucalossi, U. Veronesi and N. Cascinelli (Eds), 1974, Vol. 33, 743-753

(12) Milham S. : Methods in occupational mortality studies. J. Occup. Med., 1975, 17, 581-585

(13) Pedersen E., Hogetveit A.C. and Andersen A. : Cancer of respiratory organs amor workers at a nickel refinery in Norway. Int. J. Cancer, 1973, 12, 32-41

(14) Fox A.J., Lindars D.C. and Owen R. : A survey of occupational cancer in the rubber and cablemaking industries : results of five-year analysis, 1967-71. Brit Industr. Med., 1974, 31, 140-151

(15) Matanowski G., Seltser R., Sartwell P.E., Diamond E.L. and Elliott E. : The current mortality rates of radiologists and other physician specialists: speci* causes of death. Amer. J. Epidemiol. 1975, 101, 199-210

(16) International Agency for Research on Cancer. IARC Monographs on the Evaluation of Carcinogenic Risk of Chemicals to Man.

Volume 1 : 1972

Volume 2 : Some inorganic and organometallic compounds, 1973

Volume 3 : Certain polycyclic aromatic hydrocarbons and heterocyclic compounds, 1973

Volume 4 : Some aromatic amines, hydrazine and related substances, N-nitroso compounds and miscellaneous alkylating agents, 1974

Volume 5 : Some organochlorine pesticides, 1974

Volume 6 : Sex hormones, 1974

Volume 7 : Some anti-thyroid and related substances, nitrofurans and industrial chemicals, 1974

Volume 8 : Some aromatic azo compounds, 1975

(17) Agthe C. and Wilbourn J.D. : IARC Programme on the Evaluation of Carcinogenic Risk of Chemicals to Man. Proceedings XIth International Cancer Congress, Florence, 1974. Vol. 3 : Cancer Epidemiology, Environmental Factors. P. Bucalossi, U. Veronesi and N. Cascinelli (Eds), 1975, Excerpta Medica, Amsterdam.

(18) Court-Brown A.M. and Doll R. : Leukaemia and aplastic anaemia in patients irradiated for ankylosing spondylitis. Special Report Medical Research Council No. 295, London : HMSO, 1957

(19) Doll R.and Kinlen L. : Immunosurveillance and Cancer : Epidemiological Evidence. Brit. Med. J., 1970, 4, 420-422

(20) Clemmesen J. : Are anticonvulsants oncogenic? Lancet, 1974, I, 705-707

(21) Royal College of General Practitioners : Oral Contraceptives and Health, 1974, Pitman Medical, London

(22) Johansson S., Angervall L., Bengtsson P.U. and Wahlquist L. : Urothelial tumors of the renal pelvis associated with abuse of phenacetin containing analgesics. Cancer (Philad.) 33, 1974, 743-753

(23) Cutler S.J., Schneiderman M.A. and Greenhouse S.W. : Some Statistical Considerations in the Study of Cancer in Industry. Amer. J. Pub. Health, 1954, 44, 1159-66

(24) Anon. : Rauwolfia and Breast Cancer. Lancet, 1975, II, 312-313

(25) Herbst A.L., Uldfelder H. and Poskanzer D.C. : Adenocarcinoma of the vagina. Association of maternal stilbestrol therapy with tumor appearance in young women. New Engl. J. Med., 1971, 284, 878-881

(26) Hakulinen T., Hovi L., Karkinen-Jaaskelinen M., Penttinen K. and Saxén L.: Association between influenza during pregnancy and childhood leukaemia. Brit. Med. J., 1973, 4, 265-267

(27) Neutel C.I. and Buck C. : Effect of Smoking During Pregnancy on the Risk of Cancer in Children. J. Nat. Cancer Inst., 1971, 47, 59-63

(28) Haenszel W. and Kurihara M. : Mortality from cancer and other diseases among Japanese in the United States. J. Nat. Cancer Inst., 1968, 40, 43-69

(29) Haenszel W., Berg J.W., Segi M., Kurihara M. and Locke F.B.: Large Bowel Cancer in Hawaiian Japanese. J. Nat. Cancer Inst., 1973, 51, 1765-1779

(30) Staszewski J., Slomska J., Muir C.S. and Jain D.K. : Sources of Demographic Data on Migrant Groups for Epidemiological Studies of Chronic Diseases. J. Chron. Dis., 1970, 23, 351-373

VII

MEASURING POTENTIAL CARCINOGENS IN THE ENVIRONMENT

MESURE DES CANCÉROGÈNES POTENTIELS DANS L'ENVIRONNEMENT

Chairman / *Président :* Pr. R. TRUHAUT

Environmental Pollution and Carcinogenic Risks
Pollution de l'environnement et risques cancérogènes

ANALYSIS OF ATMOSPHERIC CARCINOGENS AND THEIR COFACTORS

E. SAWICKI (*)

Environmental Research Center
Research Triangle Park, North Carolina 27711, USA

I. Introduction

II. Hydrocarbons
   A. Benzo(a)pyrene (BaP)
      1. Introduction
      2. Sampling
      3. Extraction
      4. Cleanup
      5. Analytical finish
   B. Polynuclear aromatic hydrocarbons (PAH)
      1. Introduction
      2. Sampling
      3. Analysis
   C. Alkanes

III. Aldehydes
   A. Introduction
   B. Analysis

IV. Alkylating Agents

V. Amines

VI. Carcinogenic Vapors
   A. Bis-chloromethyl ether and analogous compounds
      1. Sampling
      2. Analysis
   B. Epoxides
   C. Nitrosamines
      1. Introduction
      2. Sampling
      3. Analysis
   D. Halogenated compounds

VII. NO_x
   A. Nitrogen oxides
   B. Nitrates

VIII. Sulfate and Sulfite
   A. Introduction
   B. Sampling
   C. Analysis

XI. Screening Tests
   A. Analytical
   B. Bioassay procedure

References   page 325

Tables I to VII   page 337

(*) Sampling and Analysis Methods Branch Atmospheric Chemistry and Physics Division Environmental Sciences Research Laboratory U.S. Environmental Protection Agency.

I. Introduction

A fairly large amount of evidence has been given in the first paper of this series (1) indicating that the composition of any mixture suspected of being carcinogenic should be characterized completely in terms of precarcinogens, carcinogens, cocarcinogens, anticarcinogens, mutagens, etc. The composition of the atmosphere was discussed in that paper in terms of gases, vapors and particulates. For many of the constituents concentrations were given in terms of background, average urban, and maximal levels. Especially striking was the relative lack of data on concentrations of the majority of the pollutants in our highly polluted areas. The genotoxic effects (i.e. carcinogenicity, mutagenicity, cocarcinogenicity, precarcinogenicity, etc.) of the various air pollutants were also thoroughly discussed in that paper.

In this paper we will discuss some of the problems in the analysis of a few of the air pollutants which have been postulated as playing important roles in human aging, cancer, mutation and possibly in atherosclerosis. The environmental concentrations of benzo(a)pyrene (BaP) have been used to show the wide prevalence of some of these carcinogens in the human environment. In addition a variety of chemical and bioassay screening tests have been gathered together; these indicate the presence or concentration of some key pollutant or family of pollutants or whether potential carcinogens or mutagens are present. Quick, reliable, non-instrumental indicator tests are needed also for the detection of dangerous one-shot initiating concentrations of key carcinogens and mutagens. With such a quick warning, personnel in the area can leave quickly so as to minimize any physiological damage (readily apparent or apparent many years later after a long latent period). Non-instrumental devices of this type could be carried by an individual or be kept adjacent to a tank car, a key industrial operation, or a storage area. Such a device should be able to give a quick reliable warning of danger due to accident or leakage.

Continuing improvement in the performance of analytical instruments is steadily lowering the limits of detectability. Many instrumental methods are capable of microgram and nanogram limits and occasional methods have demonstrated picogram (10^{-12}g) and femtogram (10^{-15}g) limits. The detection limits in instrumental analysis have been reviewed in several articles (2). It is important to remember that each method has a range of sensitivities that depends on the compound being studied, the mixture being analyzed, and the procedure employed.

Accuracy is a perfection toward which we strive and never seem to attain. Alas all too often the "specific" method with no interferences or problems acquires with the passing of years wrinkles, aches, and pains and eventually it too limps along with the older, less sophisticated methods of yesteryear. Accuracy data reported in the literature are usually based upon the recovery of the added test-standard or the recovery of an added radioactive standard; it is based upon an incomplete knowledge of (i) the composition of the atmosphere in terms of interferences and (ii) the effects of sampling and analysis on the concentration of the test substance. Thus, we have no guarantee that what is reported for the atmosphere is the same as what is actually there. The magnitude of the problem is seen by the fact that in the Intersociety Manual (3) every hand-picked method (supposedly the best of the lot) for the analysis of an air pollutant(s) contains a section on accuracy and precision virtually distinguished by the absence of data on the accuracy of the methods. However, in spite of the shortcomings (known and unknown) of our "best" state of the art methods, the data obtained with the methods can be extremely useful in deciding the necessity or direction of action especially when used in a relative or comparative manner.

Before closing this section I would like to mention multiple factor etiology. Judging from animal experiments and what we know of the human condition the evidence indicates that a large proportion of human cancers are due to a combination of factors rather than a single factor. Consider that exposure of men to asbestos dust and cigarette smoke (4) or to radiation (uranium miners) and cigarette smoke (5) results in an extremely high incidence of lung cancer.

In reinforcement of the concept of multifactor etiology is the sensitivity of the cytochrome P-450 mixed-function oxidase system to air pollutants and other exochemicals. This system has a central role in the body's defense against chemical agents, whether they are normal body constituents or are introduced from the environment. The system is responsible for the detoxification of many of the potentially harmful environmental pollutants, but unfortunately the system can be overwhelmed so that toxic and lethal syntheses can also take place especially when a promutagen or procarcinogen is involved. The system is highly inducible, that is, its activity can be greatly increased by exposure to environmental pollutants and drugs that act as substrates for the system. More than 200 environmental pollutants, carcinogens, mutagens, drugs, steroid hormones, insecticides and other foreign chemicals are now known to stimulate exochemical metabolism in experimental animals; similar effects have been shown in man. A few substances are known which inhibit the mixed-function oxidase system. The P-450 system is also the site of competitive interaction among the various environmental chemicals and their metabolites. The consequences of this inducibility, inhibition and competition have important implications in genotoxicity studies and in the meaningful analysis for the environmental chemicals which could cause cancer or mutation phenomena in man.

II. Hydrocarbons

Of all the families of compounds in the environment, the hydrocarbons are the easiest to characterize and to assay. And of these, the aliphatic (C_1 to C_5), the monocyclic aromatic (C_6 to C_8), and the polynuclear aromatic (4 to 6 ring) have been the most amenable to analysis.

A. Benzo(a)pyrene (BaP)

1. <u>Introduction</u>. Since much of our effort has been concentrated on BaP because of its carcinogenicity, ubiquity and distinctive chromatographic and spectral properties, I will concentrate my efforts on reporting the environmental concentrations of this important pollutant in urban atmospheres, Table I, in various effluents and highly polluted atmosphere, Table II, and in various environmental mixtures, Table III.

In reference to the carcinogenicity of BaP, every member of the PAH family has metabolic and carcinogenic pathways similar to that of BaP's and is metabolized to detoxified materials, toxicants and mutagens similar to those obtained from BaP. Concerning the ubiquitous nature of BaP and other carcinogens it is more than likely that every individual has a lurking chemical memory of some of his (her) contacts with carcinogens covalently attached to the genetic system. If this is so, then synergistic effects of other carcinogenic contacts, the enhancing effects of the various cocarcinogenic contacts, and the inhibitory effects of various anticarcinogenic factors become determining factors in the time of appearance and the malignancy of a tumor.

BaP is considered an indicator substance for the PAH family. Thus, with an increase in some type of pollution, e.g. coal combustion, auto exhaust fumes, coal tar pitch fumes, petroleum pitch volatiles, forest fire, etc., an increase in the concentration of BaP is an indication of an increase in the concentrations of other PAH found in that type of pollution. This does not mean that the same relative proportions of arenes are present no matter what the type of pollution, for the amounts of benzo(ghi)perylene and coronene are greater relative to the amount of BaP in auto exhaust effluent as compared to coal combustion effluents. Similarly, the amounts of the fluoranthenic hydrocarbons compared to the benzpyrenes is very much less in petroleum pitch volatiles than it is in coal tar pitch volatiles while the relative amounts of alkylated PAH to parent PAH is very much greater in petroleum pitch volatiles than in coal tar pitch volatiles.

The little experimental work performed on human beings with BaP is only suggestive of its malignant properties. The application of a 1% solution of BaP in benzene 120 times over a period of 4 months in apparently healthy areas of the body resulted in the development of regressing verrucae (154). Similar changes were found on accidental exposure to BaP (155). A man experimenting with mice developed a persistent nodule diagnosed histologically as squamous epithelioma after a 3 weeks exposure to BaP (156).

The respiratory tract takes up as much as 98% of the benzpyrene fraction of tobacco smoke (157). The penetration of BaP into the organism through the respiratory tract has been proved by the demonstration of the presence of BaP in pulmonary lymph nodes (158).

The background level of BaP in the human organism is in the order of magnitude of 0.5 ug/100 g dry substance (159). Considerable amounts of BaP in the liver, spleen, kidney, heart, and skeletal musculature of newborn babies were detected. This carcinogenic level decreases gradually during growth to reach another maximum due to exogenous BaP in adult age.

The application of BaP alone is half as active as a mixture of carcinogenic and non-carcinogenic hydrocarbons containing an equivalent amount of BaP as found in automobile exhaust gas and cigarette smoke (160). It has been postulated that an increase of 1 ng of BaP per cubic meter can result in a 5% increase in the pulmonary cancer death rate (161).

In the Soviet Union the maximum permissible concentration of BaP was established as 0.1 ng/m^3 in the ambient atmosphere and 150 ng/m^3 in work places (162).

Some studies have been reported on the formation of BaP through pyrolysis. Essentially these precarcinogens can form BaP only by combustion at fairly high temperatures (~600 - 1000°). At lower and higher temperatures the yield of BaP falls off. Pyrolysis of toluene, ethylbenzene, propylbenzene and butylbenzene gave yields of 0.002, 0.065, 0.17, and 0.92% in the tar (163). The pyrolysis of a petrol at high temperatures gave a 37.3% yield of tar with 1.42% naphthalene in the tar (164), while the pyrolysis of anthracene at 950° gave 1.6% BaP, 42% phenanthrene and 25% pyrene in the tar (165). Pyrolysis of tobacco constituents at 700° such as dotriacontane, stigmasterol and aliphatic hydrocarbons gave yields of 0.13, 0.6 and 0.003% BaP in the tar (166). Other tobacco constituents, such as the phytosterols are claimed to be specific precursors for PAH such as BaP, benz(a)anthracene and chrysene (166, 167). The importance of acetylene as a precursor of BaP is shown by the report that its pyrolysis gives a tar containing 2% BaP (168). At 600 - 800° a yield of 0.1% to 0.15% BaP has been obtained through the pyrolysis of isoprene (169). The curing of tobacco at 650° gives up to 800 ng BaP/kg (170). The amount of BaP found among the pyrolysis products of materials, such as agar-agar, glues, humectants, logwoods, natural dyes and starches, varied between 6 ug/kg of starting material for extract of logwood to 470 ug/kg for agar-agar (171). The pyrolysis of dicetyl at 800° gave 0.04% BaP (172). The importance of temperature is shown by the 0.003% and 0.034% yields of BaP from aliphatic hydrocarbons at pyrolysis temperatures of 700° and 800°; respectively (172) and the 0.000014% and 0.00012 - 0.0089% yields of BaP from carbohydrates, amino acids and fatty acids at pyrolysis temperatures of 500° and 700°; respectively (173).

One of the factors working against the formation of PAH during pyrolysis is the presence of nitrates. Tobaccos with elevated alkali nitrate content inhibit the pyrosynthesis of PAH (124).

2. *Sampling*. Of all the methods utilized for the collection of BaP and other PAH from the atmosphere, the filtering methods have been the most thoroughly investigated. And of all the filtering methods the high volume method utilizing glass fiber filters has seen the most thorough study. Other filters such as quartz, cellulose, teflon and cellulose esters have seen some use.

The high volume method utilizing glass fiber filters (174) has been collaboratively tested (175). The relative standard deviation (coefficient of variation) for single analyst variation (repeatability of the method) is 3.0%. The relative standard deviation for interlaboratory variation (reproducability of the method is 3.7%). The minimum amount of particulate matter detectable by this method is 3 mg (95% confidence level). This is equivalent to 1 - 2 ug/m^3 for a 24-hr sample.

A variety of samplers are available to collect particles of various size ranges. The best known are the Anderson size-fractionating cascade impactors. Particles collected on stages 2 through 7 of an Anderson sampler representing aerodynamic sizes from 0.43 to 7.0 um are purported to be deposited in the pulmonary, tracheobronchial and nasopharyngeal compartments of the respiratory system (176). Cascade impactors are subject to error because of particle bounce-off, re-entrainment, wall losses, and particle size changes either due to chemical reaction or moisture content changes. Dichotomous (cutoff at about 2 to 3.5 um) and trichotomous (cutoffs at 1.7 um and 3.5 um) samplers are also available.

Some of the important factors in any of the sampling methods are collection efficiency, practical usability of the filter, lack of contamination on the filter or from the sampler and sampling rate. The sampling rate can vary from ~93 m^3/hr for a high volume filter (400 cm^2 filter), 0.6 m^3/hr for a low volume filter (5 cm^2 filter surface), 1-2 m^3/hr for a large impinger, and ~5 m^3/hr for a typical cascade impactor. An example of contamination from a sampler is the production of small amounts of carbon aerosol by high volume samplers (177).

The comparison of the performance of a membrane-filter personal air sampler with that of a high volume sampler showed good correlation when suitable care was taken in weighing the membrane filters and in compensating for absorbed water (178).

Investigations on the size distribution of BaP - containing particles in urban air has disclosed that more than 75% by weight of BaP is in the "respirable" range in Pittsburgh air, especially in the winter (r<2.3 um) (179), about half of the atmospheric BaP in Budapest is in the range of r<0.15 um (180), the majority of PAH content in Toronto, Ontario is associated with particles below 3.0 um diameter (181), and 85% of the particulate PAH collected in a large Canadian city during the autumn-winter-spring of 1972 - 1973 are "respirable" with 60% of the total associated with submicron particles which could reach the alveoli (182).

It would appear from these studies that there is no real advantage in determining PAH according to particulate size since the greatest part are "respirable." On the other hand, it has been reported that "respirable" dust fractions from coking plants contained somewhat less BaP than the "non-respirable" fractions (183).

One of the main reasons for knowing the size of the particles in which genotoxicants are contained is that the physical pathway of the particle in the body could be estimated.

Reliable data about penetration into the airways of most air pollutants alone or in combination is still lacking. In the case of particulates the pattern of initial deposition in the tracheobronchial tree or other regions of the respiratory tract is dependent upon the particle size, composition, shape, and density; the pattern of air flow; and the physical dimensions and structure of the airways (184). Particles with aerodynamic diameters greater than 10 - 20 um are trapped in the nasal passages when breathed through the nose, and retained in the pharynx and larynx when breathed through the mouth. Particles with diameters of 2 - 10 um are largely trapped in the tracheo-bronchial tree while particles from 2 - 0.1 um generally penetrate deeply into the lungs, depositing in the bronchioles or penetrating into the alveoli. Particles that are deposited in the nasopharyngeal and tracheobronchial regions of the respiratory tract are normally removed to the pharynx, often by cilial action, and swallowed within minutes to hours; those which penetrate into the alveoli may remain from hours to years. The clearance mechanisms resulting in removal of particles from the various compartments of the respiratory system via the circulatory and lymphatic systems and the alimentary canal widen internal bodily contact of the particles and/or their extractable components and metabolites (185).

The importance of particle sizing in the effect of exochemicals on the lung is also shown by the following reports. Following intratracheal instillation of BaP in gelatin - 0.9% NaCl solution suspension (0.3 to 5 um with 50% of the particles in the 0.5 to 1.0 um range) the BaP is deposited and retained primarily in the alveolar region and taken up by alveolar cells as shown by ultraviolet fluorescence spectroscopy (186). In line with this is the report that intratracheal instillations of BaP suspended in 0.9% NaCl and gelatin resulted in a high percentage of malignant tumors, primarily adenocarcinomas, in the peripheral lung (187). These results are in contrast to those using BaP - ferric oxide (large particles, originally with a diameter less than 0.98 μm but clumped together into larger aggregates) wherein the carcinogen is deposited in the larger airways on the mucous ciliary escalator and enters epithelial cells after elution from the ferric oxide particles (188). Using these types of ferric oxide particles as the carrier for BaP resulted in a high incidence of upper airway lesions, i.e. tracheobronchial lesions primarily of the epidermoid carcinoma type (189, 190).

The rate and time of sampling can affect the collection efficiency for the more volatile PAH and even affect the stability of material already collected on a filter (191). When 2 samples were collected from polluted air at 1.2 m^3/min for 2 hrs. on 8 x 10 inch glass fiber sheets and one of these had filtered air drawn through for 2 hrs. at 1.2 m^3/min, the concentrations of the individual tetracyclic, pentacyclic and hexacyclic arenes were approximately the same. On the other hand a sample was collected from the air inside the gas works over a period of 5 hours; analysis immediately and after 3 weeks exposure to the warm air of the gasworks showed unchanged values for the pentacyclics (e.g. BaP analyzed immediately 200 ng/m^3 and after 3 weeks 191 ng/m^3) and decreased values for the tetracyclic arenes (191). An air sample was analyzed immediately and a year later. The year-old sample had been kept in a sealed envelope in the dark. The losses in PAH ranged from 92% for fluoranthene, 88% for pyrene, 32% for BaP, 21% for anthanthrene, 10% for benzo(ghi)perylene and 1% for coronene. Even with pure BaP (20 ug on a paper sheet) only a 25% loss of BaP was obtained after air was drawn through for 3 days (192). Where samples contain carbon and even some pitch, the PAH are much more stable on the filter. Cyclohexane solutions of PAH are not very stable (193). Cyclohexane extracts of airborne particulate samples analyzed immediately and after 6 months storage in the dark showed essentially no change in the concentrations of tetracyclic, pentacyclic and hexacyclic arenes (191).

3. <u>Extraction</u> of PAH from particulates is usually accomplished by a 5 - 12-hour Soxhlet extraction with benzene or cyclohexane at the boiling point. The disadvantages of this method are (i) benzene is a human leukemogen, (ii) during the extraction and evaporation steps some of the polar material decomposes and some polymerizes, (iii) the extraction time is too long, (iv) benzene is a more efficient extractant than cyclohexane, but contains much more extraneous polar material in the extract than does cyclohexane, and (v) the reproducibility of the method is not as good as obtained with some of the more recently developed methods.

Other methods of "extraction" include sublimation at an elevated temperature (194 - 196) or with temperature programming (197 - 198), ultrasonic extraction (199, 200) and polytron extraction.

Homogeneous glass fiber samples (16 cm^2) containing airborne particulates were analyzed by Soxhlet and ultrasonic extraction with cyclohexane (199). Sonification of the mixture at room temperature or below takes about 12 min. The relative standard deviation for 6 ultrasonic extracts was $\pm$ 1.3% and for 4 Soxhlet extracts $\pm$ 26%. An independent investigation of the benzene-soluble extraction method pointed out many of its faults and emphasized its poor reproducibility (201). In this method a personnel sampler containing a glass fiber filter (without organic binder) ahead of a silver filter in a cassette was used. Weight loss of fiber or silver chips from the filter and adipate ester contamination from the Millipore Tenite cassettes were major problems in the gravimetric analytical procedure. In a comparative study of Soxhlet and ultrasonic extraction of atmospheric particulate PAH the optimum condition for the ultrasonic extraction was found to be 30 min in 60 ml of solvent (200). The Soxhlet and ultrasonic extraction procedures were compared for seven different solvents with boiling points ranging from 34 to 81.4°C. For benzo(a)pyrene and aromatic compounds, the ultrasonic method using methanol as solvent proved to be best. Both methods were applicable for ether, dichloromethane, chloroform, and benzene. The ultrasonic method has the advantages of being rapid, requiring low temperatures (0-10°C), and causing no changes in the spectra.

One of the best low boiling solvents for the extraction of polynuclear compounds may be acetone (202).

Sublimation of PAH from airborne particulates at about 250 - 300° has several advantages over the solvent extraction methods. These are:

(i) Potentiality for automatic extraction of many samples over a short period of time.

(ii) Solvents are not necessary. Thus, solvent purification and toxicity are not problems in the method.

(iii) The method discriminates for the hydrocarbons since these compounds are the lowest boiling compounds in the mixture and are somewhat more stable than the polar compounds.

(iv) To ward off decomposition sublimation was done under vacuum or under a small flow of nitrogen (203).

Possible difficulties in the sublimation type of extraction are (i) decomposition of PAH, (ii) decomposition of the polar compounds. The polar residue after sublimation of the hydrocarbons has undergone some decompositions and some condensations, so it may be unusable for further study. (iii) the time involved in the whole extraction procedure is probably much longer than the time utilized in Polytron extraction.

Temperature programming essentially consists of extraction and separation. This is its biggest advantage. The disadvantage is artifact formation, the problem of "synthesizing" compounds which were not present originally in the air.

Polytron extraction has its share of advantages and disadvantages. The extraction is fairly efficient since it involves micromechanical and ultrasonic agitation. The extraction of a large filter (500 cm^2) can be accomplished below room temperature within 5 - 10 min. Cyclohexane or water work well as the extractant. The disadvantages are that a pure solvent has to be used and ultrasonic treatment should be kept to a minimum so as to minimize decomposition. Of course, after extraction the residue can be extracted further with other solvents.

To minimize decomposition the solutions can be concentrated at very low temperatures under freeze drying conditions (204). An evaporated benzene extract of airborne particulates has been found to be stable for 3 years when kept in the refrigerator in the dark (205).

The various solvents used in the extraction of airborne particulates have been enumerated and thoroughly discussed (206).

4. <u>Cleanup</u>. The relationship of cleanup to the other steps in the analysis of BaP and other PAH is shown in Table IV. Liquid-liquid extractions have been used (21,207,209). Essentially they involve extraction of a methanol-water (4:1) solution of the dry organic particulate extract with cyclohexane and the final extraction of the purified PAH fraction in cyclohexane into nitromethane.

Thin layer and column chromatography can be used to separate BaP and the PAH as one entity from other pollutants.

5. <u>Separation</u>. Temperature programming cannot separate BaP from benzo(e)pyrene or perylene. TLC cannot separate BaP completely from other PAH. However, with 2-dimensional alumina - cellulose acetate TLC BaP is easily separated from the other PAH and the other fluorescent pollutants in organic extracts of airborne particulates (42,209).

Manual column chromatography has seen extensive use in the past for the separation of BaP and other PAH obtained from air samples (210 - 215). BaP is usually separated out with benzo(e)pyrene, perylene, benzo(b)fluoranthene and benzo(k) fluoranthene. With poor separation this mixture can even contain benzo(ghi)perylene and anthanthrene. With a long enough column BaP can be separated from the benzofluoranthenes. Spectral techniques (to be discussed) are available to take advantage of these types of separations. Unfortunately this type of separation can take half a day. The fairly good resolution of a longer column is tempered with the longer times necessary for a separation and for a spectral examination.

High performance liquid chromatography has the potential to separate BaP from other PAH within minutes. For example, with a tunable UV detector set at 383 nm, a Lichrosorb column and cyclohexane, BaP was determined in the presence of pyrene, fluoranthene, benz(a)anthracene and benzo(e)pyrene (216). The method is sensitive to 1 ng and has an analysis time of 20 min. The method has potential.

Using a 40% cellulose acetate column BaP is readily separated from other PAH by HPLC (217,218). With columns of Durapak OPN and cellulose acetate a much larger number of PAH are resolved (217). The excitation and emission spectra of PAH separated by HPLC can be recorded during the separation by stopping the elution at peak maxima (219). The main problems of cellulose acetate separation in HPLC is that (i) the cellulose acetate packs down during separation and stops the flow and (ii) cellulose acetate washes off the column.

Gas chromatography is probably the best way to separate the PAH but it presents problems in the complete separation of BaP from other PAH. BaP and benzo(k)fluoranthene are most readily separated by most GC columns. BaP and perylene are difficult to resolve by gas chromatography. Benzo(e)pyrene (BeP) is very difficult to separate from BaP by GLC and is the major interference plaguing most GC methods for the analysis of BaP and the PAH. Although some researchers have reported separating these two by GLC, the resolution is usually unsatisfactory for routine analysis. Some papers have reported very good resolution for the separation of BaP and BeP using sodium chloride-chromosorb columns (220,221). However, other workers attempting to reproduce these results have had difficulties. Excellent separation of BaP from BeP has been obtained with a nematic liquid crystal column (222), but the column bleeds so badly at the recommended elevated temperature that the resolution disappears after a few runs.

These various results indicate that after a proper cleanup BaP can be separated from the other PAH. This means that eventually a simple routine GC method for the analysis of atmospheric BaP should be possible. However, the main difficulty to overcome is column reproducibility over extended periods of time, for at the high temperature required for BaP, and especially PAH analysis, column performance may not be sufficiently constant to guarantee, for extended periods of time, the extremely high resolution needed.

6. <u>Analytical finish</u>. In the past ultraviolet - visible absorption spectrophotometry has been used in the analysis of BaP and other PAH following conventional alumina column chromatography (223,224). Pentane is the ideal solvent in the analysis of BaP and the PAH, especially after column chromatographic separation. It has the following advantages.

(i) It is the solvent of lowest polarity. The PAH show more fine structure in the solvent than they do in cyclohexane, methanol, benzene, etc.

(ii) It is ideal in quantitative analysis because of the sharpness of the bands. Consequently, there is less interference (in pentane as opposed to other solvents) in the analysis of BaP and the other individual PAH from interfering compounds present in this solution.

(iii) With a good separation BaP can be characterized unequivocally by its triplet band near 380 nm. This particular spectrum has not been found in any other compound. As yet, no methyl BaP has been found that has these bands at the exact wavelength maxima shown by BaP. This band system changes to a broader based doublet in cyclohexane and to a singlet with a shoulder in more polar solvents. The characterization of BaP can be further reinforced by evaporation of the pentane and solution of the residue in concentrated sulfuric acid. The fluorescence excitation and emission spectra of BaP in H_2SO_4 are very distinctive.

(iv) The volatility of pentane is an advantage in that the volume of the solution can be adjusted readily in either direction. If further investigation is necessary, the solution can be readily evaporated so that other types of investigations can be easily performed.

The main problem with pentane is its volatility and flammability, but these are problems easily handled by the usual careful work. With a stoppered cuvette evaporation is negligible. Purification of pentane through distillation is a quick, simple process.

The main interference in the UV spectral analysis of BaP is benzo(k)fluoranthene, BkFT. If poor separation of BaP from benzo(ghi)perylene is achieved, this latter compound is a serious interference for its ultraviolet spectra is closely similar to that of BaP, especially in cyclohexane. However, this problem should not occur under the usual separation conditions. Analysis for BaP at 401 nm is not possible if BkFT is present, for this latter compound absorbs very much more strongly at 401 nm than does BaP. Analysis for BaP in pentane at the main band, λs 377, 379, 382 nm, can present a problem if substantial amounts of BkFT are present. But the presence of BkFT is readily ascertained by the relative heights of the two outer triplet bands and the relative height of the 382 nm band to the 401 nm band. The amount of BkFT can be estimated from its absorption at 401 nm or by quenchofluorimetry (225) or by the fluorescence method of Dubois et al (226). Utilizing the base line method with analysis at λ 377, 379, 382 for BaP and at 401 nm for BkFT (corrected for weaker absorption of BaP at this wavelength) a small correction factor is utilized to estimate the correct amount of BaP in the sample.

A more selective and faster method of analysis for BaP is the GC - ultraviolet absorptiometric method which has been used in the analysis of PAH in auto exhaust (227, 228) and coke oven (229, 230) effluents.

Fluorescence spectroscopy has seen much use (231). The problems in this type of analysis are :

(i) Instrumental reproducibility. Whereas a half dozen different ultraviolet visible absorption spectrophotometers can give the same absorbance reading for a standard over 10 years, one spectrophotofluorimeter can give widely varying readings for a standard during one 8-hour day. Thus, standards have to be used continuously.

(ii) Quenching phenomena. If the solutions are too concentrated, spurious excitation maxima would be obtained and the intensity of the emission band would be much less than it should be (232 - 234). This is not a problem in gas phase fluorescence measurements. It is a serious problem in solid phase fluorescence measurements.

(iii) Excimer formation. In solid phase fluorescence measurements if the amount of BaP (or a polynuclear aromatic compound) on a thin layer plate is fairly high, excimer formation can take place as shown by a decrease in the intensity of the usual emission spectral bands and the development of a broad featureless band at longer wavelengths derived from the excimer of BaP (or a polynuclear aromatic compound) (235,236). This phenomenon does not take place in solution nor in the gas phase.

(iv) Sensitized fluorescence. On dry TLC plates energy transfer processes can take place more readily so that sensitized fluorescence becomes possible. Thus, the fluorescence spectrum of anthracene or benz(a)anthracene containing picogram amounts of naphthacene consists of the excitation spectrum of anthracene or benz(a)anthracene and the emission spectrum of naphthacene (235).

(v) Photooxidation on a solid chromatogram or pherogram. Working with fluorescent spots on a thin layer plate the process of photooxidation can be retarded by faster analysis, use of an adsorbent that discourages photooxidation, impregnation of the plate with caffeine (237) or spraying the spots of interest with plastic (238).

(vi) Base-line assay at 3 wavelengths. Where a base-line method of assay (readily accomplished in UV-Vis adsorption spectrophotometry) is used in fluorimetric assay, the instrument has to have excellent stability and the analyst has to know how to cope with possible quenching problems and with the problem of shifts in emission wavelength maxima with changes in the quantity and type of other compounds in the test solution.

The use of thin layer chromatography (239) and fluorescence (231) in air pollution analysis has been reviewed. The advantages of fluorescence analysis in the analysis of BaP and other PAH are the greater sensitivity of the method and sometimes the increase in selectivity as compared to absorption spectral methods of assay. Other advantages have been previously enumerated (231,239).

One of the most specific and most sensitive methods for BaP is through its fluorescence in concentrated sulfuric acid (39, 205, 240, 241). Preliminary separation can be by PC, TLC, GC or liquid-liquid extraction, dependent on the type of sample being analyzed. It has been recommended that in the TLC method alumina-coated aluminum foil sheets be used, the sulfuric acid solvent be purged with nitrogen before fluorimetric measurement and that analysis be at F470/550 (241). The benzo(a)pyrene standard utilized in the procedure should be pure. Air samples and coal tar pitch samples can be analyzed by the TLC procedure. Petroleum pitch volatile samples have interference in them for they give values that are 2 to 3 times too high (216). The liquid-liquid extraction procedure (essentially sulfuric acid extraction of the cyclohexane extract of air particulates) has even more interference problems ; many types of highly polluted samples cannot be used with this non-chromatographic method. This method needs further improvement for it could prove to be a very useful screening test. Probably with further cleanup (e.g. other preliminary liquid-liquid extractions or a fast pass through a microcolumn) the method could be made more selective.

A method utilizing paper chromatography has been described (39). This method consists of column chromatographic cleanup, separation of BaP from the other PAH on acelylated paper, elution of the BaP spot, evaporation, solution of residue in sulfuric acid and fluorimetric analysis.

Low temperature fluorescence methods are also available for the analysis of BaP and other PAH (242, 243). Quasilinear spectra are obtained at liquid nitrogen temperatures. Since much fine structure is obtained the method is capable of unequivocal characterization of BaP.

By combining the high resolution of the spectrofluorimeter with that of the gas chromatograph highly selective methods for BaP are possible. Thus, although it is possible to separate BaP from BkFT with most GC columns these compounds cannot be distinguished by fluorimetry with any great accuracy since their excitation and emission wavelengths are closely similar. On the other hand, while BaP and perylene and BaP and BeP are very difficult to resolve by gas chromatography, BaP can be determined readily in the presence of both BeP and perylene. Utilizing separation by GLC with the optical resolution of the fluorimeter, complicated mixtures should be capable of being analyzed for BaP with some degree of accuracy. Thus, GC separation followed by fluorimetric analysis of a hydrocarbon or sulfuric acid solution or in the gas phase can be used. Even petroleum pitch volatiles could be analyzed in this fashion. A GC - gas phase fluorescence instrument has been developed for such analyses (244) and utilized in the analysis of a variety of mixtures (245). BeP, BkFT and perylene do not interfere in this method. Until such an instrument is perfected its main problem is downtime and the common problem of most GC methods - use of 2-3 ul of test sample and the consequent 1000 - fold loss in sensitivity.

A variety of some of the older methods for the determination of BaP have been compared (209).

Mass spectrometry is another technique that has seen much use lately. It has been interfaced with temperature programming (197, 198), thin-layer chromatography (246) and gas chromatography (247) for the analysis of the PAH. The problem with temperature programming as a separation step is that isomers and other compounds with closely similar sublimation or boiling points would not be separated. For example, compounds like BaP and BeP would not be separated. In addition there would be possibilities of decomposition and artifact formation.

The main advantage of the TLC - MS setup is the high order of sensitivity obtained. About 5×10^{-14} g of BaP can be detected. This integrated ion current method has not yet been applied to analysis of polluted air.

The most successful method for a thorough analysis of atmospheric PAH is the one using the GC-MS-COMP procedure (247). Laboratories utilizing this method could, of course, use this setup for the analysis of BaP, if they were able to separate BaP from BeP and other PAH in the GC step. However, because of the importance of a BaP analysis, a simple screening procedure for BaP would be preferred. The glories and the shortcomings of the GC-MS-COMP approach will be discussed in the PAH section

A very large number of analytical approaches for the determination of BaP (and the PAH) are available in the literature. Many of them are excellent but have not been discussed due to the lack of time. They all involve some combination of the principals shown in Table IV.

B. Polynuclear Aromatic Hydrocarbons (PAH)

1. Introduction. In the first paper of the series we have given genotoxic data showing why air pollution analysts should be interested in a wide variety of hydrocarbons and in other chemical components of the environment besides the primary toxicants. Even the tetracyclic PAH which are usually considered to be non-carcinogenic can cause malignant tumors under appropriate conditions. For example, consider anthracene which is considered to be inactive. When it (pure?) is applied topically as an ointment in conjunction with non-carcinogenic long wavelength UV irradiation, malignant tumors are obtained with high efficiency in mice (248). It is postulated that the tumors arise from covalent binding of anthracene to DNA (249). Anthracene can be bound covalently in vitro to calf thymus DNA by means of long wavelength UV irradiation.

It is most likely that quite a few components of a carcinogenic mixture affect the activity of the mixture. Thus, in the comparison of the coal tar pitches with the asphalts, the PAH content of the coal tar pitches is several orders of magnitude greater than that of the asphalts while the tumorigenic activity of coal tar pitch and asphalt on mouse skin are strong and weak, respectively (64). The fluoranthenic family of hydrocarbons are much more prominent in coal tar and its effluents than in the asphalts and their effluents (64,250), as is BaP in relation to BeP and benz(a)anthracene in relation to chrysene. The coal tar pitches are also relatively poor in alkylated PAH compared to the parent PAH in contrast to the asphalts. However, it should be emphasized that some petroleum pitch volatiles contain up to 0.4% BaP and 5.5% PAH (total of 9 hydrocarbons) (250). Even in incinerator effluents it has been reported that the benzofluoranthenic family of hydrocarbons are about 16 times as high in concentration as are the benzpyrenes in the gaseous effluents and about twice as high in the incinerator residues (251). Interestingly enough about 91% of the emitted PAH were in the residues, 8.7% in the stack gases and less than 0.3% in the water effluent.

The important factors in the analysis of atmospheric PAH are given in Table IV For example, a recent analytical scheme for the separation and analysis of PAH in cigarette smoke condensate (252) contains many of the factors listed in Table IV This method could be modified and applied to air pollution analysis. The procedure consists of separation of the condensate into acids, bases and neutrals by liquid-liquid extraction, cleanup of the PAH fraction by chromatography of the neutral fraction on activated silicic acid, further concentration and cleanup of the PAH fraction by gel filtration, and finally separation of the PAH by gas chromatography.

2. Sampling. The big problem in any analysis for pollutants present in trace amounts in the air is whether the collected chemicals truly represent what's in the air. The questions are how do you collect an accurate sample and how do you prevent interaction in the collected sample, decomposition and artifact formation. Thus, the accuracy of PAH measurement is not known and would be difficult to determine.

Another difficult problem is the sublimation cutoff point for the PAH. Pentacyclic and larger PAH are probably collected quantitatively on the filter; tetracyclic hydrocarbons are partially collected while tricyclic and smaller lower boiling hydrocarbons are usually not collected unless the air has a substantial amount of tarry material. In line with this is the report that tetracyclic arenes are very

much more volatile than the pentacyclics which are much more volatile than benzo(ghi)
perylene (253). Coronene is the least volatile of these compounds as one would expec
However, it has been reported that the tetracyclic arenes can be collected with a 100
collection efficiency by sampling onto glass filters treated with glycerin tricapryla
(254). The method involves the additional step of hydrolyzing the ester before
extraction with cyclohexane.

   3. Analysis. Probably the most promising of the methods of analysis for
the PAH is the GC-MS-COMP method. This is the method that has the most potential
of analyzing for the extremely large number of arenes present in many environmental
mixtures. For example, PAH in tobacco and marijuana smoke condensate (255) and in
airborne particulates (247) have been analyzed by this technique.

   Let us examine the state of the art of this technique, especially as it applies
to air particulates. More than 70 major PAH having from two to seven rings were
separated and identified in an air sample. Samples smaller than 0.1 mg could be
analyzed for nanogram amounts of individual compounds. The potential is fantastic.
However, let's look at some of the problems we face. Of the more than 70 PAH identified in the particulate fraction approximately 26 would be found mainly in the
vapor phase and would go through the filter. Of the 54 new compounds identified by
GC-MS not one was unequivocally identified as to structure primarily due to lack of
standards. The best that could be done with this technique is possible identification of the parent arene ring. But here again we have the isomer problem which
plagues mass spectrometry. The best hope in this case is a preliminary separation
of the isomers. The same problems are found in the separation and analysis of the
150 polynuclear components in tobacco and marijuana smoke condensates (255).

   Another technique which shows promise for the analysis of PAH and other fluorescent pollutants consists of sublimation, gas chromatography and gas phase fluorescence analysis (256). However, the 3 main components of this instrument need further
development before the method can realize its full potential.

   C. Alkanes

   In the first paper of the series the optimum size of the molecule with and with
out aromatic rings necessary for optimum cocarcinogenic activity was discussed. Ana
ysis based on such findings have never been attempted. Some of the available method
of analysis are shown in Table V. Analysis by GC-MS of the ether extract of automobile exhaust fine particulates showed the presence of hundreds of compounds, including about 50% saturated aliphatic, 5% polynuclear aromatic, and 30% oxygenated
hydrocarbons (257). Analysis of the volatilized material from temperature-programme
automobile exhaust particulates by high resolution mass spectra showed that aliphati
and aliphatic substituted benzenes constituted 99% by volume of the total volatile
organic matter associated with the particulates (258). Bioassay tests are needed to
test for the cocarcinogenic and mutagenic activity of these various aliphatic fracti
The main problem here is to determine the genotoxic activity of the types of aliphat
fractions for which we have assay methods. Once the activity of these materials is
known, then the analytical methods can be refined, if necessary.

III. ALDEHYDES

A. Introduction

There are ten families of aldehydes of interest to researchers in the field of genotoxicity. These include (i) the gaseous aldehyde, formaldehyde, (ii) low molecular weight aldehydes present in the gaseous state in the atmosphere, e.g. acetaldehyde, propionaldehyde, acrolein, crotonaldehyde, acetaldol, etc. (iii) the larger unsaturated aldehydes, e.g. 3, 4, 5 - trimethoxycinnamaldehyde, some of which are present in sawdust and other wood dusts, mixtures playing important roles in human nasopharyngeal tumorigenesis, (iv) epoxyaldehydes, the only well-known representative being carcinogenic glycidaldehyde, (v) PAH aldehydes formed in the metabolism of PAH and methylated PAH, (vi) nitrosamine aldehydes, postulated, ulticarcinogens formed metabolically from the nitrosamines, (vii) by-product endogenous aldehydes formed in the metabolism of procarcinogens, e.g. malonaldehyde following skinpainting with BaP, (viii) endogenously formed aldehydes following irradiation of tissue, e.g. cholesterol-α-oxide from tissue cholesterol, (ix) nucleic acid, lipid, carbohydrate and protein (amino acid) aldehydes formed during histochemical tests, and (x) DNA aldehydes derived from depurination and possibly depyrimidination processes resulting from contact of cellular DNA with alkylating and other destructive agents. The physiological properties of these various aldehydes have not as yet been adequately investigated.

Many of the precursors of these aldehydes may be just as important as the aldehydes from at least two aspects. The precursor, unlike the reactive aldehyde, may be able to reach a 'hot spot' in the genetic apparatus, be transformed to the aldehyde and cause a mutagenic effect. In addition some of the precursors could form the aldehyde during the analysis and thus give a spurious value for aldehyde concentration.

B. Analysis

The first four families of aldehydes will be discussed from the aspect of atmospheric sampling and analysis. The present system of analysis for the aldehydes is inadequate. Either too many procedures and instruments are used or methodology is not available. Present state of the art methodology includes impinger collection followed by colorimetric (259-267) or fluorimetric (268) analysis for formaldehyde, colorimetric analysis for acrolein (269), GC analysis for the C_2-C_5 aldehydes (270) and colorimetric analysis for total aliphatic aldehydes (271).

The problem with the chromotropic acid method (247, 259-261) is that although other aldehydes do not give positive results, formaldehyde precursors can be hydrolyzed or oxidized to formaldehyde (272) thus causing spurious results. The formaldehyde values obtained in the atmosphere and in some atmospheric pollution sources are probably too high because of this phenomenon.

The 2,4-pentanedione colorimetric procedure is more selective for formaldehyde since it is done under mild conditions and no other aldehydes react in the procedure. Of the many compounds studied as possible interferences in the method NO_2 gives the maximum interference; 530 μg/m^3 gives a 2.86% interference (264). The 2,4-pentanedione fluorimetric procedure has the potential to be much more sensitive for atmospheric formaldehyde than the other formaldhyde methods.

The present system of analysis for the aldehydes of polluted atmospheres is inadequate. This is too bad because many of these aldehydes are mutagenic and/or carcinogenic and many aldehydes play very important roles in the physiological activity of living tissue. One possible method of polypollutant analysis of the aldehydes would involve collection and stabilization on a solid device and then analysis by HPLC, GC, or GC-MS-COMP.

IV. ALKYLATING AGENTS

These types of compounds are one of a family of electrophilic agents. All ulticarcinogens appear to be electrophilic agents and somatic mutagens. Thus, an alkylating agent is essentially a raw somatic mutagen or a possible direct carcinogen. Such a molecule does not need metabolism for activation. If it can survive competing reactions with the various cell constituents of less vital importance in cancer initiation and alkylate DNA, then the first step in potential carcinogenesis has been initiated. It has been emphasized that direct alkylating agents have a high likelihood of altering intercelluar DNA and producing mutations, with such interactions the initiating step in the induction of cancer (273). However, it has been emphasized that living tissue is potentially immortal, and that the mere existence of the immortality of the germ cell line from generation to generation indicates that the problems imposed by recurrent mutation, error-catostrophe, simple wear-and-tear processes and so on, were solved at an early stage in the origin of life on Earth (274).

Most alkylating agents are potentially direct carcinogens. They do not need to be metabolized to be activated. However, some of them may be so reactive that before they can get into a cell to react with DNA, they are used up in reactions with exocellular chemicals. Over 200 compounds have been tested with 4-(p nitrobenzyl) pyridine (275). The following types of alkylating reagents reacted: alkyl halides, esters of strong acids, lactones and sultones, epoxides, aziridines, diazo compounds, nitrosamides, geminal-disubstituted compounds, onium compounds, activated double bonds, azoalkanes and hydrazones, and 1-aryl-monoalkyl- and -dialkyltriazenes. The quantitative reactivity of 90 alkylating agents with 4-(p-nitrob enzyl) pyridine has been reported (276). In that paper, many methods of analysis for the alkylating agents have been described. A variety of alkylating agents giving positive results in the 4-p-nitrobenzylpyridine test have been shown to be carcinogenic (277-279). e.g., methyl iodide, benzyl chloride, dimethyl sulfate, diethyl sulfate, methyl methanesulfonate, methyl p-toluenesulfonate and trimethylene oxide produced local sarcomas near the point of injection (277). Inhalation of 3 ppm dimethyl sulfate for one hour five times per week produced squamous carcinomas of the nasal cavity or neurogenic tumors in 8 out of 27 exposed rats.

Many other reagents have been described which can be used in screening tests for environmental alkylating agents (276). Some of them are 4-pyridinecarboxaldehyde 4-nitrophenylhydrazone, 4-pyridinecarboxaldehyde 2-benzothiazolyl hydrazone, 4-acetylpyridine 4-nitrophenylhydrazone and 4-acetylpyridine 2-benzothiazolylhydrazone (279a), 4-picoline and o-dinitrobenzene (279b) and pyridoxal 4-nitrophenylhydrazone. Unfortunately, these screening tests have seen little use in screening environmental mixtures for potential problem pollutants. Tests of the aliphatic fractions of particulate samples obtained from submarine air, auto exhaust fumes, and urban atmospheres gave positive results (279a). Other mixtures giving positive results include aromatic, acidic, and neutral oxygenated fractions obtained from urban airborne particulates, a neutral fraction from cigarette smoke tar and a composite benzene-soluble fraction.

The importance of the alkylating agents and their precursors in mutagenesis and carcinogenesis has been discussed (280). 4-p-Nitrobenzylpyridine has been used more than any other reagent in the analysis of alkylating agents in biological fluids (281-284). Methylation, ethylation, propylation and butylation products of this reagent can be separated by TLC (285). Arene oxides which are potential ulticarcinogens react readily with p-nitrobenzylpyridine (286-290) as do aziridines (291).

The methods have seen some use in the detection of alkylating agents in tobacco smoke with 4-p-nitrobenzylpyridine (292,293) and in fumes from auto exhaust and the limited aeration smoldering of plastics, paper and cellulose at temperatures ranging between 300 and 500° (294). In the latter case 4-p-nitro-benzylpyridine and 4-pyridine carboxaldehyde 2-benzothiazolylhydrazone were used as reagents.

It should be emphasized that one should not expect alkylating activity of a compound to correlate with its carcinogenic activity since there are others factors involved in carcinogenicity, e.g., metabolism, reactions with other components of living tissue, etc.

V. AMINES

Although sampling and analytical methods for amines are available, they have never been adequately evaluated. Sampling can be on silica gel (295). Two problems in long term storage are migration of the sample to the backup section and oxidation of the sample. Amines can be analyzed directly by GC with some difficulty or they can be perfluoracylated and then assayed by GC. Sampling is the main problem. Analysis of atmospheric amines has had low priority in the past so that thoroughly tested methods for the various types of amines have not yet been developed.

The importance of the amine stems from the fact that some of the anilines are carcinogenic (296), many of the polynuclear aromatic amines are carcinogenic and many of these -- in addition to the primary, secondary, tertiary and some quarternary aliphatic amines -- are precarcinogens. Thus, the primary amines could react in vivo with nitrous acid to form the potentially genotoxic diazonium salts, stability ranging from fairly stable for many of the polynuclear aromatic digzonium salts to highly unstable for the alkyl diazonium salts. The aliphatic secondary and tertiary amines could react with nitrous acid to form nitrosamines in vivo and probably in highly polluted atmospheres. This latter type of reaction has been reported for atmospheres containing fairly high concentrations of dimethylamine and moderate amounts of nitrogen dioxide (297). Peak amounts of dimethylnitrosamine were found when concentrations of dimethylamine and NO_2 were 0.1 ppm and 45 $\mu g/m^3$, respectively. These concentrations of amines and nitrogen dioxide would be exceeded in highly polluted atmosphere near some highly industrialized areas. The reaction between atmospheric NO, NO_2 and a secondary amine to form a nitrosamine was postulated in 1969 (276) and probably even before that date. Saturated imino heterocyclic compounds and arylalkylamines could also be precarcinogens. Thus, phenylmethylamine in the presence of NO_x could form phenylmethylnitrosamine whose ulticarcinogen is probably the phenyldiazonium salt with formaldehyde as a possibly genotoxic by-product.

The biggest problem we face in this area is the need for reliable methods of sampling and analysis for the atmospheric amines and nitrosamines. Dependent on the situation, derivative formation and gas chromatography, HPLC, fluorimetry and colometry could be used.

VI. CARCINOGENIC VAPORS

There are a fairly large number of carcinogens which in pure bulk are liquid or solids but are gaseous when present in the atmosphere in trace amounts. We will discuss a few types.

A. Bis-chloromethyl ether and analogous compounds

1. *Sampling* has usually been on a solid device, e.g., Tenax GC (298), Porapak Q (299-301) and Chemosorb 101 (302). A method of sampling is desired where (i) the collection and desorption efficiencies are quantitative, (ii) atmospheric humidity does not affect the results, (iii) repeatable use of the sampling cartridge is possible (iv) breakthrough volumes are relatively high and are not affected by humidity, (v) the sampled material is stable for at least a week in the cartridge, (vi) the sampling adsorbent is free of contamination and does not decompose to interfering byproducts during storage and desorption, and (vii) artifact formation from atmospheric precursors does not take place.

Tenax GC has been selected by Pellizzari et al as the best available sorbent for general use (298). It has been thoroughly studied in the laboratory and the field and has been applied to the sampling of carcinogenic vapors in the atmosphere. It has a high operating temperature in that even up to 350° C very little background bleeding is observed. Another attractive feature is that it has a low retentive index for water. Too great a retention of water as in Chromosorb 104 might complicate recovery and analysis of adsorbed compounds by GS-MS. The effects of transportation and storage of bis-chlormethyl ether were investigated (298). Immediate analysis gave 100% recovery; following transportation of a cartridge from Research Triangle Park, N.C. to San Francisco and back (total transportation time of 6 days) the recovery dropped to $58 \pm 5\%$ after 1 week and $41 \pm 4\%$ after 2 weeks. Without transportation the recovery was $65 \pm 5\%$ in 1 week.

An alternative method of sampling (wherein chloromethyl methyl ether and bis-chloromethyl ether can be collected) consists of collection in glass impingers containing a methanolic solution of the sodium salt of 2,4,6-trichlorophenol (303). Reaction of the chloromethyl ethers with phenol forms stable derivatives with significantly increased detector sensitivity, the chloromethyl methyl ether and the bis-chloromethyl ether being detected at the 3.3 and 4.7 $\mu g/m^3$ levels, respectively. Two impingers have to be used in series to trap the compounds in approximately 90% yield.

2. *Analysis*. Since bis-chloromethyl ether is a human carcinogen and causes lung cancer in rats on exposure to levels of 100 ppb in air for periods of several months (304), the presence of this carcinogen in the atmosphere is of some concern. The data for Frankel et al (305) has established an order of magnitude for bis-chloromethyl ether formation under non-catalytic conditions; in one of a series of experiments 0.4% formaldehyde plus 4% HCl, in the presence of water vapor, yielded 5 ppm of bis chloromethyl ether. A possible source of bis-chloromethyl ether was found in the thermosetting of acrylic polymers (306).

In the high resolution mass spectrometric procedure the collected organic material is desorbed into the reservoir of the mass spectrometer heated inlet system and then examined at high resolution for the presence of the ion m/e 78.9950 ($C_2H_4OCl^+$) which is the most intense ion in the bis-chloromethyl ether mass spectrum (301). One part per billion of BCME can be determined. The difficulties with the method are (i) chloromethyl methyl ether gives the same ion, (ii) control of adsorption and regeneration effects, and (iii) possible decomposition of BCME in the reservoir and its associated valves.

Greater specificity and high sensitivity of detection can be achieved using GC-MS. In one method after collection, desorption and GC separation, MS monitoring at m/e 79 and its chlorine isotope peak at 81 is utilized (302). Positive detection of BCME at the 1 ppb level is dependent upon there being simultaneous response of m/e 79 and m/e 81, in the 3:1 ratio needed for a chlorine isotope pair, at the retention time for BCME. A difficulty would arise if the ratio of m/e 79 to m/e 81 deviated markedly from the expected 3:1 due to the presence of an interferent at the m/e 79 and/or m/e 81 regions.

Given clean conditions, a detection limit of less than 10^{-2} ppb can be reached using single ion detection with the mass fragment of mass number 79 ($ClCH_2OCH_2^+$) (300). However, better selectivity is achieved with multiple ion detection of the four characteristic fragments with mass numbers 49, 51, 79 and 81. With an electron energy of 70 ev their intensity ratio is 9:3:18:6. With this system of analysis there is a loss of sensitivity of about 10. If all 4 fragmentograms display peaks at the retention time of BCME, but the intensity ratios do not conform to the fragmentation pattern, the relatively lowest peak can be taken as an upper limit of the BCME concentration (300).

Greater specificity can be obtained by monitoring the base ion of BCME, m/e 78.9950 ($ClCH_2OCH_2^+$), exclusively at high mass spectra resolution during the chromatogram. Other possible interferences, such as the ions C_6H_7 (79.0547µ), $^{13}CC_5H_6$ (79.0503), $^{13}C_2C_4H_5$ (79.0458), ^{79}Br (78.91839), $C_2HF^{35}Cl$ (78.9749), and especially C_4H_3Si (79.0047µ) would not be a problem due to the GC step. The size of the hydrocarbon background peaks m/e 79.0458 - 79.0547 would vary considerably depending upon the operating state of cleanliness of the spectrometer (299). The lower limit of the method is placed at 0.01 ppb when a 10 l. air sample is taken.

In the impinger method (303) after a known volume of air is drawn through glass impingers containing a methanolic solution of the sodium salt of 2,4,6-trichlorophenol, the sample is heated on a steam bath for 5 min., cooled and diluted with an equal volume of distilled water and hexane for extraction. Stable derivatives are produced with BCME and chloromethyl methyl ether. An aliquot of the hexane is analyzed by electron capture gas chromatography.. The sensitivity of the method is 0.5 ppb (v/v) when a 10 liter air sample is used. Interferences can be expected from highly halogenated organic compounds or compounds that could produce the same derivatives. The 2,4,6-trichlorophenol should be pure. The disadvantages of the method are the handling of liquids, extractions and the dilution of samples by the hexane extraction. Good sensitivity could also be obtained by using phenol as the reagent, the entire 2 ml of hexane test solution and HPLC with a UV detector. A much more highly sensitive method could be developed by using a fluorogen-forming reagent, HPLC and a fluorescence detector.

B. <u>Epoxides</u>

Very little work has been done on the characterization and determination of atmospheric epoxides. They are present but outside of that fact very little is known about these air pollutants. Large numbers of olefins are present in the atmosphere which could under proper conditions be oxidized to alkene oxides. The importance of these latter compounds stems from the fact that some of them could be carcinogenic while many of the others could be stable and resistant to enzymic degradation and thus in some cases could be potent epoxide hydrolase inhibitors. In this latter sense they could be important carcinogens.

Arene oxides (286) and alkene oxides (275) are active alkylating agents towards p-nitrobenzylpyridine, a widely used model acceptor for alkylating agents (307).

Some of the epoxides which have been tentatively identified in air samples collected on Tenax GC include ethylene oxide, propylene oxide and styrene oxide. Ethylene oxide and propylene oxide react with guanosine at the N-7 position (308) while ethylene oxide reacts with adenine nucleotides predominantly at the N-1 position (309). Propylene oxide is also known to bind covalently with DNA (310). Styrene oxide has been shown to be carcinogenic. These epoxides were collected on Tenax GC and then characterized by mass spectrometry.

Further study is needed, for the identification needs more certain confirmation, a method for quantitative analysis needs to be developed and other atmospheric epoxides need to be identified. One would expect many more of these compounds in polluted atmospheres because of the large number and quantity of unsaturated hydrocarbons emitted daily into the atmosphere.

C. Nitrosamines

1. Introduction. Environmental relevance is fundamentally an expression of priorities or an ordering of rank. Where a "firefighting" situation involving emergency health problems arises, a new priority is established. If this involves a new pollutant, a highly expensive monopollutant monitor cannot be used. Instrumental techniques involving a greater freedom of movement, a very broad potential for polypollutant analysis and an ongoing analytical setup are needed for such firefighting situations. Thus, when problems arise due to the presence of genotoxic materials in the atmosphere, the fast strike capability of rapid sampling followed by rapid characterization and assay by an ongoing polypollutant analytical setup is ideal for the emergency problem. However, this type of setup must be thoroughly evaluated ahead of time. The cruel fact is that the generation of unevaluated and artifact-influenced numbers and of repetitive data which is not needed and not used raises the expense of environmental protection without contributing significantly to our knowledge of our polluted environment.

The nitrosamines and nitrosamides are of special importance for the following reasons: (i) hundreds of them have been shown to be carcinogenic, (ii) dependent on structure they have a wide-ranging organotropic carcinogenic effect in rats as demonstrated with 67 different nitrosamines (311); (iii) their precursors and the catalytic agents accelerating the lethal synthesis of the nitrosamine are in our environments and to a varied extent in human living tissue, (iv) Nitrosamines and their precursors have been reported in the atmosphere, and (v) Quite a few hypotheses have been postulated indicating a relationship between minute amounts of nitrosamines in foods or beverages or large amounts of nitrosamine precursor, nitrate, in water and human cancer. Much data has been given in a recent monograph (312).

Dimethylnitrosamine concentrations in the range of 0.43 -- 0.001 ppb range were found in atmospheres containing dimethylamine and NO_2 (297). In Baltimore, Maryland on August 23, 1975, concentrations of 0.96 (GC) and 0.84 (HPLC) ppb of dimethylnitrosamine were reported by Fine et al (313).

2. Sampling. Adsorbents used to collect dimethylnitrosamine include activitated charcoal, type AS (297) and Tenax GC (298). An alternative method is to sample in a cold trap at -78°C (313); before analysis the sample is thawed and extracted with methylene chloride.

The main problems in sampling are decomposition of the nitrosamines and artifact formation. Thus, it has been shown in the collection of cigarette smoke condensate that under the conditions of cold or liquid trapping nitrosamines are produced in the trapped liquid (314).

The lack of artifact formation has to be conclusively demonstrated in any sampling method. Nitrosamines could not be detected in activated charcoal tubes containing 10 μl of 40% dimethylamine through which 10 m^3 of air containing 28 μg NO_2/m^3 had been passed.

3. <u>Analyses</u>. Fine and coworkers (315-318) have developed a highly selective system of analysis utilizing gas chromatography for separation, catalytic pyrolysis of the nitrosamine to the nitrosyl radical and detection of the NO following reaction with ozone by a chemiluminescence detector. Instead of gas chromatography HPLC can be used (319). With this latter system both volatile and non-volatile nitrosamines can be determined. Compounds which give positive results with the detector and thus could be interferences include alkyl nitrites, metallic nitrites, 2, 2', 4, 4', 6, 6'-hexanitrodiphenylamine, alkali nitrates and nitric acid (316). Other possible interferences could be alkyl nitrates, nitrous acid and peroxyacetyl nitrate. Any others compounds that react with ozone to give luminescence can be eliminated by interposing a -154°C cold trap between the catalytic pyrolyzer and the ozone reaction chamber. In the case of the NO precursors positive results could be obtained that could be due to a non-nitrosamine, e.g., a nitrite or a nitrate.

A wide range of analytical techniques have been used in the past for the determination of the nitrosamines, e.g., TLC, GLC, colorimetry, fluorimetry, polarography, all in a wide range of variations. The methods have been criticized and several recent studies indicate that many, if not all, of these methods may lack the necessary specificity and/or sensitivity to give reliable results (320-324).

The method of Bretschneider and Matz (297) utilizes GC for the analytical finish. Past experiences with this type of method have shown that care must be used in the routine assay because of the as yet unknown interferences.

In the GC-MS method for low levels of nitrosamines it is possible to detect 25 ng of a nitrosamine (325). A high resolution mass spectrometer should be used. A compound eluting from a gas chromatograph is identified as a nitrosamine if it has the correct retention time and possesses the characteristic ion in its spectrum. High resolution mass spectrometry monitoring for NO^+ ions will detect a minimum of 0.1 - 0.2 μg of individual nitrosamines (320). Five times better sensitivity can be obtained for dimethylnitrosamine by recording only the mass corresponding to the molecular ion. The main interferences for NO^+ (29.9980) are $C^{18}O$ (29.9992) and $^{15}N_2$ (30.0002) and for the molecular ion (74.0480) the ^{13}C isotope of the trimethylsilyl ion (74.0502) and the ^{29}Si isotope (74.0469) (326). These latter ions could arise from contamination from silanized GC columns or GC-MS interfacing lines. Dimethylnitrosamine can be characterized by the ions with m/e (relative intensity) : 30(52.4), 42(100), 43(51.2) and 74(87.8).

D. Halogenated Compounds

It is unfortunate that only a little time and effort can be spent on these important groups of compounds. Some of the families of compounds include alkylhalides (e.g. methyl iodide), halogenated methanes (e.g. CCl_4 and $CHCl_3$), Freons, halogenated alkanes, halogenated olefins (e.g. vinyl chloride), halogenated pesticides (e.g. DDT), polychloro polynuclear compounds (e.g. PCBs), etc. Many of these compounds are carcinogenic and many can affect carcinogenesis in various ways.

Systems of polypollutant analyses for these various families of compound need to be developed. The various available sampling methods and a GC-MS-COMP separation -- analytical finish need to be quantitated, evaluated and applied.

Using Tenax GC cartridges for collection, thermal desorption, capillary gas-liquid chromatography coupled to a mass spectrometer with an on-line computer recording data on magnetic tape and generating normalized mass spectra and mass fragmentograms, 21 halogenated hydrocarbons have been detected, including the carcinogens vinyl chloride, chloroform, carbon tetrachloride and trichloroethylene, and numerous oxygen, sulfur, nitrogen, and silicon compounds (327). Much further work needs to be done to quantitate this procedure.

Some of the halogenated pollutants are present in both the vapor and particulate forms, e.g. some pesticides and polyhalopolynuclear compounds; the systems of sampling that are available would have to be evaluated on that basis.

Sampling and analysis are simpler when only one toxic pollutant is being determined under the mistaken impression that other pollutants play no role in the genotoxicity of the main toxicant. Even then, there are many problems as shown in the determination of atmospheric vinyl chloride by an unevaluated procedure (327a).

VII. NO_x

The important pollutants of this type include N_2O, NO, NO_2, nitrous acid, nitric acid inorganic nitrates, alkyl nitrites, inorganic nitrates, alkyl nitrates, inorganic nitrates, aryl nitrates and PAN and analoguus compounds. In addition, nitroalkanes, nitroarenes, nitroamines, nitrosamines, nitrosoalkanes and nitrosoarenes could also exist in some polluted atmospheres. To keep the discussion within bounds we will briefly discuss NO_2 and inorganic nitrates only.

A. Nitrogen oxides

In the previous paper the presence of substantial amounts of nitrous acid in polluted atmospheres was discussed. The evidence for the presence of this compound in air was, at most, meager. Confirmation through infrared spectroscopy and other means is needed before the analytical methdology for the compound and the problems its presence would cause in the analysis of NO_2 could be taken seriously. If nitrous acid is present in some polluted atmospheres, then all methods for NO_2 would have to be modified.

We will briefly discuss the NO_2 methods since NO_2 is an oxidizing agent and is the main nitrosamine precursor in the air (assuming substantial amounts of nitrous acid are not present in heavily polluted atmospheres).

The main problems with the sodium hydroxide collection method for NO_2 are (i) the collection efficiency varies drastically with the concentration and (ii) NO, CO and CO_2 are interferences as are formaldhyde plus NO and phenols plus NO.

The manual colorimetric Griess-Saltzman method is acceptable under carefully controlled conditions; it is not recommended for more than 1 hr of collection time. Storage of samples is not recommended if substantial amounts (levels exceeding NO_2 concentration level) of strong oxidizing or reducing agents are present. Ozone and peroxyacyl nitrates interfere only if present in large amounts. High ratios of sulfur dioxide to nitrogen dioxide (30:1) can cause bleaching of the color and result in low values of NO_2. This method cannot be used in 24-hour integrated sampling procedures.

Automated versions or modifications of the Griess-Saltzman reaction have been used for about 20 years. The static calibration procedure for the method is not reliable since NO_2 concentrations can be misestimated by as much as 20%. Another problem is the breakdown of the reagent in light to yellow compounds; this coloration could have significant consequences at low NO_2 concentrations. The reliability of these analyzers below 0.1 ppm has been questioned (328). Significant operational problems are the control of liquid flow preparation and replenishment of reagents and contact of gas and liquid. The method does suffer negative interference from ozone (329). Dynamic calibration is required in the method. Collaborative testing has shown that the method has a significant positive bias. Comparison of this method with the manual TGS-ANSA, automated arsenite and chemiluminescence methods showed that, on the average, the arsenite and the TGS-ANSA methods gave the same NO_2 values while the chemiluminescence method values were slightly lower (330). The automated Griess-Saltzman methods were subject to the greatest interference giving significanlty lower readings at the 353 and especially at the 667 $\mu g/m^3$ level. The chemiluminescence, arsenite and TGS-ANSA methods did not seem to be affected by changes in the O_3 level.

The lower detectable limit in the automated Griess-Saltzman method are approximately $19 \mu g/m^3$.

The collection efficiency in the sodium hydroxide - sodium arsenite procedure is approximately 82% over a recommended range of use of 20 to 75° $\mu g/m^3$. Nitric oxide is a positive interferent while carbon dioxide is a negative interferent. Over the range of 50 to 310 $\mu g/m^3$ NO and 200 to 500 ppm CO_2 the average effect of these 2 interferents is to increase the NO_2 response by only 10 $\mu g/m^3$ over the range 50 to 250 $\mu g/m^3$ NO_2 (331). One of the problems in the method is that arsenite may be a human carcinogen. Other possible problems are that NO in the presence of formaldehyde and/or phenols could be a greater interference than NO by itself.

The TGS procedure has a collection efficiency of 92% over the recommended range of 20 to 700 $\mu g/m^3$ and does not suffer interferences from nitric oxide, nitrates, nitric acid, sulfur dioxide, ozone, ammonia, carbon monoxide, formaldehyde and phenol (332, 333). It has given a satisfactory collaborative test. The problem with this 24 hr method is that it involves impinger sampling and is a manual procedure. Although the collecting solution is stable for about 3 weeks and is biodegradable, it is not as stable as the sodium hydroxide -- sodium arsenite sampling solution. The lower detectable limit of the TGS method is approximately $15/\mu g/m^3$.

The continuous chemiluminescent method for NO (334) is a subtractive method, e.g. $NO + NO_2 - NO = NO_2$. The NO_2 is measured (with NO after reduction to NO) by the light emitted on its reaction with ozone. Any compounds which will be converted to NO in the thermal converter will interfere with the measurement of the NO_x concentration. The interference caused by nitric acid, while showing considerable variation, is often close to quantitative. Ethyl nitrate at low concentrations gives a 100% response. Peroxyacetyl nitrate (PAN) also interferes. However, where alkyl nitrate, alkyl nitrite, nitric acid and PAN levels are relatively low the interference would be negligligible. In areas like St. Louis or Los Angeles where PAN and HNO_3 concentrations are high during the afternoon hours, up to one-half of the observed NO_2 might actually be PAN and HNO_3 on the average. However concentrations of NO_2 do not fluctuate in the same manner as do PAN and HNO_3, so that the interference may not be too large. The variability in the HNO_3 response factor could only tend to decrease the potential interference. Common air pollutants such as O_3, NO_2, CO, NH_3 and SO_2 do not interfere in the chemiluminescent detection of NO with O_3. The lower detectable limit is 22 μ g/m^3 for a one-hour sampling period.

Personal sampling methods are also available for the determination of NO_2, e.g. sampling into triethanolamine on a suitable support, desorption and colorimetric estimation (335). The various methods have not as yet been satisfactorily evaluated.

B. <u>Nitrates</u>

Reliable sampling and analytical methods are needed for the analysis of atmospheric nitrate. The usual sampling method for nitrate is by high volume filtration through a glass fiber filter. A huge variety of analytical methods are available; they include atomic absorption, colorimetry, fluorimetry, polarograph ion selective electrodes, chemiluminescence, etc. Other methods have been developed based on reduction of nitrate to NO, NO_2, or NH_3. The analytical methods are not completely satisfactory or have not as yet been evaluated thoroughly.

Consider the potential probelms in sampling e.g, the possible synthesis of artifact nitrate on the filters from atmospheric NO and NO_2, the trapping and neutralization of gaseous nitric acid, the formation of inorganic nitrate from PAN degradation products, the reaction of nitrate precursors with the filter material or with the collected particulate matter, the fixation of collected nitrate by substances in and on the filters so that it cannot be recovered quantitavely, loss of nitrate by reduction, formation of stable nitrite compounds on the filter, nitration of material on the filters, etc. In addition, the various variables would have to be investigated, e.g. meteorological factors, sampling conditions and filter composition.

Space and time are not available to discuss the many problems of the nitrate analytical methods. Suffice it to say that new possibilities are being looked into. For example, nitrate in the presence of sulfuric acid could nitrate benzene, thiophene etc. The nitrobenzene or nitrothiophene formed in this way could be determined with electron capture GC or with HPLC. Nitrate could be thermally decomposed to NO_x which would then be determined with a chemiluminescent monitor. However, all new approaches need a thorough evaluation. Many of these studies are being pursued. Artifact formation has to be carefully looked into for it can take place readily some time. As for example the discovery that the unexpectedly high concentrations of PAN over the Atlantic were spurious since the compound was formed by reaction of precursors on the glass surface of the sampling syringe (336).

VIII. SULFATE AND SULFITE

A. Introduction

The genotoxic aspects of sulfates and sulfites have been discussed in the first paper of the series. The simplest organic sulfate, dimethyl sulfate, is inactive when administered to animals orally or intravenously (277) but shows distinct carcinogenic activity when administered by inhalation and may be associated with cancer of the lung in humans exposed industrially to it (278). On the other hand, Ellgehausen has stated "Since dimethyl sulfate has been manufactured and processed on a large scale for over 50 years - the present annual turnover in Europe is approximately 25,000 tons - it is highly improbable that dimethyl sulfate has a cancerogenic effect on humans" (300).

The main problem in the analysis of sulfate and sulfuric acid lies in the sampling procedures. With all the advances that have taken place in automation, computer technology, and nucleic acid chemistry, we still do not have reliable methods of analysis for atmospheric sulfate and sulfuric acid.

B. Sampling

A vast amount of data on atmospheric sulfate has been accumulated over the years by many groups. The problem is are we collecting data which is inaccurate and, if so, is there any means for improving the accuracy of the data. The problems in the sampling procedure must be examined closely and corrected by appropriate modifications.

There are three main problems in sampling for atmospheric sulfate. These are possible high blank background, artifact formation, and sulfate modification.

Some filters used for collecting airborne particulates have relatively higher concentrations of sulfate (337). In the Gelman Spectro Quality Type "A" glass fiber sheet sulfate and nitrate were less than 100 µg and 50 µg, respectively, per 8 x 10 sheet.

The oxidation of SO_2 to sulfate on a filter can give erroneously high sulfate values. Some of the factors leading to this type of artifact formation are high pH glass filter (338), filter particulate (339), soot (340), trace metal catalysts, particulate sulfite oxidation, SO_2 concentration, relative humidity, sampling time/rate, and sample storage conditions.

Sulfate modification can take the form of a change of water-soluble sulfate to insoluble sulfate by reaction of the sulfate with particulate salts or with ions present in the filter, e.g., barium ions. Other modifications would involve sulfuric acid reactions.

In sampling for H_2SO_4 the same problems are found as for sulfate, but there are additional ones due to the acidic and oxidative nature of sulfuric acid. A diminution of the concentration of sulfuric acid found in air is observed for increases in flow rates and sampling times (341). It is postulated that this phenomenon is due to the neutralization of H_2SO_4 at the surface of the glass fiber sheet. Some of the factors that could affect the accuracy of H_2SO_4 determination include filtering media composition (at elevated temperatures, even Teflon reacts), organic vapors, gaseous pollutants, collected particulates, air flow rates, relative humidity, sampling time, sampling volume, particle sizes, storage time, conversion of SO_2 to H_2SO_4 on the filter, reduction of H_2SO_4 to SO_2 on the filter, sulfonation of pollutants on the filter, etc. Many of these factors are to be investigated in a project concerned with developing a reliable procedure for the sampling and analysis of sulfuric acid. Since reliable methods are not available for the determination of sulfuric acid, much controversy surrounds the subject of atmospheric sulfuric acid. Thus, it has been stated recently that the weak acid nature of rain casts doubt on the concepts that the acidity of rain is increasing and that these increases are due to strong acids such as sulfuric acid (342).

For the collection of sulfite large concentrations of appropriate metallic compounds are necessary to stabilize the sulfite in a complex. Thus, sulfite is found around copper smelters.

C. Analysis

Volatile organic sulfates can be collected on an adsorbent such as Tenax GC and then following desorption the sulfate can be separated and assayed by GC-MS. This has been done with atmospheric dimethyl sulfate (300). Since organic sulfates would be highly reactive, storage stability could be a problem. After dimethyl sulfate has been collected the adsorber tubes are washed at room temperature with purified helium to remove the residual water and then stored dry at -20°C. In this way samples can be stored for several days.

At the present stage it looks like the best way to determine H_2SO_4 is directly in the air with a real time monitor or by forming a stabilized derivative of H_2SO_4 as it is being collected. As far as I know, reliable systems of this type are not available.

IX. SCREENING TESTS

A. Analytical

A variety of screening test values are given for some air pollutant families in Table VI. Table VII lists a variety of additional useful screening tests. Other screening tests that could be used in the air pollution field include the nuclear magnetic resonance determination of the percent aromatic hydrogens in an organic particulate fraction (250), colorimetric determination of total pyridine bases near tarring operations (348), estimation of PAH by length of fluorescent zone on column (349), carbon number distributions, CH, CHN, and CHNO ratios of various air particulate fractions (350), and high resolution GLC profiles of air pollutant PAH (351, 352).

The importance of a screening test can be shown with BaP. The following are the reasons BaP has been picked as the key analyte in the study of carcinogenic mixtures containing the PAH.

(i) Several studies involving skin painting of humans with BaP have shown precarcinogenic effects. As a result of one of these studies, Cottini and Mazzoni (154) stated in 1939 "benzpyrene, if applied to human skin for protracted periods, would be carcinogenic as it is in animals."

(ii) BaP is carcinogenic in all species examined; it has been administered through oral, skin, and intratracheal routes.

(iii) Every member of the carcinogenic PAH family has metabolic and 'carcinogenic' pathways similar to that of BaP's, and is metabolized to detoxified materials, toxicants and mutagens similar to those obtained from BaP.

(iv) The human and animal enzyme systems are closely similar. For example, the liver mixed function oxidase system, the lung enzyme systems and the formation of powerful alkylating agents (carcinogenic to animals and mutagenic to bacteria) through epoxidation is common to animal and man.

(v) Other things being equal, higher and lower levels of BaP in atmospheres containing coke oven effluents would mean higher and lower levels, respectively, of other PAH.

(vi) Wherever humans have been in extended contact with mixtures containing fairly high concentrations of BaP and other PAH, cancer has resulted. And the larger the concentration of BaP and the PAH, the larger the percentage of cancer cases.

The lists of screening tests are incomplete, but they do give some idea of what's available. Their usefulness? Well, suppose one wished to check an organic fraction from incinerator effluents for the presence of direct-acting carcinogens one would test for the presence of alkylating agents with one of the reagents listed in Table VII.

If an individual wished to monitor a high risk coke oven area where BaP concentrations would be expected to be high and the cancer risk would be considerably increased, one of the BaP methods described in Table VII could be used. If large numbers of samples are to be assayed, then the GC - gas phase fluorescence method could be used. The instrumentation presently available is a successful prototype model but needs sophisticated mechanical improvement before it can be used routinely.

When the organic fraction of a particulate sample is needed either to measure the TpAH or other organic compounds, then ultrasonic or, better yet, a polytron extraction of particulates is desirable. The use of a benzene-soluble fraction quantity as a screening test is undesirable because a Soxhlet extraction takes about 6 to 12hr, benzene is a leukemogen, some thermal decomposition could take place, and the reproducibility of the method is bad. Much more reliable and reproducible results are obtained quicker with the TpAH test. This test will give some idea of the quantity of non-polar organic particulate material present in the air. In addition, with this HPLC procedure the penta-, tetra-, tri-, and dicyclic PAH cuts as well as the monocyclic fraction could be determined. This separation could be followed with further separation and GC-MS-COMP analysis to give a much more thorough analysis for hydrocarbons than has yet been possible.

There are over a dozen methods for determining the possible carcinogenicity of a chemical to a human individual. The ones that show the most promise in more rapidly developing an understanding, a prevention and a cure of cancer are the recently developed rapid prescreening assays (353). There are about 8 main types, some of which could be developed into rapid reliable tests for carcinogens eventually. Some of these methods could prove useful in air pollution studies. Obviously, these types of studies would have to be coordinated with animal and epidemiological investigations.

One of the most popular systems has been the Ames' system (354) which is essentially a plate test utilizing the bacterium Salmonella Typhimurium which is a histidine auxotroph (done on histidine-free media containing a dimethylsulfoxide extract of the test mixture, the bacteria and the supernatant extract of a rat or human liver homogenate fraction). With this simple back mutation-test, the reversion from histidine requirement to growth on a histidine-free media is measured. Essentially, carcinogens are metabolized to mutagens which then revert the bacteria to histidine prototrophy and the colonies then grow and can be counted or measured. A large number of these strains have been developed, they have a low spontaneous reversion rate and a high sensitivity to certain types of carcinogens, dependent on the strain. There is a strain which detects mutagens that cause base substitution; it reverts either by direct mutation or by suppressor mutations, a second strain capable of detecting frameshift mutagens and a third strain capable of detecting large deletions. These strains have 3 mutations - loss of hystidine synthesis capability, loss of the excision repair system and the loss of the lipopolysaccharide barrier that coats the surface of the bacteria. These test strains have been used in detecting mutagenic activity in cigarette smoke condensates and also to detect mutagens in the urine of rats given small amounts of carcinogens. With this system many halo-alkanes widely used in industry and the home have been shown to be mutagenic.

A biologic method of classification of mutagens is available which supplements the Ames method of classification by chemical specificity for inducing reversions. E. coli is used as the test system. Five types of mutageneses are proposed. The mutagens are classified into the two large groups involving misrepair and misreplication by the kind of DNA repair involved in transforming mutagen-induced damage to mutation.

A large number of spot tests can be done on one plate for air samples and their appropriate concentrations. A large number of other types of methods are available. The background, the faults, advantages and the various ramifications of the chosen bioassay methods should be known before use. The sample to be assayed should be a clean, reproducibly-obtained sample. Samples should be taken in high-risk areas and in pollution-controlled areas. Everything possible should be known about the sample, where taken and chemical composition.

Summary

Problems in the sampling and analysis of a variety of key air pollutants have been considered. The pollutants of primary interest were those with carcinogeni mutagenic or cofactor activity. These include benzo[a]pyrene, polynuclear aromatic hydrocarbons, aldehydes, alkylating agents, amines, chloromethyl ethers, epoxides, nitrosamines, nitrogen dioxide, nitrates, sulfate and sulfite. Screening tests of the analytical and bioassay types were also discussed; a large variety of these tests were summarized in tables.

Concentrations of benzo[a]pyrene in urban atmospheres, in highly polluted atmospheres and effluents, and in a large variety of environmental mixtures were reported.

Résumé

Les problèmes que posent l'échantillonnage et l'analyse de divers importants polluants de l'air ont été examinés. Les polluants présentant un intérêt primodial sont ceux qui ont une activité cancérogène, mutagène ou de cofacteur : benzo-[a]pyrène, hydrocarbures aromatiques polynucléaires, aldéhydes, agents alcoylants, amines, chlorométhyl-éthers, époxydes, nitrosamines, peroxyde d'azote, nitrates, sulfate et sulfite. Les tests de détection analytiques et les épreuves biologique ont été également discutés; une large gamme de ces tests est résumée dans des tableaux.

Il a été fait état de concentrations de benzo[a]pyrène dans les atmosphères urbaines, les atmosphères et les effluents très pollués et dans des mélanges environnementaux très divers.

Analysis of atmospheric carcinogens and their cofactors

E. SAWICKI

REFERENCES

(1) E. Sawicki, Composition and genotoxic aspects of the atmosphere, presented at the Workshop for the Investigation of the Carcinogenic Burden by Air Pollution in Man in Hannover, West Germany on October 22-24, 1975.

(2) F. W. Karasek, Research/Development, 25, 6, 36 (1974); 26, 20 (1975).

(3) Intersociety Committee, Methods of Air Sampling and Analysis, American Public Health Association, Washington, D. C. 1972.

(4) I. J. Selikoff, E. C. Hammond and J. Churg, J. Am. Med. Assoc. 204, 106 (1968).

(5) F. E. Lundin, Jr., J. W. Lloyd, E. M. Smith, V. E. Archer and D. A. Holaday, Health Phys., 16, 571 (1969).

(6) G. J. Cleary, Int. J. Air Water Pollution, 7, 753 (1963).

(7) D. Rondia, Arch. Belg. Med. Soc., 18, 220 (1960).

(8) G. E. Moore, M. Katz, and W. B. Drowley, J. Air Poll. Control Assoc., 16, 492 (1966).

(9) V. Bkramovsky, Acta Unio Int. Cancr., 19, 733 (1963).

(10) V. Masek, Cesk. Hyg. (Prague), 10, 86 (1965).

(11) J. M. Campbell and J. Clemmersen, Danish Med. Bull., 3, 205 (1956).

(12) B. T. Commins and R. E. Waller, Atmos. Env., 1, 49 (1967).

(13) R. E. Waller, Brit. J. Cancer, 6, 8 (1952).

(14) R. L. Cooper, Analyst, 79, 573 (1954).

(15) P. Stocks, B. T. Commons, and K. V. Audrey, Int. J. Air Water Pollution, 4, 141 (1961).

(16) P. Stocks, Brit. J. Cancer, 20, 595 (1966).

(17) G. Chatot. et. al., Atm. Env., 7, 819 (1973).

(18) H. O. Hetiche, Staub, 25, 365 (1965).

(19) H. O. Hetiche, Staub, 23, 136 (1963).

(20) W. Kutscher and R. Tomingas, Staub, 29, 18 (1969).

(21) H. O. Hetlche, Int. J. Air Water Poll., 8, 185 (1964).

(22) M. Zamfirescu and I. Ardelean, Wiss. Z. Humboldt Univ. Berlin Math. Maturw. Reihe, 19, 509 (1970).

(23) M. Kertesz-Saringer, J. Morik, and Z. Morlin, Atm. Env., 3, 417 (1969).

(24) M. Kertesz-Saringer and Z. Morlin, Atm. Env., 9, In press (1975).

(25) A. M. Mohan-Rao and K. G. Vohra, Atm. Env., In press (1975).

(26) Y. Aldoh, N. Agndaie, M. R. Darvich, and M. H. Khorgami, Atm. Env., 6, 949 (1972).

(27) A. D'Ambrosio, Mikrochim. Acta, 927 (1961).

(28) A. D'Ambrosio, F. Pavelka, E. Calderoni, and F. Ciardo, Centro Provinciale per lo studio sugli inquinamenti atmosferici (1958).

(29) F. C. Petrielli and S. Karitz, Giorn. Ig. Med. Prev., 3, 83 (1962).

(30) P. Valori, cited in W. Kutscher and R. Tomingas, Staub, 29, 18 (1969).

(31) H. Sakabe, H. Matsushita, H. Hayashi, K. Nozaki, and Y. Suzuki, Ind. Health (Japan), 3, 126 (1965).

(32) S. Abe, Air Pollution News (Japan), 63, 5 (1971).

(33) K. Fujie, Proc. Symp. Japan Soc. Air Pollution, 252 (1974).

(34) P. J. Blokzijl and R. Guicherit, TNO-News, 653 (1972).

(35) J. Just, S. Maziarka, and H. Wyszynska, Polskiego Tygodnika Lekarskiego, 23, 1553 (1968).

(36) C. W. Louw, Am. Ind. Hyg. Assoc. J., 520 (1965).

(37) J. de la Serna, B. Sanchez-Fernandez-Murias, and T. F. Caballero. Rev. San. Hig. Pub., 46, 87 (1972).

(38) P. Anechina, J. M. Romero, B. Sanchez-Fernandez-Murias, and J. de la Serna, Rev. San. Hig. (Spain), 44, 885 (1970).

(39) G. Lindstedt, Atm. Env., 2, 1 (1968).

(40) T. Müller, Z. Praventimed., 11, 157 (1966).

327

(41) E. Sawicki, W. C. Elbert, T. R. Hauser, F. T. Fox, and T. W. Stanley, Am. Ind. Hyg. Assoc. J., 21, 443 (1960).

(42) E. Sawicki, T. W. Stanley, W. C. Elbert, and J. D. Pfaff, Anal. Chem., 36, 497 (1964).

(43) R. B. Faoro, J. Air Poll. Control Assoc., 25, 638 (1975).

(44) R. J. Gordon and R. J. Bryan, Env. Sci. Technol., 7, 1050 (1973).

(45) M. N. Bolotowa, Gig. Sanit, 4, 96 (1966).

(46) R. P. Hangebrauck, D. J. von Lehmden, and J. E. Meeker, PHS Publication No. 999-AP-33, 1967.

(47) R. Gladen, Chromatographia, 5, 236 (1972).

(48) W. Schrodter, P. Studt and P. Voigtsberger, Z. Bakt. Hyg., B158, 50 (1973).

(49) R. P. Hangebrouck, R. P. Lauch and J. E. Meeker, Am. Ind. Hyg. Assoc. J., 27, 47 (1966).

(50) D. Hoffmann and E. L. Wynder, Cancer, 15, 93 (1962).

(51) P. J. Lawther, B. T. Commins, and R. E. Waller, Brit. J. Ind. Med., 22, 13 (1965).

(52) J. O. Jackson, P. O. Warner and T. F. Mooney, Jr., Am. Ind. Hyg. Assoc. J., 35, 276 (1974).

(53) E. Sawicki, T. R. Hauser, W. C. Elbert, F. T. Fox and J. E. Meeker, Am. Ind. Hyg. Assoc. J., 23, 137 (1962).

(54) E. Sawicki, F. T. Fox, W. C. Elbert, T. R. Hauser, and J. E. Meeker, Am. Ind. Hyg. Assoc. J., 23, 482 (1962).

(55) I. W. Davies, R. M. Harrison, R. Perry, D. Ratnayaka and R. A. Wellings, Env. Sci. Technol., In press (1975).

(56) M. N. Bolotova, Y. S. Davydov and N. G. Nikishina, Med. Zh. Uzb., (11), 51 (1967).

(57) P. J. Lawther, Roy. Soc. Health J., 91, 250 (1971).

(58) V. Galuskinova, Neoplasma, 11, 465 (1964).

(59) L. P. Elliott and Dr R. Rowe, J. Air. Poll. Control Assoc., 25, 635 (1975).

(60) R. I. Larsen and V. J. Konopinski, Arch. Env. Health, 5, 597 (1962).

(61) J. Bonnet, in E. Sawicki and K. Cassel, Jr., Eds., Symposium: Analysis of Carcinogenic Air Pollutants, NCI monograph No. 9, Superintendent of Documents, Washington, D. C., 1962, pp. 221-223.

(62) L. M. Shabad and G. A. Smirnov, Atm. Env., 6, 153- (1972).

(63) J. S. Harington, Nature, 193, 43 (1962).

(64) L. Wallcave, H. Garcia, R. Feldman, W. Lijinsky and P. Shubik, Toxicol. Appl. Pharmacol. 18, 41 (1971).

(65) E. Sawicki, Arch. Env. Health, 14, 46 (1967).

(66) P. P. Dikun and I. I. Nikberg, Vopr. Onkol., 4, 669 (1958).

(67) G. Bittersohl, Arch. Geschwulstforsch, 38, 198 (1971).

(68) R. Tye, A. W. Horton and I. Rapien, Am. Ind. Hyg. Assoc. J., 27, 25 (1966).

(69) W. Lijinsky and A. E. Ross, Food Cosmetic Toxicol., 5, 343 (1967).

(70) V. Masek, Z. Jach, and J. Kandus, J. Occ. Med, 14, 548 (1972).

(71) W. Lijinsky, I. Domsky, G. Mason, H. Y. Ramahi and T. Safavi, Anal. Chem., 35, 952 (1963).

(72) L. M. Shabad, A. Y. Khesina, A. B. Linnik and G. S. Serkovskaya, Int. J. Cancer, 6, 314 (1970).

(73) W. Lijinsky, V. Saffiotti and P. Shubik, J. Natl. Cancer Inst., 18, 687 (1956).

(74) H. G. Maier and W. Stender, Deut. Lebensmitt Rdsch., 65, 341 (1969).

(75) K. Soos, Z. Lebensm. Unters.-Forsch., 156, 344 (1974).

(76) G. Grimmer, Erdoel Kohle, 19, 578 (1966).

(77) B. S. Ruchkovskey, V. V. Konstantinov, M. A. Tiktin, M. M. Morgulis, G. K. Babanov, V. O. Gorodiskaya and N. V. Tkach, Voprosy Pitaniya, 31, 74 (1972).

(78) Y. K. Masuda and M. Kuratsune, Gann., 62, 27 (1971).

(79) D. Schmahl, Deutsche Med. Wochenschr., 99, 2424 (1974).

(80) G. Grimmer and G. Wilhelm, Dtsch. Lebensmittel-Rdsch., 62, 19 (1966).

(81) B. P. Dunn, personal communication.

(82) H. J. Cahnmann and M. Kuratsure, Anal. Chem., 29, 1312 (1957).

(83) W. Lijinsky and C. R. Raha, Toxicol. Appl. Pharmacol., 3, 469 (1961).

(84) R. H. White, J. W. Howard, and C. J. Barnes, J. Agr. Food Chem., 19, 143 (1971).

(85) H. Weiss, A. Brockhaus, and G. Koern, Z. Bakteriol. Hyg., 155, 142 (1972).

(86) CRC Project No. CD-9-61, Progress Report on Composition, Odor, and Eye Irritation for Period May 1, 1962 to February 15, 1963.

(87) K. S. Rhee and L. J. Bratzler, J. Food Sci., 33, 626 (1968).

(88) I. Berenblum and R. Schoental Brit. J. Exp. Path., 24, 232 (1943).

(89) G. Grimmer and H. Bohnke, J. Assoc. Off. Anal. Chem., 58, 725 (1975).

(90) N. D. Gurelova, P. P. Dikun, L. D. Kostenko, O. P. Gretskaja, and A. V. Emshanova, Novost. Onkol., 8 (1971).

(91) J. W. Howard and T. Fazio, J. Agric. Food Chem., 17, 527 (1969).

(92) D. Luks, Rev. Ferment. Ind. Aliment, 28, 111 (1973).

(93) D. Luks and J. Lenges, personal communication.

(94) H. Elmenhorst and W. Dontenwill, Z. Krebsforsch., 70, 157 (1967).

(95) J. M. Campbell and A. J. Lindsay, Chem. Ind., 951 (1957).

(96) J. M. Campbell and A. J. Lindsay, Chem. Ind., 64 (1955).

(97) G. Grimmer, J. Jacob and H. Hildebrandt, Z. Krebsforsch., 78, 65 (1972).

(98) M. Blumer, Science, 134, 474 (1961).

(99) W. Fritz and R. Engst, Z. Ges. Hyg., 17, 271 (1971).

(100) L. M. Shabad, A. Keesina, and S. S. Khitroro, Vestn. Akad. Med. Navk SSSR, 23, 6 (1968).

(101) L. M. Shabad, Z. Krebsforsch; 70, 204 (1968).

(102) M. Blumer, Science, 134, 476 (1961).

(103) L. M. Shabad, Cancer Research, 27, 1132 (1967).

(104) W. Fritz and R. Engst, Enfassing Auswirkengen Luft-verunre Iningungen, vortr. Lufthyg. Kolloq, 3rd; 97-106 (1972).

(105) J. Zdrazil and F. Picha, Neoplasma, 13, 49 (1966).

(106) A. Y. Tsipenyuk and I. Donina, Gig. Sanit., 38, 95 (1973).

(107) M. Kuratsune, J. Natl. Cancer Inst., 16, 1485 (1956).

(108) M. Kuratsune and W. C. Hueper, J. Nat. Cancer Inst., 20, 37 (1958).

(109) H. Falk, P. Kotin and I. Markul, Cancer, 11, 482 (1958).

(110) L. J. Wood, J. Appl. Chem., 11, 130 (1961).

(111) T. Hirohata, Y. Masuda, A. Horie, and M. Kuratsure, Gann., 64, 323 (1973).

(112) A. M. Manzone and C. Rossi, Tumori, 59, 295 (1973).

(113) Yu. L. Kogan, Voprosy Onkol., 19, 84 (1973).

(114) L. M. Shabad and Y. L. Cohan, Arch. Geschwnlstforsch., 40, 237 (1972).

(115) Y. Shiraishi, T. Shirotori and Y. Sakagami, J. Food Hyg. Soc. Japan, 13, 41 (1972).

(116) H. R. Bentley and J. G. Burgan, Analyst, 83, 442 (1958).

(117) B. L. Van Duuren, J. Natl. Cancer Inst., 21, 1 (1958).

(118) J. Bonnet and S. Neukomm, Helv. Chim. Acta., 39, 1724 (1956).

(119) J. Bonnet and S. Neukomm, Oncologia (Basel), 10 124 (1957).

(120) Z. Cordon, E. T. Alvord, H. J. Rand, and R. Hitchcock, Brit. J. Cancer, 10, 485 (1956).

(121) R. L. Cooper and A. J. Lindsey, Brit. J. Cancer, 9, 304 (1955).

(122) R. Latarjet, J. L. Cusin, M. Hubert-Habart, B. Muel, and R. Royer, Bull. Cancer, 43, 180 (1956).

(123) H. Elmenhorst and G. Grimmer Z. Krebsforsch., 71, 66 (1968).

(124) D. Hoffmann, G. Rathkamp, K. D. Brunnemann, and E. L. Wynder, Sci. Total Environ., 2, 157 (1973).

(125) W. Dontenwill et al, as reported in (124).

(126) E. L. Wynder and D. Hoffmann, in Environment and Cancer, 24th Annual Symposium on Fundamental Cancer Research, 1971, Williams and Wilkins Co., 1972, pp. 118-141.

(127) A. Savino and G. Scassellati Sforzolini, Riv. Ital. Igiene, 29, 265 (1970).

(128) L. Mallet, Congr. Expertise Chim. Vol. Spec. Conf. Commune 4th Athens, 301 (1964); through Chem. Abstr., 66, 84240b (1967).

(129) M. Repetto and D. Martinez, Env. J. Toxicol. Env. Hyg., 7, 234 (1974).

(130) W. Graf and H. Diehl, Arch. Hyg. (Berlin), 143, 405 (1959).

(131) L. M. Shabad, Vestnik Akad. Med. Nauk SSSR, 27, 35 (1972).

(132) N. Y. Yanysheva, Y. I. Kostnetskiiand Z. P. Fedorenko, Gig. Sanit., 7, 71 (1974).

(133) J. Borneff and H. Kunte, Arch. Hyg. Bakteriol., 153, 220 (1969); through Chem. Abstr. 71, 73890r (1969).

(134) J. Borreff and H. Kunte, Arch. Hyg. (Berlin), 148, 585 (1964).

(135) J. Borneff, Arch. Hyg. (Berlin), 148, 1 (1964).

(136) I. Siddiqui and K. H. Wagner, Chemosphere, 1, 83 (1972).

(137) A. P. Il'nitskii, L. G. Rozhnova, and T. V. Drozdova, Hyg. Sanit., 36, 816 (1971).

(138) J. Borneff and H. Kunte, Arch. Hyg., 149, 226 (1965).

(139) P. P. Dikun and A. I. Makhinenko, Gig. Sanit., 28, 10 (1963) through Chem. Abstr. 59, 3640g (1963).

(140) Z. P. Fedorenko, Gig. Sanit., 29, 17 (1964).

(141) P. Wedgwood and R. L. Cooper, Analyst, 79, 163 (1954).

(142) P. Wedgwood and R. L. Cooper, Analyst, 81, 45 (1956).

(143) I. A. Veldro, L. A. Lakhe, and I. K. Arro, Gig. Sanit., 30, 104 (1965).

(144) L. Mallet, A. Perdriau and J. Perdriau, Compt. Rend., 256, 3487 (1963).

(145) J. Bourcart and L. Mallet, Compt. Rend., 260, 3729 (1965).

(146) L. Mallet and J. Sardou, Compt. Rend., 258, 5264 (1964).

(147) L. Mallet, M. Tendron, and V. Plessis, Ann. Med. Leg., 40, 168 (1960).

(148) A. Depuis, Tech. Eau., 14, 25 (1962).

(149) G. Eglinton, B. R. T. Simoneit and J. A. Zoro, Proc. Roy. Soc. London, B189, 415 (1975).

(150) J. Borneff and R. Fischer, Arch. Hyg. (Berlin), 146, 183, 334 (1962); 146, 572 (1963).

(151) W. Giger and M. Blumer, Anal. Chem, 46, 1663 (1974).

(152) H. Hellmann, Deutsche Gewasserkundliche Mitteilung, 18, 155 (1974).

(153) J. B. Andelman and M. J. Suess, Bull. World Health Org., 43, 479 (1970).

(154) G. B. Cottini and G. B. Mazzone, Am. J. Cancer, 37, 186 (1939).

(155) C. P. Rhoads, W. E. Smith, N. S. Cooper and R. D. Sullivan, Proc. Am. Assoc. Cancer Res., 1, 40 (1954).

(156) E. Klar, Klin. Wschr., 17, 1279 (1938).

(157) D. Schmal, H. Consbruch and H. Druckrey, Arzneim. Forsch., 3, 403 (1953).

(158) J. Sula, Cas. lek. ces., 91, 1029 (1952).

(159) W. Graef, Society for Ecology, Proc. Conf. Load Loadability Ecosystems, Giessen, West Germany, 1972, p. 115-122.

(160) D. Schmahl and K. G. Schmidt in E. Karbe and J. F. Park, Eds. Experimental Lung Cancer, Carcinogenesis and Bioassays, Springer-Verlag, Berlin, 1974, 139-145.

(161) B. W. Carnow, Arch. Environ. Health, 27, 207 (1973).

(162) L. M. Shabad, Kazan, Med. Zh., 5, 92 (1973).

(163) G. M. Badger and T. M. Spotswood, J. Chem. Soc., 4420 (1960).

(164) G. M. Badger, J. K. Donnelly and T. M. Spotswood, Aust. J. Chem., 16, 392 (1963).

(165) G. M. Badger, J. K. Donnelly and T. M. Spotswood, Aust. J. Chem., 17, 1147 (1964).

(166) G. M. Badger, J. K. Donnelly and T. M. Spotswood, Aust. J. Chem., 18, 1249 (1965).

(167) E. L. Wynder, G. F. Wright and J. Lham, Cancer, 11, 1140 (1958).

(168) G. M. Badger, G. E. Lewis and I. M. Napier, J. Chem., Soc., 2825 (1960)

(169) E. Gil-Av and J. Shabtai, Nature, 197, 1065 (1963).

(170) J. S. Gilbert and A. J. Lindsey, Brit. J. Cancer, 11, 398 (1957).

(171) E. Kroeller, Dtsch. Lebensmitt.-Rdsch., 61, 16 (1965); 150 (1965).

(172) J. Lam, Acta. Path. Microbiol. Scand., 39, 198, 207 (1956).

(173) Y. Masuda, K. Mori and M. Kuratsune, Gann., 58, 69 (1967).

(174) Reference Method for the Determination of Suspended Particulates in the Atmosphere (High-Volume Method). Appendix B, "National Primary and Secondary Ambient Air Quality Standards," Federal Register 36: (84), Part II, April, 1971.

(175) H. C. McKee, R. E. Childers, O. Saenz, Jr., T. W. Stanley and J. M. Margeson, J. Air Poll. Control Assoc., 22, 342 (1972).

(176) Anderson 2000 Inc., "Anderson Sampler Simulates Human Respiratory System", Atlanta, Ga., April, 1971.

(177) R. J. Countess, J. Air Poll. Control Assoc., 24, 605 (1974).

(178) P. M. Duvall and R. C. Bourke, Env. Sci. Technol., 8, 765 (1974).

(179) L. DeMaio and M. Corn, J. Air Poll. Control Assoc., 16, 67 (1969).

(180) M. Kertesz-Saringer, E. Meszaros and T. Varkonyi, Atm. Env., 5, 429 (1971).

(181) R. C. Pierce and M. Katz, Arch. Env. Sci. Technol. 9, 347 (1975).

(182) A. Albagli, H. Oja, and L. Dubois, Env.Letters, 6, 241 (1974).

(183) V. Masek, Z. Arbeitsmed., 24, 213 (1974).

(184) R. B. Schlesinger, Bio-Science, 23, 567 (1973).

(185) Task Group on Lung Dynamics, Health Physics, 12, 173 (1966).

(186) A. R. Kennedy and J. B. Little, Cancer Res., 35, 1563 (1975).

(187) M. C. Henry, C. D. Port, R. B. Bates and D. G. Kaufman, Cancer Res., 33, 1585, 1592 (1973).

(188) A. R. Kennedy and J. B. Little, Cancer Res., 34, 1344, 1352 (1974).

(189) J. B. Little and W. F. O'Toole, Cancer Res., 34, 3026 (1974).

(190) U. Saffiotti, F. Cefis and L. H. Kolb, Cancer Res., 28, 104 (1968).

(191) B. T. Commins, in E. Sawicki and K. Cassel, Eds., Analysis of Carcinogenic Air Pollutants, National Cancer Institute Monograph No. 9, Superintendent of Documents, Washington, D. C., 1966, pp. 225-233.

(192) N. H. Ketcham and R. W. Norton, Arch. Env. Health, 1, 194 (1960).

(193) J. Von Borneff and R. Knerr, Arch. Hyg. Bakt., 143, 405 (1959).

(194) W. L. Ball, G. E. Moore, J. L. Monkman and M. Katz, Ind. Hyg. J., 222 (1962).

(195) H. Matsushita and Y. Suzuki, Bull. Chem. Soc. Japan, 42, 460 (1969).

(196) H. Arito, R. Soda and H. Matsushita, Ind. Health, 5, 243 (1967).

(197) D. Schuetzle, A. L. Crittenden and R. J. Charlson, J. Air. Poll. Control Assoc., 23, 704 (1973).

(198) D. Schuetzle, D. Bronn, A. L. Crittenden and R. J. Charlson, Env. Sci. Technol., 9, 838 (1975).

(199) C. Golden and E. Sawicki, Int. J. Anal. Chem., 4, 9 (1975).

(200) G. Chatot, M. Castegnaro, J. L. Roche, R. Fontanges and P. Obaton, Anal. Chim. Acta., 53, 259 (1971).

(201) H. J. Seim, W. W. Hanneman, L. R. Barsotti and T. J. Walker, Am. Ind. Hyg. Assoc. J., 35, 718 (1974).

(202) T. W. Stanley, J. E. Meeker and M. J. Morgan, Env. Sci. Technol. 1, 927 (1967).

(203) J. L. Monkman, L. Dubois and C. J. Baker, Pure Appl. Chem., 24, 731 (1970).

(204) E. Wittgenstein, E. Sawicki, and R. L. Antonelli, Atm. Env., 5, 801 (1971).

(205) E. Sawicki, et al, Health Lab. Sci., January Supplement, 7, 56 (1970).

(206) R. E. Schaad, Chromatog. Rev., 13, 61 (1970).

(207) D. Hoffman and E. L. Wynder, Anal. Chem., 32, 295 (1960).

(208) A Liberti, G. P. Cartoni and V. Cantuti, J. Chromatog., 15, 141 (1964).

(209) E. Sawicki, T. W. Stanley, W. C. Elbert, J. Meeker and S. McPherson, Atm. Env., 1, 131 (1967).

(210) F. Weigert and J. C. Mottram, Cancer Res., 6, 97 (1947).

(211) P. Wedgewood and R. L. Cooper, Analyst, 78, 170 (1953).

(212) P. Kotin, H. L. Falk and M. Thomas, Arch. Ind. Hyg., 9, 164 (1954).

(213) B. T. Commins, Analyst, 83, 386 (1958).

(214) E. Sawicki, W. C. Elbert, T. W. Stanley, T. R. Hauser and F. T. Fox, Int. J. Air Poll., 2, 273 (1960).

(215) E. Sawicki, W. C. Elbert, T. W. Stanley, T. R. Hauser and F. T. Fox, Anal. Chem., 32, 810 (1960).

(216) H. J. Seim, Private communication.

(217) N. F. Ives and L. Giuffrida, JAOAC, 55, 757 (1972).

(218) H. Klimisch, Anal. Chem., 45, 1960 (1973).

(219) H. Hatano, Y. Yamamoto, M. Saito, E. Mochida and S. Watanabe, J. Chrom., 83, 373 (1973).

(220) R. M. Duncan, Am. Ind. Hyg. Assoc. J., 30, 624 (1969).

(221) B. H. Gump, J. Chrom. Sci., 7, 755 (1969).

(222) G. M. Jamini, K. Johnston and W. L. Zielinski, Jr., Anal. Chem., 47, 670 (1975).

(223) G. J. Cleary, J. Chromatog., 9, 204 (1962).

(224) G. Grimer and A. Hildebrandt, J. Chromatog., 20, 89 (1965).

(225) E. Sawicki, T. W. Stanley and W. C. Elbert, Talanta, 11, 1433 (1964).

(226) L. Dubois, A. Zdrojewski, C. Baker and J. L. Monkman, Air. Poll. Control Assoc. J., 17, 818 (1967).

(227) R. A. Brown, T. D. Searl, W. H. King, W. A. Dietz and J. M. Keeliher, Final Report, "Rapid Methods of Analysis for Trace Quantities of Polynuclear Aromatic Hydrocarbons and Phenols in Automobile Exhaust Gasoline and Crankcase Oil," CRC-APRAC Project CAPE-12-68. National Technical Information Service, Springfield, Va., 22151.

(228) E. Sawicki, et al., Health Lab. Sci., 11, 228 (1974).

(229) T. D. Searl, F. J. Cassidy, W. H. King and R. A. Brown, Anal. Chem., 42, 954 (1970).

(230) E. Sawicki et al., Health Lab. Sci., 11, 218 (1974).

(231) E. Sawicki, Talanta, 16, 1231-1266 (1969).

(232) E. Sawicki and H. Johnson, J. Chromatog., 23, 142 (1966).

(233) M. Zander, Z. Anal. Chem., 229, 352 (1967).

(234) E. Sawicki, T. W. Stanley and H. Johnson, Mikrochim. Acta, 178 (1965).

(235) E. Sawicki, T. W. Stanley and H. Johnson, Microchem. J., 8, 257 (1964).

(236) H. Johnson and E. Sawicki, Talanta, 13, 1361 (1966).

(237) J. Lam and A. Berg, J. Chromatog., 20, 171 (1965).

(238) E. Sawicki, T. W. Stanley, W. C. Elbert and M. Morgan, Talanta, 12, 605 (1965).

(239) C. R. Sawicki and E. Sawicki, Thin-layer chromatography in air pollution research, in G. Pataki, Ed., Advances in Thin-layer Chromatography, Ann Arbor Scientific Publishing Co., Ann Arbor, 1970.

(240) E. Sawicki, T. R. Hauser and T. W. Stanley, Int. J. Air. Poll.,

(241) R. A. Lannoye and R. A. Greinke, Am. Ind. Hyg. Assoc. J., 35, 755 (1974).

(242) G. E. Fedoseeva and A. Y. Khesina, Z. Prikl. Khim., 9, 282 (1968).

(243) G. F. Kirkbright and C. G. de Lima, Analyst, 99, 338 (1974).

(244) H. P. Burchfield, R. J. Wheeler, and J. B. Bernos, Anal. Chem., 43, 1976 (1971).

(245) J. Mulik, M. Cooke, M. Guyer, G. Semeniuk, and E. Sawicki, Anal. Letters, 8, 511 (1975).

(246) R. J. Majer, R. Perry and M. J. Reade, J. Chromatog., 48, 328 (1970).

(247) R. C. Lao, R. S. Thomas, H. Oja, and L. Dubois, Anal. Chem., 45, 908 (1973).

(248) W. Heller, Strahlentherapie, 81, 529 (1950).

(249) G. M. Blackburn, J. Buckingham, R. G. Fenwick, P. Taussing, and M. H. Thompson, J. Chem. Soc. Perkin 1, 2809 (1973).

(250) R. A. Greinke and I. C. Lewis, Presented at the Symposium on "Chemistry, Occurrence and Measurement of Polynuclear Aromatic Hydrocarbons" at the 170th ACS National Meeting in Chicago on August 8, 1975.

(251) I. W. Davies, R. M. Harrison, R. Perry, D. Ratnayaka, and R. A. Wellings, Env. Sci. Technol., in press (1975).

(252) M. E. Snook, W. J. Chamberlain, R. F. Severson, and O. T. Chortyk, Anal. Chem., 47, 1155 (1975).

(253) R. C. Lao, R. S. Thomas, L. Dubois, and J. L. Monkman, presented at the International Symposium on the Health Effects of Environmental Pollution, Paris, France, June 24-28, 1974.

(254) A. Brockhaus, Atm. Env., 8, 521 (1974).

(255) M. L. Lee, M. Novotny, and K. D. Bartle, Anal. Chem., 47, in press (1975).

(256) H. P. Burchfield, E. E. Green, R. J. Wheeler, and S. M. Billedean, J. Chromatog., 99, 697 (1974).

(257) K. W. Boyer and H. A. Laitinen, Env. Sci. Technol., 9, 457 (1975).

(258) J. B. Moran, M. J. Baldwin, O. J. Manary, and J. C. Valenta, "Effect of Fuel Additives on the Chemical and Physical Characteristics of Particulate Emissions in Automotive Exhaust", Final Report to the Environmental Protection Agency by the Dow Chemical Co., Midland, Mich., June 1972.

(259) A. P. Altshuller and S. P. McPherson, J. Air Poll. Control Assoc., 13, 109 (1963).

(260) N. A. Renzetti and R. J. Bryan, J. Air Poll. Control Assoc., 11, 421 (1961).

(261) P. W. West and B. Sen, Z. Anal. Chem., 153, 177 (1956).

(262) A. P. Altshuller, D. L. Miller, and S. F. Sleva, Anal. Chem., 33, 621 (1961).

(263) T. Nash, Biochem. J., 55, 416 (1953).

(264) T. Odaira, K. Asakuno, T. Izumigawa, M. Mori, H. Morita, and J. Miyai, Proc. Symp. Japan Soc. Air Poll., 13th, 121 (1972).

(265) V. D. Yablochkin, Lab. Delo, 719 (1966).

(266) N. Yamate, T. Matsumara, and M. Tonomura, Bull. Nat'l. Inst. Hyg. Sci., Tokyo 86, 58 (1968).

(267) N. Yamate and T. Matsumura, J. Japan Soc. Air Poll., 4, 121 (1969).

(268) S. Bellman, Anal. Chim. Acta, 29, 120 (1963).

(269) I. R. Cohen and A. P. Altshuller, Anal. Chem., 33, 726 (1961).

(270) D. A. Levaggi and M. Feldstein, J. Air Poll. Control Assoc., 20, 312 (1970).

(271) E. Sawicki, T. R. Hauser, T. W. Stanley, and W. C. Elbert, Anal. Chem., 33, 93 (1961).

(272) E. Sawicki, T. R. Hauser, and S. McPherson, Anal. Chem., 34, 1460 (1962).

(273) N. Nelson, B. Van Duuren, S. Laskin, M. Kuschner, R. Albert, B. Pasternack, and F. Mukai, Chem. Eng. News, 5, October 6, 1975.

(274) D. C. Wallace, Med. J. Australia, 829 (June 28, 1975).

(275) R. Preussman, H. Schneider, and F. Epple, Arzneim.-Forsch., 19, 1059 (1969).

(276) E. Sawicki and C. R. Sawicki, Annals N. Y. Acad. Sci., 163, 895 (1969).

(277) H. Druckrey, H. Kruse, R. Preussman, S. Ivankovic, and C. Landschutz, Z. Krebsforsch., 74, 241 (1970).

(278) N. Nashed and S. Ivankovic, Z. Krebsforsch., 68, 103 (1966).

(279) R. Preussman, H. Hengy, and H. Druckrey, J. Liebigs Ann. Chem., 684, 57 (1965).

(279a) E. Sawicki, D. F. Bender, T. R. Hauser, R. M. Wilson, Jr., and J. E. Meeker, Anal. Chem., 35, 1479 (1963).

(279b) D. F. Bender, E. Sawicki, and R. M. Wilson, Analyst, 90, 630 (1965)

(280) J. A. Miller and E. C. Millen, Lab. Investigation, 15, 217 (1966).

(281) K. Norpoth, H. Schriewer and H. M. Rauen, Arzneimittel-Forsch, 21, 1718 (1971).

(282) Y. L. Tan and D. R. Cole, Clin. Chem., 11, 58 (1965).

(283) B. Truhaut, E. Delacoux, G. Brule and C. Bohuon, Clin. Chim. Acta, 8, 235 (1963).

(284) J. Epstein, R. W. Rosenthal, and R. J. Ess, Anal. Chem., 27, 1435 (1955).

(285) H. Hengy, Z. Anal. Chem., 272, 46 (1974).

(28) L. G. Hammock, B. D. Hammock, and J. E. Casida, Bull. Env. Contam. Toxicol., 12, 759 (1974).

(287) C. T. Bedford and J. Robinson, Xenobiotica, 2, 307 (1972).

(288) A. Dipple and T. A. Slade, Eur. J. Cancer, 6, 417 (1970).

(289) A. J. Swaisland, P. L. Grover, and P. Sims, Biochem. Pharmacol., 22, 1547 (1973).

(290) P. Sims, Biochem. J., 125, 159 (1971).

(291) D. Lalka and T. J. Bardos, J. Pharm. Sci., 62, 1294 (1973).

(292) R. Norpoth and T. Papatheodorou, Naturwissenschaften, 57, 356 (1970).

(293) K. Norpoth, T. Papatheodorou and G. Venjakob, Beitrage Tabakforsch., 6, 106 (1972).

(294) K. Norpoth, G. Manegold, R. Brucker, and H. P. Amann, Z. Bakt. Hyg., B156, 341 (1972).

(295) E. E. Campbell, G. O. Wood, and R. G. Anderson, Development of Air Sampling Techniques. California Univ., Los Alamos, New Mexico, Los Alamos Scientific Lab., Atomic Energy Comm. Contract W-7405-ENG. 36 LASL Proj. R-059, Progress Rpt. LA 5484-PR, UC-41, 6pp, January, 1974.

(296) J. C. Arcos and M. F. Argus, Advances Cancer Res., 11, 305 (1968).

(297) K. Bretschneider and J. Matz, Arch. Geschwullstforsch., 42, 36 (1973).

(298) E. D. Pellizari, J. E. Burch, R. E. Berkley, and J. McRae, to be published.

(299) K. P. Evans, A. Mathias, N. Mellor, R. Silvester, and A. E. Williams, Anal. Chem., 47, 821 (1975).

(300) D. Ellgehausen, Anal. Letters, 8, 11 (1975).

(301) L. Collier, Environ. Sci. Technol., 6, 930 (1972).

(302) L. A. Shadoff, G. J. Kallos, and J. S. Woods, Anal. Chem., 45, 2341 (1973).

(303) R. A. Solomon and G. J. Kallos, Anal. Chem., 47, 955 (1975).

(304) S. Laskin, M. Kuschner, R. T. Drew, V. P. Capiello, and N. Nelson, Arch. Env. Health, 23, 135 (1971).

(305) L. S. Frankel, K. S. McCallum, and L. Collier, Env. Sci. Technol., 8, 356 (1974).

(306) M. D. Hurwitz, Am. Dyestuff Reporter, 62, (March, 1974).

(307) F. Oesch, Xenobiotica, 3, 305 (1973).

(308) P. D. Lawley and C. A. Wallick, Chem. Ind., 633 (1957).

(309) H. G. Windmueller and N. O. Kaplan, J. Biol. Chem., 236, 2716 (1961).

(310) P. D. Lawley and M. Jarman, Biochem. J., 126, 893 (1972).

(311) S. Ivankovic and D. Schmahl, Z. Krebsforsch., 69, 103 (1967).

(312) P. Bogovski and E. A. Walker, Eds, N-Nitroso Compounds in the Environment, IARC Scientific Publications No. 9, International Agency for Research on Cancer, Lyon, 1974.

(313) D. H. Fine, Personal Communication, 1975.

(314) D. E. Johnson, J. D. Millar, and J. W. Rhoades, Nitrosamines in Tobacco Smoke in Toward a Less Harmful Cigarette, Nat. Cancer Inst. Monograph 28, 181 (1968).

(315) D. H. Fine and D. P. Rounbehler, J. Chrom., 109, 271 (1975).

(316) D. H. Fine, F. Rufeh, D. Lieb, and D. P. Rounbehler, Anal. Chem., 47, 1188 (1975).

(317) D. H. Fine, F. Rufeh, and D. Lieb, Nature (London), 247, 309 (1974).

(318) D. H. Fine, D. Lieb, and F. Rufeh, J. Chromatog., 107, 351 (1975).

(319) P. E. Oettinger, F. Huffman, D. H. Fine, and D. Lieb, Anal. Letters, 8, 411 (1975).

(320) G. M. Telling, T. A. Bryce, and J. Althorpe, Agr. Food Chem., 19, 937 (1971).

(321) J. K. Foreman, J. F. Palframan, and E. A. Walker, Nature (London), 225, 554 (1970).

(322) A. Henriksen, Analyst, 95, 601 (1970).

(323) K. Heyns and H. Koch, Tetrahedron Letters, 10, 741 (1970).

(324) J. W. Howard, T. Frazio, and J. O. Watts, J. Assoc. Offic. Anal. Chem., 53, 269 (1970).

(325) T. A. Bryce and G. M. Telling, 20, 910 (1972).

(326) C. J. Dooley, A. E. Wasserman, and S. Osman, J. Food Sci., 38, 1096 (1973).

(327) E. D. Pellizzari, J. E. Bunch, R. E. Berkley and J. McRae, to be published (1975).

(327a) S. P. Levine, K. G. Hebel, J. Bolton, Jr., and R. E. Kugel, Anal. Chem., 47, 1075A (1975).

(328) Air Quality Criteria for Nitrogen Oxides, Environmental Protection Agency, Air Pollution Control Office, Publication No. AP-84, January, 1971.

(329) R. E. Baumgardner, T. A. Clark, J. A. Hodgeson, and R. K. Stevens, Anal. Chem., 47, 515 (1975).

(330) L. J. Purdue, G. G. Akland, and E. C. Tabor, Comparision of Methods for Determination of Nitrogen Dioxide in Ambient Air, U. S. Environmental Protection Agency, Research Triangle Park, N. C. 27711, June, 1975.

(331) M. E. Beard, J. C. Suggs, and J. H. Margeson. Evaluation of Effects of NO, CO_2, and Sampling Flow Rate on Arsenite Procedure for Measurement of NO_2 in Ambient Air, Research Report, Environmental Monitoring Series, Office of Research and Development, Environmental Protection Agency, EPA-650/4-75-019, April, 1975.

(332) R. G. Fuerst and J. H. Margeson. An Evaluation of the TGS-ANSA Procedure for Determination of NO_2 in Ambient Air, Research Report, Environmental Monitoring Series, Office of Research and Development, Environmental Protection Agency, EPA-650/4-74-047, November, 1974.

(333) J. Mulik, R. Fuerst, M. Guyer, J. Meeker, and E. Sawicki, Int. J. Env. Anal. Chem., 3, 333 (1974).

(334) L. P. Breitenback and M. Shelef, J. Air Poll. Control Assoc., 23, 128 (1973).

(335) E. D. Palmes, A. F. Gunnison, J. DiMattis, J. Gunnison, and C. Tomczyk, Am. Ind. Hyg. Assoc. J., 36, A-15, (1975).

(336) J. E. Lovelock and S. A. Penkett, Nature, 249, 434 (1974).

(337) C. Gelman and J. C. Marshall, Am. Ind. Hyg. Assoc. J., 36, 512 (1975).

(338) R. M. Burton et al, J. Air Poll. Control Assoc., 23, 277 (1973).

(339) J. W. Coffer, SO$_2$ Oxidation to Sulfate on a High Volume Air Sampler Filter. A thesis submitted in partial fulfillment of the degree of Master of Science in Engineering, University of Washington, 1974.

(340) T. Novakov, S. G. Chang, and A. B. Harker, Science, 186, 259 (1974).

(341) L. Dubois, A. Zdrojewski, T. Teichman, and J. L. Monkman, Int. J. Env. Anal. Chem., 1, 259 (1972).

(342) J. O. Frohliger and R. Kane, Science, 189, 455 (1975).

(343) T. B. McMullen, R. B. Faoro, and G. B. Morgan, presented at the Annual Metting of the Air Pollution Control Assoc., New York, June 22-26, 1969.

(344) L. L. Ciaccio, R. L. Rubino, and J. Flores, Env. Sci. Technol., 8, 935 (1974).

(345) D. Schuetzle, A. L. Crittenden, and R. J. Charlson, presented at the 65th Annual APCA Meeting, Ambient Air and Source Measurement, June 18-22, 1972, Miami Beach, Florida.

(346) D. Schuetzle, "Computer Controlled High Resolution Mass Spectrometric Analysis of Air Pollutants," Doctor of Philosophy dissertation, University of Washington, 1972.

(347) R. G. Smith, J. D. MacEwen, and R. E. Barrow, Am. Ind. Hyg. Assoc. J., 20, 149 (1959).

(348) V. Masek, Bitumen, Teere, Asphalta, Peche, 25, 349 (1974).

(349) P. K. Mukherjee and A. Sinha, Technology, Sindri, 10, 279 (1973).

(350) J. L. Shultz, A. G. Sharkey, Jr., R. A. Friedel, and B. Nathanson, Biochem. Mass Spectrometry, 1, 137 (1974).

(351) K. D. Bartle, M. L. Lee, and M. Novotny, Intern. J. Environ. Anal. Chem., 3, 349 (1974).

(352) M. L. Lee, K. D. Bartle, and M. V. Novotny, Anal. Chem., 47, 540 (1975).

(353) D. R. Stoltz, L. A. Poirier, C. C. Irving, H. F. Stich, J. H. Weisburger, and H. F. Grice, Toxicol. Appl. Pharmacol., 29, 157 (1974).

(354) B. N. Ames, E. G. Gurney, J. A. Miller, and H. Bartsch, Proc. Nat. Acad. Sci. USA, 69, 3128 (1972).

Analysis of atmospheric carcinogens and their cofactors

E. SAWICKI

TABLES I to VII

Table I.	Concentrations (µg/1000 m3) of BaP in Urban Atmospheres
Table II.	Concentrations of BaP in Highly Polluted Atmospheres and Effluents
Table III.	Concentrations of BaP in Environmental Mixtures
Table IV.	Atmospheric BaP (PAH) Analysis
Table V.	Atmospheric Aliphatic Hydrocarbons
Table VI.	Screening Test Values for Air Pollutant Families
Table VII.	Other Useful Screening Test for Air Pollutants

Table I

Concentrations (ug/1000 m³ air) of BaP in Urban Atmospheres

Country - City	Year	Concentration[a] Winter	Summer
Australia			
Sydney (6)	1962/63	8	0.8
Belgium			
Liege[b]	1958-1962	110	15
Canada, Ontario (8)	1961/62		
Sarnia		3.5	1.6
Windsor		15.0	7.8
Chatham		5.0	2.3
London		3.2	1.7
Kitchener		2.7	1.2
Brantford		5.2	2.2
Hamilton		9.4	5.7
St. Catharines		9.1	3.8
Toronto		5.4	6.4
Oshawa		2.6	1.7
Peterborough		10.0	1.8
Belleville		2.0	1.7
Kingston		11.0	4.0
Brockville		1.5	1.6
Cornwall		20	18.5
Ottawa		2.6	0.6
Orillia		14	1.3
North Bay		4.9	2.9
Sudbury		11	1.2
Sault Ste. Marie		3.9	4.0
Port Arthur			1.3
Czechoslovakia (9,10)			
Prague	~1964	122(high)	19(low)
100 meters away from pitch battery at Orlova - Lazy Coke Kilns		1800-3000	
400 meters away		1100-1900	
Denmark			
Copenhagen (11)	1956	17	5
England[c]			
London, St. Bartholomew's Hospital Medical College	1957-1962	1-2200	
County Hall (12)	1963-1965	1-54	
Bilston	1949/1950	46	
Bristol		27	
Burnley		13	
Cannock		27	
Hull		19	
Leicester		18	
Sheffield (13)		29	
Salford (14)		42	
Salford (15)	Feb., 1953	210	
Burnley		110	
Darwen		32	
Gateshead		35	
Lancaster		62	
Merseyside (St. George's Dock)		20	
Ripon		31	
Salford		15	
Warrington		108	
York		31	
Northern England and Wales (15)		24	
		11-108	

Table I

Concentrations (μg/1000 m³ air) of BaP in Urban Atmospheres

Country - City	Year	Concentration[a] Winter	Summer
Finland			
Helsinki (16)	1962/63	5	2
France			
Paris	1958	300-500	—
Lyon (17)	March, 1972	1.3	
Germany (18)			
Bonn	Feb., 1965	130	
Hamburg (19,20)	1961/62	340(high)	18(low)
Hamburg (21)	1961-1963	180	17
Hamburg	1961-1963	390(high)	10(low)
Dusseldorf	1963	75	4
Dusseldorf	~1964	41	
Dusseldorf	Feb., 1965	130	
Dortmund	~1964	101	
Gelsenkirchen	~1964	213	
Oberhausen	~1964	150	
Recklingharsen	~1964	62	
Bochum	~1964	415	
Bochum	Feb., 1965	240	
Wanne-Eickel	~1964	52	
Bottrop (18)	~1964	36	
Duisburg	~1964	150	
Hagen	~1964	54	
Essen	~1964	110	
Castrop - Rauxel	~1964	220	
Mannheim (20)	1965	28(high)	0.2(low)
	1966	58(high)	1.3(low)
	1967	43(high)	2.4(low)
Near vicinity of tar distillation plant (22)		700-5500	

Country - City	Year	Concentration[a] Winter	Summer
Hungary			
Budapest	~1968	1000[d]	32
	1971-1972	27[e]	
Iceland			
Reykjavik (11)	~1955		3
India			
Bombay (25)	1973		
Near gas plant - coal is fuel			170-860
Street (traffic density ~60 vehicles min⁻¹)			15-36
Near street and kiln for firing pottery			17-230
Residential suburbs			0.8-3.9
Iran			
Teheran (26)	1971	6	0.6
Ireland (16)			
Belfast	1961/62	51	9
Dublin	1961/62	23	3
Italy (27,30)			
Bologna	Feb., 1965	212(high)	6(low)
Genoa	~1964	37	1
Milan	1958-1960	610(high)	3(low)
Milan	Jan. 26, 27 1959	610	
Rome	1963-1966(winter)	20-147	
Japan (31,33)			
Muroran	~1965	110-160	
Osaka	1970	50	15
	1971	11	

Table I

Concentrations (ug/1000 m^3 air) of BaP in Urban Atmospheres

Country - City	Year	Concentration Winter	Summer
Sapporo	Feb., 1961	200	
Tokyo	Feb., 1964	15	
Netherlands^f (34)			
Amsterdam	1968	22	
	1969	18	2
	1970	5	2
	1971	8	
Delft	1968	20	3
	1969	18	1
	1970	12	3
	1971	6	3
Rotterdam	1968	15	3
	1969	23	1
	1970	19	3
	1971	23	2
Vlaardingen	1968	35	3
	1969	32	4
	1970	13	5
	1971	16	9
The Hague	1968	23	
	1969	12	4
	1970		4
	1971	13	
Norway			
Oslo (16)	1956	15	1
	1962/63	14	0.5

Country - City	Year	Concentration^a Winter	Summer
Poland(average for 10 large cities) (35)	1966/67	130	30
Gdansk	1966,1967	84,64	
Katowice		76,75	
Krakow		63,63	
Lodz		45,45	
Opole		54,47	
Poznan		48,49	
Szczecin		44,62	
Warszawa		29,29	
Wroclaw		57,64	
Zabrze		130,100	
South Africa (36)			
Durban	June, 1964		5,14,28
Johannesburg	May, 1964		49
	May, 1964		1100^g
Pretoria	1963/64	10	22
Spain			
Madrid (37)	1969/70	120	0
Madrid (38)	1969	9	0
Sweden			
Stockholm	1960	10	1
Stockholm (39)	-1967	27(high)	2(low)
Switzerland			
Basle (40)	1963-1964	S-80	

Table I

Concentrations (ug/1000 m^3 air) of BaP in Urban Atmospheres

Country - City	Year	Concentration Winter	Summer
USA[h], ~100 large urban communities	1958-1959	6.6(41)	
32 large urban stations (43)	~1962	5 (42)	
" " " "	1966-1967	3.3	
" " " "	1968	2.7	
" " " "	1969	2.9	
" " " "	1970	2	
Los Angeles (44)	June 1971-June 1972	1.1,0.5,3.5,0.03	
USSR (45)			
Leningrad	~1965	15	
Kiov	~1965	9	
Tashkent	~1965	110	
100 meters away from a pitch boiling plant of a cardboard factory in Taskent		129	
Near coke furnaces		570	
500 meters to the south of the furnaces		120	

[a] Average values. Where one value is reported it is an annual average value, unless otherwise stated. In some cases high and low values of the year are reported.

[b] In December, 1959 values ranging from 79 to 103 obtained; in February, during foggy weather values of 330-360 ug BaP/1000 m^3 were reported (7).

[c] In London during period in which smoke condensation was high - December, 1957 - a concentration of BaP as high as 2200 ng/m^3 was recorded, but after that episode winter values did not exceed 54 ug/1000 m^3 (12).

[d] Taken during a period of heavy smog (23).

[e] Winter values. Maximum value of 56 ug/1000 m^3 obtained (24).

[f] During A December fog values as high as 138 ug/1000 m^3 and 190 ug/m^3 were reported in Delft (1969) and Rotterdam (1964), respectively. In Rotterdam average winter and summer concentrations have steadily decreased from 1960 to 1971.

[g] Near a road tarring operation.

[h] The samples obtained in 1958/1959 and 1962 were analyzed by ultraviolet absorption spectrophotometry following alumina column chromatography. The remainder of the samples were analyzed by spectrophotofluorimetric analysis following alumina thin layer chromatography. Examination of composite samples (January - March, 1959) from 28 non-urban sites in various areas of the USA gave values ranging from 0.2 to 51 ug BaP/gram of particulates and ~0.01 to 1.9 ug BaP/1000 m^3 air (41).

Table II

Concentrations of BaP in Highly Polluted Atmospheres and Effluents

Area	Concentration, ng/m^3	Area	Concentration, ng/m^3
Auto exhaust (16 samples)	4000 (average)^a	Open burning	2800;4200;173,000
Coal-fired residential furnaces(46)	2200-1,500,000	Phonograph record plant (56)^c	
Coal-fired power plants (46)	~30-930	air from vinylite shop	3400-8000
Coal-fired unit, intermediate size (46)	49-7900	plant-territory	8-52
Coke oven, above gas works retorts (51)	216,000	Retort houses, maximal results (57)	2,300,000
Coke oven battery, on battery locations (52)	172-15,900	above horizontal retorts	220,000
, off battery locations	21-1200	Rubber products plant (56)^d	
battery roof	6700	shop air	43-51
larry car	6300	25 meters from shop	23-33
pusher	960	100 meters from shop	15-17
pump house	260	Sidewalk tarring operations (54)	52,110, 78,000
brick shed	380	Silicon carbide (carborundum) plant (56)^e	
cortez van	150	air in crusher shop	300-900
Garage air (Cincinnati downtown) (53)	33	air from coke ovens	400-730
Roof tarring operations (54)	90, 870, 14,000	100 meters from coke ovens	200-410
Gas works retort houses (51)	3000 (average)	500 meters from coke ovens	72-180
above the retorts	220,000	100 meters from plant	28-56
Gas-fired heat generation units (46)	~20-350	Smoky atmosphere	
Incinerators, municipal (46)^b	17, 19, 2700	Beer hall in Prague (58)	28-144^f
Incinerators, commercial (46)	11,000, 52,000	Arena (59)	0.7-22^g
Oil-fired heat generation units (46)	~20 - 1900	Tar paper plant (56)^h	
		Mass boiling shop	1100-1500
		plant territory air	230-290

Table II

Concentrations of BaP in Highly Polluted Atmospheres and Effluents

Area	Concentration, ng/m^3
100 meters from plant	125-135
500 meters from plant	38-61
Truck exhaust (46)	1500-36,000
Tunnel, Blackwell (60)	350
Tunnel, Sumner (60)	690
Wall-tarring operations (61)	520 × 10^3, 640 × 10^3 1600 × 10^3, 6000 × 10^3†

^{a}Range of values from 1700 to 20,000 ng/m^3 reported by Hangebrauck et al (46); 4300 ng/m3 reported by Gladen (47). Emission of BaP from a cold running motor reported to be 89,000 ng/m3 as compared to 4500 ng/m3 from a warm running motor (48). Output of BaP from 8 automobiles ranged around 2.9 - 33.5 ug/mile, from 4 trucks 40 ug/mile (49). In a "city driving" schedule BaP was emitted at the rate of 5 ug/1-min. run (50). In this work (46,49) the BaP/BghiPER and BaP/COR ratios were calculated by us to be 0.29 and 1.34, respectively. These low values are typical of areas where auto exhaust is the important pollutant e.g. BaP/BghiPER ratios in Los Angeles and San Francisco during the 1958 summer, 1959 winter seasons were 0.25 and 0.2, respectively. During the 1959 winter season BaP/BghiPER ratios of 1.4, 1.5, 1.5, 1.5 and 1.3 were obtained for Birmingham, Nashville, South Bend, Wheeling and Youngstown, respectively. These were coal combustion type areas. The BaP/COR ratios show similar differences; in the auto exhaust polluted areas values ranged around 0.5 in winter and 0.2 in the summer, while the coal combustion polluted areas ratios varied from 4 to 7 in the winter.

^{b}In the combustion of municipal refuse in a continuous feed incinerator it was found that about 9, 3450 and 3.5 mg of BaP and BeP were emitted per day in the stack gases, the residues and the water effluents, respectively (55).

^{c}Petroleum pitch and montan wax utilized in LP records contained 0.4 - 0.9 and 0.003 - 0.006% BaP, respectively.

^{d}Carbon blacks utilized in rubber products contained 0.00002 to 0.0038% BaP.

^{e}Petroleum coke utilized in plant contained 0.0027 - 0.0001% BaP.

^{f}Dependent on number of people smoking and ventilation. At same time urban air in Prague contained 2.8 - 4.6 ng BaP/m3.

^{g}In this air conditioned stadium BaP concentrations were 0.69, 7.1 and 21.7 for 0,9500 and 13500 people in the stadium respectively. Carbon monoxide and particulate concentrations also increased with an increase in the number of people in the stadium.

^{h}The coal tar, coal tar pitch and the boiling mass utilized in the plant contained 0.35 - 1.0, 0.4 - 2.0 and 0.3 - 1.4% BaP, respectively.

^{i}Within one hour worker could inhale ~3 mg of BaP, the amount of BaP he'd get from smoking approximately 300,000 cigarettes or breathing polluted air containing 10 ng BaP/m3 for approximately 70 years.

Table III

Concentrations of BaP in Environmental Mixtures

Material	Concentration µg/g	(Refs.)	Material	Concentration µg/g	(Refs.)
Airport runway sweepings	0.18	(62)	incinerator, vegetable matter	4500	(65)
Asbestos	up to 0.00055	(63)	motel space heater, gas	51	(65)
Asphalt	0.1 - 27	(64)	petroleum catalytic cracking catalyst regeneration - FCC	<8 - 97	(46)
Benzene - soluble fractions from gaseous effluents			pcc - cr - HCC	500 - 23,000	(46)
airborne particulates, urban	500[a]		pcc - cr - TCC (air lift)	3300 - 4600	(46)
industries emitting coal tar pitch fumes	18 - 950	(54)	auto exhaust	30 - 530	(46)
roof tarring operations	1300 - 19,000	(54)	truck exhaust	30 - 210	(46)
sidewalk tarring operations	1800 - 6,000	(54)	vicinity of pitch coke ovens of obsolete pattern	9000 - 13,000	(66)
coal-fired power plants	-7 - 330	(46)	Bitumen[b]	26	(67)
intermediate coal-fired units	5 - 3400	(46)	cracking residue	66	(67)
intermediate oil-fired units	<5 - 65	(46)	Bituminous coal	0.00004 - 0.001	(68)
intermediate gas-fired units	<17 - 170	(46)	Charcoal-broiled T-bone steaks[c]	0.05	(69)
coal-fired residential furnaces	990 - 10,400	(46)	Clothes, pitch coking plant workers		(70)
municipal incinerators	4 - 2000	(46)	trousers and blouses	19- 176	
commercial incinerators	4500, 5900	(46)	underwear, pants	0.9 - 6.3	
outdoor burning: grass, leaves	35	(46)	underwear, shirt	6.1 - 10.5	
outdoor burning: auto components	4300	(46)	Coal tar pitch[d]	up to 14,000	(54,64,71,72)
stack home heating, coal	9400	(65)	Creosote oil	0.0001 - 0.0002 (71), 50 (73)[e]	
incinerator, garbage	14,000	(65)	Foods		
incinerator, auto parts	580	(65)	barley, rye, chicory, sugar beet heated to 2400	$4 \times 10^{-5} - 41 \times 10^{-5}$	(74)
			barley, wheat and rye[b]	$2 \times 10^{-4} - 41 \times 10^{-4}$	(76)
			frankfurters[g]	6×10^{-4}	(77)

Table III

Concentrations of BaP in Environmental Mixtures

Material	Concentration μg/g	(Refs.)
kale	$13 \times 10^{-3} - 25 \times 10^{-3}$	(76)
katsuobushi	37×10^{-3}	(78)
leeks	7×10^{-3}	(76)
lettuce	$3 \times 10^{-3} - 13 \times 10^{-3}$	(76)
meat, grilled	$1 \times 10^{-3} - 50 \times 10^{-3}$	(79)
roast and fried meat	$2 \times 10^{-4} - 6 \times 10^{-4}$	(76)
roast coffee	$3 \times 10^{-4} - 5 \times 10^{-4}$	(76)
baker's yeast, dry	$2 \times 10^{-3} - 40 \times 10^{-3}$	(80)
mussels, isolated areas	$0 - 5 \times 10^{-3}$	(81)
mussels, harbor areas	$10 \times 10^{-3} - 60 \times 10^{-3}$	(81)
shucked oysters, harbor area	$10 \times 10^{-3} - 30 \times 10^{-3}$	(82)
Hexane	0.023	(83)
Liquid smoke flavors	0.03 - 3.8	(84)
Lungs, human	0-16 ug/lung	(85)
Particulates		
auto exhaust	100	(86)
diesel exhaust	5 - 37	(41)
non-urban airborne[h] - Jan - Mar. 1959	0.2 - 51	(41)
urban airborne[i] - Jan - Mar 1959	2.4 - 410	(41)
Birmingham, Ala. 1958 - 1959	31 - 230	(41)
Hamburg 1961 - 1963	32 - 2400	(41)
Los Angeles 1958 - 1959	2.6 - 28	(41)
Zurich 1965 - 1967	3 - 28	(41)
Blackwell Tunnel, England	91 - 150	(65)
stack home heating, coal	3100	(65)
incinerator, auto parts	130	(65)
incinerator, garbage	1700	(65)
incinerator, vegetable matter	70	(65)
municipal incinerator, refuse	41	(65)
outdoor burning: floor mats, auto seats, etc.	380	(65)
outdoor burning: grass, leaves	45	(65)
municipal incinerator, refuse	41	(65)
power plant stack effluents	0.5	(65)
urban, coal tar pitch polluted	7 - 1200	(66)
urban, sedimentation samples	30 - 250	(66)
Sawdust (whole wood smoke analyzed)	0.3×10^{-3}	(87)
Shale oil	1×10^{-4}	(88)
Smoked fish	$1 \times 10^{-3} - 78 \times 10^{-3}$	(89,90)
Smoked flavors, resinous condensate	0.025 - 3.8	(91)
Smoked mackerel	7×10^{-3}	(92)
Smoked meats	$0 - 0.9 \times 10^{-3}$	(92)
Smoked foods	$1 \times 10^{-3} - 50 \times 10^{-3}$	(91,93)
Smoke from grilled bacon	0.13	(94)
Snow, per m^2 in 24 hr.	1.4 - 2.2 ug	(62)
Snuff, British	7×10^{-3}	(95)
Snuff, Zulu	0.26	(96)

Table III

Concentrations of BaP in Environmental Mixtures

Material	Concentration μg/g	(Refs.)	Material	Concentration μg/g	(Refs.)
Soil, wilderness area	$1 \times 10^{-3} - 2 \times 10^{-3}$	(62)	Ichthammol	0	(112)
Soil, near airport	$0.2 \times 10^{-3} - 9 \times 10^{-3}$	(62)	Pine	5	(72)
Soil, forest[j]	$<0.3 \times 10^{-3}$	(97)	Metashal	8	(72)
Soil, near a Russian factory	650	(99)	Tar, mineral (pharmaceutical)	3300 - 6900	(72)
Soil, near oil refinery, coal tar pitch polluted	200	(100,101)	Tar, wood	0 - 340	(113)
Soil, Iceland	$0 - 5 - 8 \times 10^{-3}$	(97)	Tar, coal	5000	(115)
Soil, Iceland, near airport	0.8	(97)	Tar, Locacorten	225	(116)
Soils, forest, garden and plowed field near Cape Cod	0.04 - 1.3	(102)	Tar, nerosin[n]	70	(107)
Soil, near sidewalk of busy Moscow street[k]	21	(103)	Tea, Japanese	0 - 0.016	(117,122)
Soil, near traffic highways	2	(105)	Tobacco leaves	$4 \times 10^{-3} - 10 \times 10^{-3}$	(128)
Solvents[l]	0.6 - 1000 μg/liter	(106)	Cigarette ends and ashes	38×10^{-3}	(129)
Soot, aircraft engines	0.25 - 30	(62)	Cigarette smoke condensate (μg BaP/100 cigarettes)[o]	0.5,1.0,1.2,2.0, 7.5 - 12.5	(130)
Soot, coal	12 - 56	(107)	Tree leaves[p]	0.01 - 0.014	{(133) (134,135)}
Soot, coffee	0.2 - 0.4	(108)	Urine, smoker[q] (μg/liter)	0.55	(136)
Soot, wood	2 - 36	(107)	Vegetation[r]	0.01 - 0.02	(137)
Soot, auto exhaust[m] (from tail pipe)	220	(53)	Water[s]	ng/liter	
Soot, domestic	300	(108)	Drinkable water	<100 <2.5	
Tar, Belgian road	~5000	(110)	Rainwater	<300	
Tar, British road	~20,000	(110)	Surface waters (uncontaminated)	<0.1	
Tar, skin (drugs)		(111)	Surface waters		
Pityroll	15		Various German rivers	1 - 110	(134,138)
Glyteer	13		Rivers receiving industrial effluents		(139,140)

Table IV

Atmospheric BaP (PAH) Analysis

Sampling	High volume
	Low volume
	Dichotomous
	Trichotomous
	Anderson
Extraction	Soxhlet
	Sublimation
	Ultrasonic
	Polytron
Cleanup	Liquid-liquid extraction
	TLC
	CC
Separation	TLC
	CC
	HPLC
	GC
	Temperature programming
Measurement[a]	ECD
	FID
	SP
	SPF (g, l or s)
	LTF
	SPP
	MS

[a] ECD = electron capture detector; FID = flame ionization detector; SP = ultraviolet - visible absorption spectrophotometry, including HPLC detectors such as the λ260 UV detector and the tunable UV detector; SPFl = UV-Vis spectrophotofluorimetry, including HPLC detectors, such as tunable excitation/emission fluorescence detector, and manual fluorimetric measurements of solutions of eluents from various chromatographic separations; SPFg = GC gas phase fluorescent detectors; SPFs = direct fluorescence measurements on thin layer plates; LTF = quasilinear fluorescence measurements in one specific solvent with geometrical dimensions approximating those of the hydrocarbon being measured; SPP = spectrophotophosphorimetry; MS = mass spectrometry.

Table III

Concentrations of BaP in Environmental Mixtures

Material	Concentration µg/g	(Refs.)
at discharge site	8000 - 12,000	
at water intake	100	
~500 m downstream	2000 - 3000	
Ground water, uncontaminated	0.1 - 0.6	(134,138)
Ground water, contaminated	10 - 30	(136)
Domestic effluents	1 - 1800	(134,138)
dried humus	3,000,000	(141)
Industrial effluents		
coke or oil gas works - before discharge	0.34 x 10⁶, 1 x 10⁶	(142)
coke byproducts- after oil separation	7 x 10³, 130 x 10³ 250 x 10³, 290 x 10³	(139)
shale oil - after treatment	2000 - 320,000	(143)
Water environment	ng/g	
Algae, water (Greenland, west coast)	60	(144)
Algae, water (Italy)	2	
Fauna, marine		
codfish	15	(144)
mollusc	60	(144)
mussel	100, 500	(145)
oyster, Virginia	2- 6	(82)
sardine, Bay of Naples	70	(145)
Plankton, marine	5 - 400	(144 - 146)
Sediments, marine (mud or sand)	1 - 15,000	(144,145,147)
Sediments, river water	4 - 17,000	(139,140,148,149)
Sediments, surface water	2 - 2,000	(135,138,150)
Sediments, marine (Buzzards Bay)	75 - 370	(151)

Table III (continued)

[a] Average value for the Soxhlet extracts of particulate samples of ~100 American urban communities - 1962.

[b] Number of malignant neoplasms in some parts of this department high.

[c] Cooked close to coals for long periods.

[d] Roofing and gas works retort workers exposed to an average of 13 - 37 ug of BaP (range 0 - 154 ug) per 8 hr. day; asphalt workers <1 ug (57).

[e] Highly carcinogenic commercial sample.

[f] Close to industries wheat and barley contained 0.0003 and 0.001 ug/g of BaP, respectively, while far away from industry 0.00015 and 0.0003 ug/g of BaP, respectively (75).

[g] Russian style frankfurters 0.59 ug/kg while more modern skinless frankfurters 0.0044 ug/kg.

[h] 28 non-urban areas.

[i] 70 USA urban areas.

[j] Up to 1.3 ug BaP/g found in some forest samples (98).

[k] Several dozen meters away at control site 0.43 ug BaP/g soil. See also (104).

[l] Acetone, benzene, cyclohexane, petroleum ether, etc.

[m] Using gasoline as fuel the average of 5 sampling periods of a normal run amounted to 220 ug BaP/g of particulate matter (53). Automobile engines using gasoline as a fuel produced 330 ug BaP/g of soot under idling conditions (zero load); engines using diesel fuel on full load produced about 12 ug BaP/g of soot (109).

[n] Produced from shale oil used as a soil conditioner and to prevent soil erosion. It is carcinogenic on application to the skin of mice. In soil sampled several weeks after nerosin had been applied on the soil surface (1 - 2.1 tons per hectare) the BaP concentration reached 0.1 - 0.2 ug/g of air-dry soil in the top layer and 0.02 - 0.05 ug/g at a depth of 10 - 30 cm (114). Even with an increase of BaP in the soil, the accumulation in plants varies, e.g. it is absorbed in sunflower seed and potato but not in cotton seed and maizeseed.

[o] Average concentrations of 1.31 ug BaP/g of cigarette smoke condensate have been reported (123). It has been reported that from 1952 to 1972 BaP in the mainstream smoke of a leading U.S. cigarette without a filter tip has declined from 38 to 24 ng per cigarette or essentially decreased from 1.25 to 0.85 ug/g of dry tar (124). The analysis of tars from a wide variety of tobacco types showed that the tumorigenicity of the condensates on mouse skin increased dramatically with an increased concentration of BaP in the condensate. The carcinogenic potential of tobacco "tars" is paralleled by the carcinogenic activity of whole smoke when measured on the larynx of the Syrian hamster (125). It has been estimated that a cigarette smoker can take in 500 - 1500 mg particulates, 0.6 - 1.8 ug BaP, and 2 - 6 mg phenol per day (126). It has also been reported that sidestream smoke as compared to mainstream smoke contains 3 times as much BaP, 5 times as much carbon monoxide and 50 times as much ammonia (127).

[p] Decaying matter under the trees contained 0.043 ug BaP/g.

[q] Highest concentration. Similarly for a passive smoker 0.23 u BaP in the afternoon urine of smokers was twice as high as in

[r] Normal level in vegetation.

[s] Solubility of BaP in tap water is about 0.05 ± 0.02 ug/liter, 0.07 ± 0.02 ug/liter and in polluted river water up to 0.8 ug/l has been recommended that the maximum allowable concentration bodies should not exceed 0.0004 ug/liter (132). Background lev reported. In rural reservoirs values in bottom sand, algae and plants, and unpolluted water gave levels less than 0.002, 0.005 ug/liter (131).

[t] The values in a wide variety of waters near Koblenz, Germany for ranged from 2 to 400 ng/liter; the sediments ranged from 4 to 80 The BaP content in deposits on a river bottom have reached 420 t such deposits are demonstrable for several months. For a review water environment see (153).

[u] Values in the highly contaminated Acushnet River which runs into should be even higher since this river has a long history of massi and dumping from the Whaling and textile industries for well over

Table V

Atmospheric Aliphatic Hydrocarbons

Hydrocarbons	ug/m^3	Collection media	Analysis
Methane	1000	Grab	GC (fid)
$C_1 - C_5$		Grab	GLC or GSC (fid)
$C_5 - C_{10}$	3 - 100[b]	Cryogenic	GC (fid)
$C_6 - C_{20}$	1 - 30[c]	Wood charcoal	GLC $\rightarrow$ MS
$C_{18} - C_{34}$ (n-alkanes)	0.5[b]	Glass fiber paper (GFP)	LC $\rightarrow$ LC $\rightarrow$ GC
Alkanes ($C_{18} - C_{50}$)	1 - 4[b]	GFP	TLC $\rightarrow$ SPF
Alkanes ($C_{12} - C_{50}$)	3[b]	GFP $\rightarrow$ Chromosorb 102	△ $\rightarrow$ MS
Alkenes	2[b]	GFP $\rightarrow$ Chromosorb 102	△ $\rightarrow$ MS

[a] GC = gas chromatography, GLC = gas liquid chromatography, GSC = gas solid chromatography fid = flame ionization detector, LC = liquid chromatography, MS = mass spectrometry; SPF = spectrophotofluorimetry, TLC = thin layer chromatography, △ = temperature programming.

[b] Total

[c] Each

Table VI

Screening Test Values for Air Pollutant Families[a]

Mixture or family	Technique	Concentration ug/m^3
Suspended air particulates	Gravimetry	
1. Total		102^b
2. 3.5 - 20 um		~22
3. 1.7 - 3.5 um		~15
4. < 1.7 um		~65
Organic[c] air particulates, total (or fine)	Gravimetry	
1. Weight		6.7^d
2. Empirical formula[e]		$(C_{32.4}H_{48}O_{3.3}N_{0.16}S_{0.083}X_{0.065}(OR)_{0.12}$
3. Infrared CO/CH ratio		$(0.0)^f$
4. Estimated double bonds per formula		$(1-2 \text{ to } 4-5)^g$
5. Molecular weight		(460)
Total hydrocarbons		3000^h
Non-methane hydrocarbons		1500
Particulate alkanes		
1. Total	$TP^i \rightarrow MS$	3.2
2. $C_{12} - C_{30}$	$TP \rightarrow MS$	1.7
3. $C_{20} - C_{50}$	$TP \rightarrow MS$	1.0
Particulate alkenes		
1. Total	$TP \rightarrow MS$	2.3
2. $C_{12} - C_{30}$	$TP \rightarrow MS$	1.6
3. $C_{30} - C_{50}$	$TP \rightarrow MS$	1.1
4. C_{50} - polymer	$TP \rightarrow MS$	0.2
Total particulate alkylbenzenes	$TP \rightarrow MS$	0.02 - 4.3
Substituted styrenes	$TP \rightarrow MS$	0.07 - 7
Naphthalene and its methyl derivatives	$TP \rightarrow MS$	0.1 - 0.3
Total acids (as acetic acid)	$TP \rightarrow MS$	0.3 - 3.0
Total organic acids (C>10)	$TP \rightarrow MS$	0.6 - 1.3
Total aliphatic aldehydes	Colorimetry	3-79
Phenols	Colorimetry	5^j
PCB	GC	$0.00005 - 0.0008^k$

a
Unless otherwise stated these are ballpark numbers. Some of these values would be increased considerably near appropriate sources of the air pollutant. Other simple screening tests which have not been used extensively include total particulate aliphatic hydrocarbons (TpaH), total particulate aromatic hydrocarbons (TpAH), total particulate polynuclear aromatic hydrocarbons (TPAH), total particulate alkylating agents, total particulate aliphatic aldehydes, total particulate aliphatic aldehyde precursors etc.

b
Average value for 217 urban stations - 1966 - 1967. Average non-urban values were 21 ug/m^3 for 10 remote areas, 40 for 15 intermediate areas and 45 for 5 proximate areas (343).

c
Benzene- soluble Soxhlet extract unless otherwise stated.

d
1.1 to 2.5 ug/m^3 in non-urban areas (343).

e
This empirical formula and the molecular weight are values obtained for a composite benzene-soluble fraction of the airborne particulates from 200 widely spaced American communities. For chloroform extracts of airborne particles empirical formulae of $C_{32}H_{48.8}N_{0.5}O_{3.6}$, $C_{32}H_{50}N_{0.2}O_{4.8}$, and $C_{32}H_{53.5}N_{0.6}O_{4.7}$ have been reported. X represents Cl, Br, and I reported in terms of chloride while OR represents alkoxy reported in terms of methoxy (279a).

f
For hexane-benzene extract. Chloroform gave CO/CH ratios of 0.41, 0.50 and 0.37 (344).

g
For a hexane-benzene extract. Values as high as 12-13 are obtained for an acetone extract (344).

h
Total hydrocarbons have been reported as high as 12000 ug/m^3.

i
TP = Thermal programming (345,346).

j
Range 0.1 - 10 ug/m^3 phenols in St. Louis. In Detroit average of 4.3 and range of 1.1 - 7.8 ug/m^3 obtained. Street samples were 8.5 - 41 ug/m^3 while phenol-emitting plant areas had 255 - 988 ug/m^3 (347).

k
Sargasso sea area

Table VII

Other Useful Screening Tests for Air Pollutants[a]

Compound or family	Technique
Aliphatic aldehydes	
1. Total	Colorimetry
2. Formaldehyde	Colorimetry or fluorimetry
3. Acrolein	Colorimetry
Alkylating agents	
1. p-Nitrobenzylpyridine	
2. 4-Pyridinecarboxaldehyde 4-nitrophenylhydrazone	
3. 4-Pyridinecarboxaldehyde 2-benzothiazolylhydrazone	
Airborne particles	Mass spectrometry
1. Particulate matter	
a. H vs C distribution for hydrocarbons	
2. Benzene-methanol extract	
a. Mass spectra	
b. H vs C distribution for hydrocarbons	
3. Neutral oxygenated fraction	
a. H vs C distribution for mono-oxygenated compounds	
b. H vs C distribution for dioxygenated compounds	
4. Basic fraction	
a. N vs C distribution	
(1) for mononitrogenated compounds	
(2) for dinitrogenated compounds	
5. Acid fraction	
a. Hvs C distribution	
(1) for mono-oxygenated	
(2) for di-	
(3) for tri-	
(4) for tetra-	
Benzo(a)pyrene	
1. TLC	SPF
2. Ion-chromatographic	SPF
3. GC	Gas phase SPF

Compound or family	Technique
Component profiling to simple fingerprinting	Gas chromatography
1. Particulates	
a. Organic fraction	
b. Hydrocarbon fraction	
c. Oxygenated fraction	
2. Organic vapors	
a. Flame ionization detector	
b. Flame photometric detector	
c. Coulson electrolytic conductivity detector	
3. C_1 – C_5 hydrocarbons	
4. Benzene and its derivatives	
5. Phenols	
Carbohydrate, particulates	Colorimetry
Non-metal elements, particulate	GC, etc.
1. Total C	
2. Organic C	
3. Carbonate C	
4. Hydrogen	
5. Oxygen	
6. Sulfur	
7. Nitrogen	
8. Halogen	
9. Alkoxy	
Olefins, particulate	Colorimetry
1. MBTH oxidative	
Total particulate aliphatic hydrocarbons (TpaH)	TLC→SPF
Total particulate aromatic hydrocarbons (TpAH)	HPLC
Total polynuclear aromatic hydrocarbons (TPAH)	Colorimetry
1. Piperonal test	
2. Envelope method for 4, 5 & 6 ring compounds	GC

[a] With high resolution mass spectrometry mass elemental composition

ECHANTILLONNAGE ET TRAITEMENTS PRELIMINAIRES DES HYDROCARBURES AROMATIQUES POLYCYCLIQUES EN VUE DE LEUR ANALYSE ULTERIEURE (*)

P. CHOVIN

Laboratoire Central de la Préfecture de Police
39 bis, rue de Dantzig, 75015 Paris, France

De nombreuses études ont été faites ces toutes dernières années, sur les méthodes de caractérisation et de dosage des hydrocarbures aromatiques polycycliques présents dans l'environnement, l'intérêt se centrant sur les composés dotés de propriétés cancérogènes et mutagènes. Le présent exposé est consacré à la revue des méthodes d'échantillonnage et aux traitements qui doivent obligatoirement précéder les dosages proprement dits, lesquels seront exposés par un autre orateur (***).

ECHANTILLONNAGE

Les hydrocarbures aromatiques polycycliques présents dans l'atmosphère sont dissous dans des particules goudronneuses ou adsorbés sur des particules solides, ces dernières pouvant elles-mêmes être enrobées de goudrons. La distinction est parfois difficile à faire mais on utilise essentiellement deux méthodes d'échantillonnage :

- par recueil dans des jauges de dépôt pour les poussières sédimentables,
- par filtration pour les poussières restant en suspension.

Utilisation des jauges de dépôt

La méthode fait l'objet de la norme NF X 43006 pour la mesure des retombées de poussières (1).

Par l'intermédiaire d'un entonnoir en plastique rigide inattaquable aux agents atmosphériques, on récupère dans un flacon plastique de 10 litres les précipitations liquides et solides en un point donné pendant une période de temps fixée généralement à un mois. L'entonnoir est protégé par une grille pour éviter l'interférence d'objets gênants, gros insectes, débris végétaux importants, etc.... Les poussières insolubles sont séparées par décantation et filtration, puis séchées et pesées ; elles serviront ultérieurement à la recherche et au dosage des hydrocarbures polyaromatiques cancérogènes (benzo-3,4 pyrène, benzo-8,9 fluoranthène en particulier). Les quantités de poussières seront exprimées en grammes par mètre carré et par mois, les quantités d'hydrocarbures aromatiques polycycliques en micro-

---

(*) 2e partie de l'exposé par le Dr Fontanges, pp. 369

grammes par mètre carré et par mois.

Les résultats suivants ont été obtenus en 1972 pour une jauge placée au niveau du Laboratoire Central :

- Poussières totales :   9,0 g. $m^{-2}.mois^{-1}$
- Poussières insolubles : 5,5 g. $m^{-2}.mois^{-1}$
- Benzo-3,4 pyrène :   18,0 $\mu g.m^{-2}.mois^{-1}$

soit environ 3 ppm en masse par rapport aux poussières insolubles.

Les valeurs obtenues correspondant, en réalité, à un prélèvement d'une durée moyenne de 15 jours, le principal problème qui se pose est celui de l'oxydation progressive du benzo-3,4 pyrène en particulier pendant la période d'échantillonnage, lorsque le recueil des poussières ne se fait pas à l'abri de la lumière. Par ailleurs les résultats peuvent dépendre de la force et de la direction des vents.

D'autre part, la méthode ne faisant pas appel à la sélectivité granulométrique, il ne sera pas possible d'établir une corrélation entre la taille des particules et la quantité d'hydrocarbures aromatiques polycycliques dosée. Or, il est intéressant, pour étudier l'impact de la pollution sur l'individu, de faire appel à des prélèvements permettant de sélectionner des particules dans une zone granulométrique fixée. Et une étude faite sur des particules d'atmosphère urbaine à l'aide d'un impacteur à cascade à 5 étages Casella a permis de mettre en évidence le fait que, principalement en période hivernale, le benzo-3,4 pyrène était contenu à plus de 75 % dans les particules de diamètre inférieur à 4,6 μm et à plus de 50 % dans les particules de diamètre inférieur à 0,3 μm (2).

De tels résultats sont intéressants à connaître lorsque l'on sait que les particules inhalables ont un diamètre inférieur à 5 μm.

Nous verrons, dans la suite de cet exposé, qu'il existe un système d'échantillonnage avec filtration qui permet de sélectionner les particules correspondant à une zone granulométrique intéressante (allant de 1 μm à 10 μm de diamètre).

<u>Filtration des particules restant en suspension</u>

Le système de base est composé d'une unité filtrante, d'une pompe et d'un compteur permettant de connaître le volume de l'air filtré. Les filtres les plus généralement utilisés sont en cellulose, en esters de cellulose ou bien en fibres de verre ; ces derniers ne prenant pas l'humidité, le calcul du taux exact de poussiè recueillies est plus aisé. Les poussières récupérées ont un diamètre moyen supérieur à une valeur donnée qui dépend essentiellement de la porosité du filtre utilisé.
Le choix sera fait en ne perdant pas de vue qu'un pore de diamètre donné retient des poussières plus fines. Les couches filtrantes les plus utilisées dans ce type d'étude ont des diamètres de pore allant de 0,8 μm à 1,2 μm. A titre d'information, on notera qu'un filtre dont les pores ont un diamètre de 0,8 μm peut arrêter des particules de 0,1 μm.

L'étude d'un problème particulier peut amener à faire un choix déterminé pour le filtre, le débit d'air et le temps de prélèvement. Par exemple, une étude sur les suies inhalables a été faite à LYON par P. CHAMBON, B. MINAUD-VANPEL et R. CHAMBON afin d'avoir un ordre de grandeur de la quantité de matière déposée sur l'arbre trachéo-bronchique (3). Le captage a été fait sur filtre Millipore en ester de cellulose de porosité 1,2 μm avec un débit moyen de 1,5 $m3.h^{-1}$. Cependant, compte tenu des arrêts et du colmatage des filtres, ce débit est en pratique plus faible et la quantité d'air filtré a varié suivant les cas de 500 m^3 à 650 m^3 pour un mois, ce qui est comparable au volume d'air inhalé par un individu pendant le même temps. Par exemple, la quantité de suie recueillie pendant le mois de décembre a été d'environ 53,3 mg (209 $\mu g.m^{-3}$ d'air) dont 15,7 mg constituaient la partie organique contenant les hydrocarbures aromatiques polycycliques. Le benzo-3,4 pyrène a été dosé et la teneur décelée a été de 49,80 $ng.m^{-3}$ d'air soit environ 238 ppm en masse par rapport aux suies.

Le choix du matériel à utiliser peut être aussi guidé par les conditions du prélèvement à effectuer. Ainsi, lorsqu'on fait un prélèvement à grand volume d'air, l'utilisation du filtre en fibres de verre est préférable à celle du filtre de cellulose car la perte de charge n'augmente que lentement avec le dépôt et un grand débit peut alors être maintenu longtemps sans que les pompes chauffent trop fortement. En revanche, avec un système utilisant le filtre en cellulose, la perte de charge augmente rapidement et il faut utiliser des pompes refroidies à ailettes car les pompes à turbine risquent de subir une surchauffe (4).

Lorsqu'on veut faire une étude des cancérogènes présents dans l'atmosphère, il faut tenir compte du fait que ceux-ci peuvent atteindre des concentrations très faibles de l'ordre du nanogramme par 1000 m^3 d'air, il sera donc nécessaire de prélever une quantité importante de poussières pour faciliter les investigations ultérieures.

Diverses techniques peuvent être utilisées : filtres groupés en série (5), prélèvements simultanés regroupés pour l'analyse ultérieure, mais la méthode la plus généralement utilisée est celle des échantillonneurs à grand volume d'air.

"High_Volume_Sampler". On peut citer comme exemples les appareils Staplex et General Metal Works. La durée d'échantillonnage varie généralement de 24 h à plusieurs jours avec un débit qui doit être compris entre 1,13 et 1,70 $m^3.mn^{-1}$. On peut noter que la collecte sur 24 h est particulièrement intéressante car on peut déterminer les concentrations moyennes des échantillons en écartant des résultats les effets de phénomènes météorologiques et de la variation de l'activité humaine diurne.

La technique a été utilisée au Laboratoire Central pour une étude particulière de pollution dans la région parisienne. Les prélèvements étaient effectués sur filtre Whatman N° 1 en cellulose de dimensions 23 cm x 8 cm pendant une durée moyenne de 24 heures avec le High Volume Sampler GMWL 2000 H de General Metal Works. Les résultats furent les suivants :

- Poussières en moyenne : 150 $\mu g.m^{-3}$
- Benzo-3,4 pyrène : teneurs de 0,3 à 16,8 $ng.m^{-3}$

La quantité minimale de poussières nécessaire au dosage ultérieur du benzo--3,4 pyrène en particulier étant de 50 mg à 150 mg, il faut utiliser des volumes d'air filtré de l'ordre de 1000 m^3. D'autre part, avec cette méthode, on récupère les poussières de granulométrie supérieure à une valeur dépendant des conditions de prélèvement. Mais il existe un autre type de collecte de particules utilisant un appareillage à haut débit d'air (1200 m^3.h^{-1}) qui sélectionne également les poussières de granulométrie allant d'environ 1 µm à 10 µm. Ce collecteur a été mis au point par le Docteur FONTANGES (6, 8) et son principe est le suivant.

Collecteur à haut débit d'air. Dans ce collecteur, se produit le transfert dans un très faible volume d'eau déminéralisée et stérile, des particules de diamètre variant de 1 µm à 10 µm contenues dans un volume d'air 2.10^7 fois plus grand environ ; l'eau est ensuite éliminée.

Le fonctionnement de l'appareil est le suivant : l'air traverse successivement, après mise en circulation forcée par un motoventilateur :

- un filtre moustiquaire en toile inox qui arrête insectes, feuilles et débris,
- une batterie de chauffe mise en action quand la température descend au-dessous de + 5°C,
- un séparateur cyclogalax permettant de séparer par voie sèche les grosses et les petites particules, celles de diamètre supérieur à 10 µm étant alors stockées à part,
- un laveur solivore qui reçoit une pulvérisation d'eau en circuit fermé avec alimentation automatique en eau déminéralisée réglée pour compenser l'évaporation et les prélèvements et qui joue un double rôle : d'une part, il sature d'humidité l'air à épurer et, d'autre part, il ajoute des particules liquides à la suspension solide non arrêtée par le séparateur cyclogalax,
- un venturi où se produit une détente continue du gaz suivie d'une condensation instantanée. Les particules formant noyaux vont se trouver enrobées d'une mince couche de liquide. Il en résulte que des gouttelettes et particules enrobées forment une véritable bruine à la sortie du venturi,
- un séparateur cyclogolax travaillant par voie humide, recueillant les gouttes et gouttelettes d'eau contenant les particules. Cependant, les particules de diamètre supérieur à 1 µm ne sont pas arrêtées et sont acheminées vers l'organe suivant :
- un ventilateur qui assure la circulation de l'air dans l'ensemble des organes précédents.

Toutes les heures un prélèvement de 1 litre de suspension peut être fait. Celle-ci est stockée au noir à 4°C puis centrifugée à 55.000 tours/mn. Le culot est repris par l'eau stérilisée puis lyophilisé pour obtenir une poudre qui servira aux déterminations chimiques et bactériologiques ultérieures. Ce collecteur permet de piéger des masses de poussières allant jusqu'à 5 g environ, dont une partie peut servir à l'étude des effets des composés chimiques présents dans les particules sur les tissus vivants (un prélèvement d'air de 97 h correspondant à 40.000 m^3 a permis de récupérer 5,18 g de poussières (7). Une étude faite par FONTANGES, ISOARD et CHATOT (8) a montré que les prélèvements faits avec le collecteur à haut débit correspondaient à des concentrations de l'ordre de 150 à 170 µg de poussières inhalables par m^3 d'air. La vitesse linéaire d'aspiration (0,4 m/sec) étant proche de celle de l'air inhalé par un individu (7), on peut avoir une idée de la quantité d'hydrocarbures aromatiques polycycliques susceptibles de se déposer dans l'appareil broncho-pulmonaire (un homme inhalant environ 15 m^3.j^{-1} d'air), la quantité de microconstituants déposés dans son organisme pendant sa vie, soit 70 ans en moyenne, sera d'environ 65 g ayant véhiculé 1/10.000, soit 6.500 µg de composés du type hydrocarbures aromatiques polycycliques dont beaucoup sont cancérogènes).

Une étude a été également faite sur l'atmosphère lyonnaise par l'équipe du Docteur FONTANGES (9) pour comparer les résultats obtenus avec son propre collecteur et ceux résultant de l'emploi d'un "High volume sampler" (Staplex). Les résultats sont comparables et l'on peut en donner l'aperçu suivant :

13 hydrocarbures aromatiques polycycliques ont été étudiés quantitativement. Il est apparu que la granulométrie des particules en suspension n'avait pas d'influence sur la teneur en hydrocarbures de chaque grain de poussière. Les dérivés du pyrène et du fluoranthène étaient majoritaires. La pollution moyenne passait pour l'ensemble des 14 hydrocarbures dosés de 1.057 µg.g^{-1} en hiver à 290 µg.g^{-1} en été. On doit noter la contribution importante des chauffages domestiques en particulier au charbon dans l'augmentation de la pollution cancérogène en période hivernale.

Pour les mêmes périodes, les concentrations en benzo-3,4 pyrène varient de 53 µg.g^{-1} en hiver à 16 µg.g^{-1} en été.

On peut signaler en conclusion que, le collecteur de particules du Docteur FONTANGES n'étant pas commercialisé, la méthode utilisant le High Volume Sampler est actuellement la plus utilisée pour les études statistiques de pollution urbaine ou industrielle.

Il convient néanmoins de citer une étude de JAEGER (10) pour doser les hydrocarbures aromatiques polycycliques adsorbés sur les poussières volantes provenant d'une centrale thermique où ces dernières ont été collectées par le moyen d'un précipitateur électrostatique.

PREPARATION DES ECHANTILLONS

A partir de poussières recueillies, il faut maintenant extraire les hydrocarbures aromatiques polycycliques (HAP). Cette séparation est assez délicate car les HAP sont plus ou moins adsorbés sur les poussières organiques ou minérales.

Suivant la méthode d'échantillonnage utilisée deux procédés peuvent être mis en oeuvre.

Cas des particules recueillies dans des jauges de dépôt

On soumet les poussières insolubles séchées à une extraction au soxhlet par un mélange de 98 % de benzène et 2 % d'éthanol jusqu'à épuisement total. On conserve la solution benzénique pour une séparation ultérieure.

Cas des particules déposées sur des filtres

Plusieurs traitements des filtres ont été proposés.

Extraction par solvants. De nombreux solvants ont été utilisés :

- Benzène (11, 12, 13)
- Cyclohexane (14),
- Ethanol ou méthanol, - Ether, acétone ou acétate d'éthyle,
- Dérivés chlorés (15, 16).

Des études ont été faites sur les qualités d'extraction de ces différents solvants.

Le benzène est le plus couramment employé (16), mais en raison de sa toxicité, il est souvent remplacé par le cyclohexane, qui présente de plus l'avantage d'extraire préférentiellement les hydrocarbures aromatiques.

Les solvants hydroxylés ou polaires ont également une grande efficacité d'extraction mais sont capables de dissoudre en même temps des produits minéraux, comme les nitrates (11, 13, 17).

On doit tenir compte, pour le choix du solvant, de la méthode de détermination qui sera utilisée ultérieurement. Ainsi, lorsqu'une détermination fluorimétrique est envisagée, les solvants chlorés ne conviennent pas car ils influencent défavorablement la fluorescence des hydrocarbures aromatiques polycycliques. C'est le phénomène de "Quenching" (11). Il est dans ce cas recommandé d'utiliser des solvants de qualité spectrale dont l'absence de fluorescence est vérifiée préalablement :

- Le tétrachlorure de carbone est utilisé pour sa transparence en spectrographie infra-rouge, mais il convient de signaler sa très grande toxicité imposant de prendre des précautions dans son emploi,

- Le sulfure de carbone pour l'identification (18) en chromatographie gazeuse (19),

En revanche, le dichlorométhane est déconseillé pour les séparations en chromatographie sur couches minces car il solubilise la cellulose acétylée. D'une manière générale, il est admis que la meilleure efficacité d'extraction est obtenue quand les polarités du solvant et du composé à extraire sont les mêmes (17). L'extraction est réalisée actuellement au moyen d'un Soxhlet, mais les temps d'extraction varient suivant les auteurs de 6 heures à 22 heures (13, 17) voire plusieurs jours. On se contente souvent d'un temps d'extraction moyen de 8 h (12, 14, 20) car de toutes manières, l'extraction n'est jamais totale et ne permet de récupérer que 75 % environ des hydrocarbures fixés sur les particules (16, 21).

Il peut y avoir en effet des pertes ou décomposition par la chaleur et la lumière et surtout des rétentions dont la méthode précitée ne vient pas à bout.

__Extraction par solvants et ultra-sons__. Préconisée par l'équipe du Docteur FONTANGES, elle a l'avantage de dissocier les poussières atmosphériques en leurs constituants organiques et minéraux. De ce fait, elle est à la fois plus rapide (temps d'extraction réduit à 30 minutes) et plus efficace.

En outre, cette méthode peut être pratiquée à froid (22, 23), ce qui est bien préférable si l'on craint que l'extraction par des solvants portés à ébullition risque d'introduire une modification et peut être une dégradation irréversible des hydrocarbures aromatiques polycycliques extraits.

Une étude comparative concernant l'emploi des ultra-sons et du soxhlet a été faite pour sept solvants différents (23).

__Extraction par sublimation__. Certains auteurs effectuent une extraction par sublimation (21). Ce procédé évite les inconvénients dus à la méthode d'extraction par des solvants (longue durée d'exécution, faiblesse du rendement) et permet d'obtenir directement les composés polyaromatiques à partir des poussières. Les filtres (obligatoirement en fibres de verre) sont chauffés sous vide (10^{-2} mm de mercure à 300°C) pendant 50 minutes. Les hydrocarbures aromatiques polycycliques sont recueillis dans un tube capillaire refroidi et lavés ultérieurement avec du benzène ou du cyclohexane (26, 27, 28). Il est possible d'effectuer ensuite une chromatographie sur couches minces.

TRAITEMENTS PRELEMINAIRES DES SOLUTIONS OBTENUES

Concentration

Les extraits organiques obtenus sont ensuite concentrés par évaporation du solvant. MONKMAN et ses collaborateurs (24) ont effectué des essais relativement à la volatilité des hydrocarbures aromatiques polycycliques pendant la concentration des solutions extraites.

L'emploi de solvants d'extraction à haut point d'ébullition comme la diméthylformamide, nécessite uu chauffage à plus de 100°C pour l'évaporation. Un chauffage prolongé à cette température peut entraîner des dégradations. Certains auteurs ont donc mis au point une méthode d'évaporation sous vide à basse température (25) présentant plusieurs avantages :

- Gain de temps,
- Moins d'altérations chimiques (polymérisation - oxydation),
- Condensation des solvants toxiques.

ELIMINATION DES CONSTITUANTS GENANTS

Les extraits organiques concentrés sont des mélanges plus ou moins complexes de composés aliphatiques, aromatiques et hétérocycliques.

TABLEAU I

Schéma de traitement des solutions chlorométhyléniques
d'extraction des poussières atmosphériques
(selon NOVOTNY & Coll., 15)

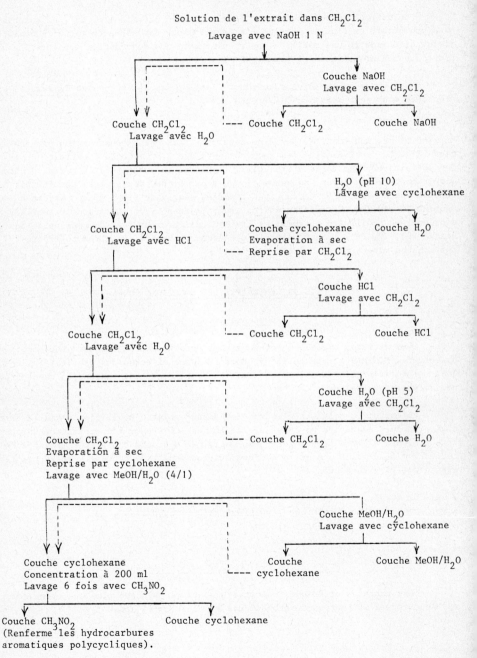

Ils doivent donc être traités chimiquement pour éliminer les constituants gênants qui sont :

- Les hydrocarbures aliphatiques,
- Les composés hydrophiles,
- Les phénols et bases pyridiniques facilement éliminés par traitement acido-basique.

On effectue le plus souvent une suite d'extractions liquide - liquide au moyen de solvants de polarités différentes qui éliminent sélectivement les composés indésirables.

Divers types de traitements peuvent être proposés.

Technique de CLOSSON et Coll. (22) : l'extrait cyclohexanique est traité par le mélange méthanol/eau qui sépare les composants hydrophiles de la fraction organique.

La solution cyclohexanique restante est ensuite traitée par le nitrométhane qui dissout les composés aromatiques en laissant dans la phase cyclohexanique les dérivés aliphatiques.

Le nitrométhane, présentant des dangers d'inflammabilité, est souvent remplacé par le diméthylsulfoxyde ou la diméthylformamide.

Technique de NOVOTNY et Coll. (15) : L'extrait organique en solution dans le dichlorométhane est traité successivement par la soude diluée, l'eau, l'acide chlorhydrique dilué pour éliminer les phénols et les bases pyridiniques.

La solution de dichlorométhane résultante est évaporée à sec et le résidu est repris par le mélange méthanol/eau qui sépare les fractions hydrophiles.

Le cyclohexane est ensuite repris par le nitrométhane comme précédemment.
Le TABLEAU I ci-après donne une représentation du schéma de principe du traitement.

La présence d'anthracène en grandes quantités peut masquer les spectres de fluorescence. On peut l'éliminer en le combinant à l'anhydride maléique par une synthèse diénique et éliminant l'excès de réactif par un lavage à la soude diluée et à l'eau.

SEPARATION DES HYDROCARBURES POLYCONDENSES ENTRE EUX

Une fois isolée, la fraction aromatique polycondensée est ensuite séparée par des procédés chromatographiques. Plusieurs types sont utilisés.

Chromatographie sur colonne

Cette méthode très ancienne est de moins en moins utilisée, car elle nécessite un temps d'exécution de plusieurs heures pour un rendement assez faible.

Différents adsorbants peuvent être mis en oeuvre :

- Oxyde d'alumine,
- Cellulose acétylée,
- Gel de silice,
- Résine de séphadex (15),
- Mélange de ces adsorbants,

La chromatographie sur colonne peut être considérée comme un traitement préliminaire pour la séparation des hydrocarbures aliphatiques.

Le choix du solvant éluant dépend de la polarité et de la nature de l'adsorbant.

Chromatographie de partage sur papier

Elle est surtout utilisée pour la séparation microanalytique. A côté de papiers filtres normaux (type Whatman), on peut utiliser des papiers acétylés (29) ou des papiers imprégnés de paraffine (5) (méthode utilisée au Laboratoire Central).

Si elle semble moins sélective que d'autres, cette méthode donne de meilleurs résultats car la récupération des produits adsorbés est pratiquement totale (29).

Chromatographie sur couches minces

Elle utilise les mêmes adsorbants et les mêmes solvants que la chromatographie sur colonne. Beaucoup plus rapide et plus sélective que cette dernière, cette méthode est de plus en plus employée dans les laboratoires.

Choix des différents supports :

- La cellulose, adsorbant organique est utilisée pour séparer les composés aliphatiques,

- le silicagel, adsorbant minéral, pour les composés phénoliques.

La diversité des adsorbants et des solvants utilisés en fait une technique relativement spécifique et aux possibilités multiples :

- Séparation des hydrocarbures appartenant à une classe donnée,

- Recherche à l'intérieur d'une classe, de composés d'intérêt particulier, pa exemple le benzo-3,4 pyrène.

Cette méthode peut donc servir également à éliminer des constituants gênants (aliphatiques ou hétérocycliques).

On effectue dans ce cas une chromatographie sur couche mince préparative (gel de silice - benzène) sur l'extrait benzénique lui-même.

On sépare ainsi les hydrocarbures polycycliques en deux zones :

- celle dont les indices de rétention vont de 0,1 à 0,8 et qui comprend les composés polycycliques hétérocycliques,

- celle dont les indices de rétention sont compris entre 0,8 et 1 et qui contient des hydrocarbures polycycliques aromatiques.

Elle est suivie d'une chromatographie bidimensionnelle sur mélange alumine/cellulose acétylée (2/1) qui sépare les hydrocarbures polycycliques aromatiques en spots individualisables. Elle élimine également les quelques hydrocarbures paraffiniques qui restaient dans la fraction aromatique.

Dans le cas de mélanges complexes ou difficilement séparables, on peut effectuer des séparations doubles en utilisant des chromatographies sur couches minces successives, par exemple, une sur alumine puis une sur cellulose acétylée (30).

Les solvants utilisés sont fonction de l'adsorbant choisi. On a généralement les couples adsorbant - solvant ci-dessous :

- Cellulose - Diméthylformamide/eau (22),
- Silicagel - Benzène/hexane,
- Alumine - Pentane/éther,
    Ethanol/toluène/eau (31).

D'un point de vue quantitatif, des auteurs ont cherché à mettre en évidence les pertes subies au cours de l'élution des plaques (32).

On surcharge l'échantillon avant séparation par une faible quantité connue de benzo-3,4 pyrène marqué au ^{14}C. La comparaison des radioactivités par spectrométrie à scintillation liquide donne le rendement global des différentes étapes opératoires.

L'intérêt de la connaissance précise des pertes est d'autant plus grand que les quantités à analyser sont plus faibles.

On a également étudié le problème de la récupération quantitative des hydrocarbures aromatiques à l'aide de deux solvants (benzène et cyclohexane) afin de déterminer un facteur de correction pour le dosage.
Les deux solvants donnent sensiblement les mêmes résultats (31).

CRHOMATOGRAPHIES EN PHASE LIQUIDE ET EN PHASE GAZEUSE

Ces méthodes ne sont mentionnées ici que pour mémoire car en réalité elles donnent direcement les résultats qualitatifs et quantitatifs attendus. Ce sont donc

En résumé, il résulte de ce qui précède que l'extraction, de même que la purification des hydrocarbures aromatiques polycycliques présents dans les poussières atmosphériques peut s'effectuer suivant plusieurs méthodes :

- Extraction par solvants avec ou sans ultra-sons,
- Sublimation,
- Distribution entre deux phases,
- Traitements chimiques,
- Chromatographies.

Certaines de ces méthodes de séparation peuvent servir en même temps de méthodes d'identification par groupes en tenant compte de la fluorescence, des indices de rétention, etc...

Toutes ne sont pas systématiquement nécessaires, l'on peut en utiliser seulement quelques unes d'entre elles, comme la chromatographie sur couches minces par exemple.

En fait, l'analyste dispose à l'heure actuelle d'un vaste arsenal de techniques qui doivent lui permettre, avec un peu de métier et d'expérience, de choisir celles qui seront le mieux adaptées au but poursuivi, compte tenu des circonstances.

Il lui restera ensuite à pratiquer les dosages individuels par des méthodes que le Docteur FONTANGES va maintenant exposer.

ABSTRACT

This review deals with the different methods of sampling of particulate matter, either suspended or sedimentable, the extraction of the contained carcinogenic hydrocarbons and the pretreatment necessary to allow a good quantitative determination of these potentially hazardous substances.

The sampling may be done by filtering a certain volume of air or collecting the sedimentable materials in deposit gauges. In the first case, filters of various nature and porosity are used and the volume of air to be filtered may vary between large limits ; but the trend is now to use high volume samplers, two types of which are described.

To extract the organic matter from the collected particulates, various solvents may be used, at their boiling point in a Soxlhet apparatus, or at ambiant temperature with the aid of ultra-sonic waves. The solutions, which contain many undesirable products, have to be subjected to a preliminary treatment in order to obtain a purified sample of the compounds to be determined. These treatments include washing with alcaline and acidic aqueous solutions to remove the impurities having phenolic or basic character. Finally, the prepurified solutions may be subjected to various chromatographic procedures, in order to make group separations which simplify the final identification and determination of the individual components present in the initial mixture.

BIBLIOGRAPHIE

(1) POLLUTION ATMOSPHERIQUE : Mesure des "retombées" par la méthode des "jauges de dépot". NORME FRANCAISE NFX 43.006, Juillet 1967.

(2) KERTESZ-SARINGER M., MESZAROS E. et VARKONYI T., : Technical note on the size distribution of benzo (a) pyrene containing particles in urban air. Atmos. Environm., 1971, 5, 429.

(3) CHAMBON P., MINAUD-VANPEL B. et CHAMBON R. : Analyse des suies atmospheriques lyonnaises. Tribune du CEBEDEAU, 1974, 365, 225.

(4) DAMS R. et HEINDRYCKX R. : A high volume air sampling system for use with cellulose filters. Atmos. Environm., 1973, 7, 319.

(5) HLUCHAN E., JENICK M. et MALY E. : Determination of airborne polycyclic hydrocarbons by paper chromatography. J. Chromatog., 1974, 91, 531.

(6) FONTANGES R. et ISOARD P. : Etude des microconstituants de l'atmosphère de la région lyonnaise à l'aide d'un nouveau collecteur à grand volume. Tribune du CEBEDEAU, 1968, 301, 674.

(7) ISOARD P., MARCOTTE F., LEMERCIER G. et FONTANGES R. : Etude de l'action expérimentale de polluants atmosphériques sur le poumon de la souris Balb/C. Poll. Atmos., 1972, p. 43.

(8) FONTANGES R., ISOARD P. et CHATOT G. : Prélèvement, analyse et effets biologiques des microconstituants atmosphériques de la région Rhône-Alpes. Lyon Pharmaceutique, 1972, 23, 253.

(9) CHATOT G., DANGY-CAYE R., FONTANGES R. et OBATRON P. : Etude de la pollution atmosphérique par les arènes polynucléaires dans la région lyonnaise à l'aide de deux collecteurs de principe différent. Atmos. Environm., 1973, 7, 819.

(10) JAEGER J. : Comportement des hydrocarbures aromatiques polycycliques adsorbés sur des corps solides - I - Extraction des hydrocarbures aromatiques polycycliques des corps solides. Ceskosl. Hyg., 1969, 14, 135.

(11) CHAD R.E., : Chromatography of carcinogenic polycyclic aromatic hydrocarbons. Chromato. Rev., 1970, 13, 61.

(12) CUKOR P., CIACCIO L.L., LANNING E.W. et RUBINO R.L. : Some chemical and physical characteristics of organic fractions in airborne matter. Environm. Sci. Technol., 1972, 6, 633.

(13) GORDON R. : Solvent selection in extraction of airborne particulate matter. Atmos. Environm., 1974, 8, 189.

(14) LAO R., THOMAS R.S., OJA H. et DUBOIS L. : Application of gas - Chromatograph - mass spectrometer - data processor ; combination to the analysis of the polycyclic aromatic hydrocarbon content of airborne pollutants. Anal. Chem., 1973, 45, 908.

(15) NOVOTNY M., LEE M.L. et BARTLE K. : The methods for fractionation, analytical separation and identification of polynuclear aromatic hydrocarbons in complex mixtures. J. Of Chromato. Sciences, 1974, 12, 606.

(16) SAWICKI E., STANLEY T.W., ELLERT W.C, MEEKER J. et PHERSON S.Mc : Comparison of methods for the determination of benzo(a)pyrene in particulates from urban and other atmospheres. Atmos. Environm., 1967, 1, 131.

(17) GROSJEAN D. : Solvent extraction and organic carbon determination in atmospheric particulate matter : The organic extraction - Organic carbon analyzer (OE - OCA) technique. Anal. Chem., 1975, 47, 797.

(18) GOLDBERG M.C., DE LONG L. et SINCLAIR M. : Extraction and concentration of organic solutes from water. Anal. Chem., 1973, 45, 89.

(19) GROB K. et GROB G. : Gas-liquid chromatographic - mass spectrometer investigation of C_6 - C_{20} organic compounds in an urban atmosphere. An application of ultra trace analysis on capillary colums. J. Chromatogr., 1971, 62, 1.

(20) CHATOT G., CASTEGNARO M., ROCHE J.L. et FONTANGES R. : Etude des hydrocarbures polycycliques aromatiques de l'atmosphère de Lyon. Chromatographie, 1970, 3, 507.

(21) HEROS M. et MALLET L. : Dosage du pyrene par spectrophotométrie d'absorption U.V. dans les vapeurs émises directement par les poussières atmosphériques et divers produits. Extension à la détermination d'autres hydrocarbures polycondensés. Analusis, 1973, 2, 176.

(22) CLOSSON A., NICOTRA C., PERILHON P. et CORNU A. : Etude des hydrocarbures polycycliques fixés sur les poussières atmosphériques. Poll. Atmos., 1970, p. 239.

(23) CHATOT G., CASTEGNARO M., ROCHE J.L. et FONTANGES R. : Etude comparée des ultrasons et du soxhlet dans l'extraction des hydrocarbures polycycliques atmosphériques. Anal. Chim. Acta., 1971, 53, 259.

(24) PUPP C., LAO R., MURRAY J. et POTTIE R. : Equilibrium vapour concentrations of some polycyclic aromatic hydrocarbons, $As_4 O_6$ and SeO_2 and the collection efficiencies of these air pollutants. Atmos. Environm., 1974, 8, 915.

(25) WITTGENSTEIN E., SAWICKI E. et ANTONELLI R.L. : Freeze-drying in the recovery of organic material from extracts of air particulate matter. Atmos. Environm., 1971, 5, 801.

(26) SCHULTZ M.J., ORNHEIM R.M. et BOVEE H.H. : Simplified method for determination of benzo(a)pyrene in ambiant air. J. Am. Ind. Hyg. Assoc., 1973, 34, 404.

(27) TOMINGAS R. et BROCKHAUS A. : Anwendung der Sublimationsmethode bei der Bestimmung von Benzo(a)pyren in Grosstadtaerosolen. Staub, 1973, 33, 481.

(28) MATSUSHITA H., ESUMI Y. et HANDA T. : Dosage rapide du benzo(a)pyrene dans des particules en suspension dans l'air. Jap. Analyst., 1972, 21, 772.

(29) D'ARRIGO V. et CAVANA M.R. : Identification and determination of aromatic polycyclic hydrocarbon by circular paper chromatography. Rass. Chim., 1974, 26, 169

(30) ABDOH Y., AGHDAIE N., DARVICH M. et KHORGAMI M. : Detection of some polynuclear aromatic hydrocarbons and determination of benzo(a)pyrene in Teheran atmosphere Atmos. Environm., 1972, 6, 949.

(31) CHATOT G., DANGY-CAYE R. et FONTANGES R. : Determination du facteur de correcti à apporter dans le dosage des hydrocarbures aromatiques polycycliques séparés par chromatographie sur couche mince. Chromatographia, 1972, 5, 460.

(32) DE WIEST F., RONDIA D. et DELLA FIORENTINA H. : Dosage du benzo(a)pyrene atmosphérique par couplage des techniques fluorimétrique et spectrophotométrique à scintillation liquide. J. Chromatogr., 1975, 104, 399.

Environmental Pollution and Carcinogenic Risks
Pollution de l'environnement et risques cancérogènes

MESURE DES CANCEROGENES POTENTIELS DANS L'ENVIRONNEMENT
II – METHODES DE DOSAGE PROPREMENT DITES

R. FONTANGES

Division de Microbiologie
Centre de Recherches du Service de Santé des Armées
108, boulevard Pinel, 69272, Lyon Cedex 1, France

I - INTRODUCTION

Monsieur le Professeur CHOVIN vient de vous décrire les différentes techniques d'échantillonnage utilisées pour prélever ce que nous désignons sous le terme de microconstituants atmosphériques.

Dans notre exposé, nous vous présenterons les méthodes que nous avons mises au point pour analyser les hydrocarbures polycycliques. A cette occasion, nous reprendrons quelques points particuliers de l'exposé précédent pour bien comprendre la succession des différentes opérations suivies.

Cette identification entre dans le cadre plus général d'une étude sur les effets biologiques des particules inhalées.

II - PRELEVEMENT

Le type d'appareillage utilisé est le collecteur à grand volume d'aspiration qui vient de vous être présenté. J'ajouterai qu'il a l'inconvénient d'être encombrant, mais qu'il a l'avantage de concentrer les particules en suspension dans l'air, dans une phase liquide. Ses grandes capacités d'aspiration lui permettent de fournir des échantillons dosables, à fréquence de prélèvements élevée.

Un collecteur mobile double la station fixe, permettant ainsi des études comparatives en fonction du lieu, des conditions météorologiques.

III - CONCENTRATION

Elle est obtenue, soit par centrifugation des suspensions recueillies toutes les heures suivie d'une lyophilisation de type classique, soit, directement, par lyophilisation rotative. Cette dernière technique, que nous avons mise au point, consiste à congeler le produit à analyser sur tambour rotatif refroidi à - 40°C, puis à dessécher les paillettes obtenues (glace + microconstituants) dans un appareil également rotatif qui lui est couplé et qui utilise le principe du lit fluidisé. L'avantage de ce dispositif est double. D'une part, les températures sont toujours très bien contrôlées, notamment au moment de la phase de sublimation, l'ensemble des opérations s'effectuant sous vide. Les phénomènes d'oxydation éventuels sont donc supprimés. D'autre part, le rendement est très important.

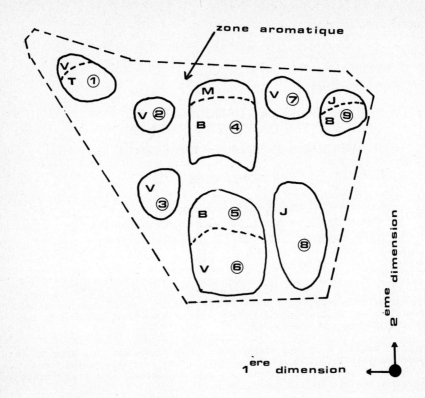

B : bleu
T : turquoise
V : vert
J : jaune
M : marron

CHROMMATOGRAMME A DEUX DIMENSIONS SUR COUCHES MINCES ALUMINE/CELLULOSE ACÉTYLÉE (2

<u>Identification des spots</u> : ① fluoranthène, pyrène ; ② benzo(a) anthracène, triphénylène ; ③ chrysène ; ④ benzo(e) pyrène, benzo(k) fluoranthène, pérylène ; ⑤ benzo(b) fluoranthène ; ⑥ benzo(a) pyrène ; ⑦ benzo(ghi) pérylène ; ⑧ ortho phénylène pyrène ; ⑨ coronène.

IV - EXTRACTION

La poudre obtenue est constituée par deux fractions quantitativement très différentes :
- la partie minérale, qui représente environ 99 % de l'ensemble,
- la partie organique, elle-même divisée en sous-fractions comprenant,
  . les bactéries et les champignons,
  . les hydrocarbures à chaînes longues saturées ou non, dits aliphatiques,
  . les hydrocarbures hétérocycliques dont un atome de carbone de certains de leurs cycles benzéniques est remplacé par un azote ou un oxygène,
  . enfin, les hydrocarbures polycycliques aromatiques qui résultent de l'accolement de différents noyaux benzéniques non substitués et où parmi lesquels se situeraient les molécules à potentialité carcinogène. Nous avons utilisé la technique d'extraction de la phase organique par les ultrasons pour continuer, d'une part, de travailler à des températures peu élevées et, d'autre part, pour réduire le temps de réponse.

Au terme d'une étude comparative utilisant comme solvant l'éther éthylique, l'acétone, le chloroforme, le méthanol, le benzène, le cyclohexane et le dichlorométhane, c'est ce dernier que nous avons retenu. Après filtration sur verre fritté de porosité n° 4, la fraction organique soluble est évaporée sous pression réduite jusqu'à un volume de 0,5 cm3 environ en vue de l'analyse chimique.

V - ANALYSE CHIMIQUE

A) Etude QUALITATIVE des arènes polynucléaires

- Nous utilisons la chromatographie sur couches minces contrôlée par la chromatographie en phase gazeuse. Ce couplage est réalisable à condition de n'employer que des adsorbants et des solvants chimiquement purs. Du fait de la grande quantité d'air aspirée par le collecteur, nous pouvons suivre, dans des temps très courts, l'évolution de la pollution de l'air pour un site donné. Ainsi, après cryodessication du produit il suffira d'une dizaine d'heures pour identifier les arènes polynucléaires (30 mn d'extraction à l'aide des ultrasons, 6 heures pour la purification de la fraction aromatique polycyclique par chromatographie sur couches minces, 2 heures pour la chromatographie en phase gazeuse de cette fraction).

- Les systèmes chromatographiques sur couches minces sont les suivants :
. système n° 1 : chromatographie préparative (adsorbant : gel de silice, solvant : benzène)
. système n° 2 : chromatographie bidimensionnelle (adsorbant : alumine-cellulose acétylée, 2:1; solvant 1ère dimension : pentane-éther, 98,5:1,5; solvant 2ème dimension : éthanol-toluène-eau, 17:4:4). Avant utilisation, la plaque est activée à 130° pendant 45 minutes.

La zone fluorescente du 1er système chromatographique, correspondant aux hydrocarbures aromatiques est raclée, éluée par le benzène et redéposée sur le deuxième système. Les spots individualisés sur ce second système sont raclés et élués par le benzène. Cet éluat, ramené à un volume connu, est injecté dans le chromatographe en phase gazeuse.

- La technique de chromatographie en phase gazeuse utilise un support Chromosorb G. La phase stationnaire dissoute dans le chloroforme est constituée par du SE-52. L'imprégnation est réalisée par évaporation de la solution chloroformique de phase, à température ambiante, sous vide. Le taux d'imprégnation est de 4,5 en poids par rapport au support sec. Le remplissage a lieu dans des colonnes de 1/8 de pouce et de 2 m de long. Les conditions de chromatographie sont les suivantes : température initiale 100° (maintenue pendant 3 minutes puis la montée en température est affichée à raison de 3°/minute) température finale, 290°; gaz vecteur, azote; débit 60 ml/mn; température du bloc d'injection 250° et température du bloc de sortie 290°.

A l'aide de ce couplage on arrive ainsi à identifier facilement et rapidement 13 arènes polynucléaires, comme l'indiquent la figure 1 et le tableau I.

Spot	Température de rétention (°C)	Hydrocarbure présent	Poids en microgrammes par g. de poussières	
			CPG	UV
1	208 202	pyrène fluoranthène	48 38	51 39
2	230 230	benz(a)anthracène triphénylène	22 10	20 0
3	253	chrysène	80	80
4	250 250	benzo(e)pyrène benzo(h)fluoranthène	80 10	78 10
5	250	benzo(1)fluoranthène	25	non dosé
6	250 253	benzo(b)fluoranthène benzo(a)pyrène	260 40	252 42
7	272	benzo(g,h,i)pérylène	66	65
8	268	O-phénylène pyrène	92	90
9	284	coronène	12	13

TABLEAU I - Identification des Hydrocarbures aromatiques polycycliques sur plaque et en phase gazeuse.

B) Etude QUANTITATIVE des arènes polynucléaires

- Dans le cas d'un dosage TOTAL, la zone aromatique délimitée sur la figure 1 est éluée par le benzène et injectée en phase gazeuse en même temps qu'une quantité connue d'un étalon R. Nous obtenons alors le chromatogramme de la figure 2.
Nous avons utilisé la technique de l'étalon interne, le coefficient de réponse relatif des arènes polycycliques ne variant que de ± 5% par rapport à l'unité.

- Dans le cas du dosage PARTICULIER d'un hydrocarbure, la même technique est adoptée, mais l'élution et l'injection se font à partir d'un des neuf spots de la figure 1. Chacun de ces spots contient un ou plusieurs hydrocarbures préalablement identifiés.

Nous remarquons, sur le tableau I, que les résultats obtenus par cette méthode sont similaires à ceux qui proviennent de l'utilisation de la spectrophotométrie ultraviolette, préconisée dès 1960 par SAWICKI.

VI - RESULTATS

La mise au point de ces méthodes de dosage des arènes polynucléaires nous a permis d'effectuer une étude comparative entre plusieurs points de prélèvements de la région lyonnaise - en choisissant des zones urbaines, suburbaines, rurales et forestières - en fonction des saisons. Il semble, d'une façon générale que les dérivés du pyrène et surtout du fluoranthène soient majoritaires (tableau II). Leur taux ne sont pas pourtant supérieurs à ceux relevés dans la majorité des grandes villes, en particulier en ce qui concerne le benzo(a)pyrène, plus fréquemment dosé. On remarque également sur ce tableau que la pollution d'une grande ville est moins importante à mesure que la consommation des différentes sources d'énergie diminue, ce qui est logique. Enfin, l'éloignement du centre de pollution aboutit à une baisse considérable du taux des arènes polynucléaires.

Hydrocarbures aromatiques dosés en µg g^{-1} de poussières	Saisons			Zones		
	Lyon (Hiver)	Lyon (Printemps)	Lyon (Eté)	Suburbaine industrialisée	Rurale	Forestière
Fluoranthène	64	27	23	11	9	1
Pyrène	51	11	9	5	4	1
Chrysène	106	21	21	8	3	1
Benzo(a)anthracène	29	0	7	1	1	1
Triphénylène	13	0	0	0	0	0
Benzo(a)pyrène	53	32	16	5	4	1
Benzo(b)fluoranthène	350	214	138	27	11	1
Benzo(k)fluoranthène	46	17	14	3	1	0
Benzo(e)pyrène	106	24	16	8	5	1
Pérylène	13	0	0	1	0	0
Orthophénylène pyrène	122	57	20	1	1	1
Benzo(ghi)pérylène	88	28	19	11	5	1
Coronène	16	7	7	5	3	1
Total	1057	439	290	86	47	10

TABLEAU II - Récapitulatif des résultats.

CONCLUSION

Au cours d'une étude générale sur la recherche des effets biologiques des microconstituants atmosphériques nous avons été amenés à mettre au point des méthodes qui conduisent au dosage rapide et reproductible des arènes polynucléaires. L'utilisation de deux collecteurs sélectifs, l'un fixe, l'autre mobile, à grand volume d'aspiration, d'un lyophilisateur rotatif à grand rendement, d'un couplage chromatographie sur couches minces-chromatographie en phase gazeuse a permis de dresser la carte de pollution de la région lyonnaise, pour 13 arènes polynucléaires, en fonction du temps et du lieu de prélèvement.

SUMMARY

MEASURING POTENTIAL CARCINOGENS IN THE ENVIRONMENT
II - Final methods of determination

As part of a general study of the biological activity of atmospheric micro constituants, we havè perfected the methods for the rapid determination of polycyclic aromatic hydrocarbons.

The method involves the use of two selective collectors, a high output rotatory freeze-drier, two-dimensional thin-layer chromatography and gas chromatography. A pollution map has been prepared of the Lyons area showing the atmospheric concentrations of 13 polycyclic aromatic hydrocarbons and their variation with weather and sampling point.

LES COMPOSES N-NITROSÉS
ANALYSE ET CARCINOGENICITE EVENTUELLE
CHEZ L'HOMME

L. GRICIUTE

Service des Cancérogènes de l'Environnement
Centre International de Recherche sur le Cancer
150, cours A. Thomas, 69008 Lyon, France

Malgré les connaissances considérables acquises ces 20 dernières années sur l'action cancérogène des composés N-nitrosés chez l'animal, le rôle que jouent ces substances en pathologie humaine reste à élucider.

Bien qu'on ne puisse extrapoler directement à la situation humaine les données expérimentales recueillies sur les cancérogènes, deux raisons principales nous conduisent à penser que les résultats de l'expérimentation animale sont applicables à l'homme :

1) Tous les cancérogènes connus induisent des tumeurs chez certaines espèces d'animales ([1,2,3]).

2) La cancérogènese est un phénomène cellulaire. Dans une large mesure, les cellules des différentes espèces d'animaux se ressemblent; l'ADN constitue la base génétique de toutes les cellules ([1]).

Nous avons maintenant des preuves indirectes indiquant que les composés N-nitrosés pourraient être cancérogènes chez l'homme :

1) On a observé que les nitrosamines induisent les mêmes lésions aiguës chez l'homme et chez l'animal : par exemple, la nécrose du foie observée chez le rat après l'administration d'importantes doses de nitrosamines a également été constatée chez l'homme ([4]).

2) Les taux de métabolisme *in vitro* de la diméthylnitrosamine sont très semblables dans le foie du rat et de l'homme ([5]).

3) Des lésions prolifératives, considérées comme précancéreuses, peuvent apparaître dans les cultures tissulaires de poumon d'embryon humain après l'administration de N-nitroso-urée ([6]).

La preuve ultime de la cancérogénicité de certaines substances a été jusqu'ici la morbidité exceptionnelle par cancer enregistrée dans certains environnements ou constatée lors d'études de cas particuliers ([7]).

Bien qu'elles soient cancérogènes, les nitrosamines diffèrent considérablement des autres substances cancérogènes présentes dans l'environnement (hydrocarbures aromatiques polycycliques, amines aromatiques, amiante, etc.). Par exemple, la nitrosation peut se produire dans l'environnement, mais également *in vivo* - aussi les nitrosamines sont-elles à la fois exogènes et endogènes.

En outre, bien que certaines nitrosamines soient organo-spécifiques, ces substances dans leur ensemble induisent maintes formes différentes de cancer chez l'animal ([8]). Si le même phénomène se produisait chez l'homme, on ne pourrait pas distinguer la morbidité générale par cancer de celle due aux nitrosamines. On ne peu donc espérer faire des études de cas particuliers sur la cancérogénicité des nitrosamines chez l'homme. Une autre méthode d'étude des nitrosamines chez l'homme consisterait à établir une corrélation entre la morbidité par cancer dans certaines régions et les quantités de composés N-nitrosés existant dans l'environnement. Cette méthode s'est avérée particulièrement fructueuse dans les études sur l'aflatoxine ([9]) Les effets cancérogènes des nitrosamines chez l'animal dépendent essentiellement de l dose, encore qu'une dose unique puisse être "efficace".

Quantité de N-nitroso diméthylamine	Tumeurs du foie chez le rat
2 mg/kg	1/26
5 mg/kg	8/74
50 mg/kg	>66%
	(Terracini, 1967 ([10]))

Tableau I. - Nitrosodiméthylamine dans l'alimentation

Pour estimer la charge totale de composés N-nitrosés dans l'environnement, convient de mesurer trois groupes de substances:

1) Composés N-nitrosés dans différents produits choisis dans l'environneme

2) Composés N-nitrosés *in vivo* (matériel biologique humain)

3) Précurseurs des N-nitrosamines (nitrates, nitrites et amines susceptibl d'être nitrosées) qui peuvent intervenir dans la nitrosation *in vivo* comme dans l'environnement.

On a démontré que la formation de N-nitrosamines est possible *in vitro* dans des conditions analogues à celles de l'estomac humain, à partir de substances exista dans les aliments ([11]). Ces substances peuvent être des substances naturelles (amine: secondaires ou tertiaires), ou des additifs alimentaires (nitrites, nitrates).

Un milieu acide est normalement nécessaire pour que la nitrosation se produi mais la présence d'un catalyseur - certaines bactéries par exemple - peut permettre nitrosation en milieu neutre. On pense que dans l'organisme humain, en présence d'infection, les localisations les plus favorables à la nitrosation sont la bouche, l'intestin et la vessie ([12]). La nitrosation est influencée par plusieurs facteurs, comme le pH qui, dans l'estomac humain, se modifie sans cesse, ou la présence de certaines substances qui inhibent (acide ascorbique ([13]), acide sulfanilique ([14])) ou catalysent (thiocyanate ([15]), certains complexes métalliques ([16])) la réaction. Les tanins, par exemple, peuvent à la fois catalyser et inhiber la formation de nitrosodiéthylamine *in vitro*, ce qui dépend dans une large mesure des conditions de pH et la concentration relative des produits en réaction ([17]). La nitrosation dépend aussi considérablement des taux des nitrites et des substances nitrosées ([18]). On sait peu de choses de la dégradation de nitrosamines *in vitro*.

Etant donné la complexité du problème de la formation des N-nitrosamines *in vivo*, il est pour l'instant plus réaliste d'étudier les N-nitrosamines exogènes en recourant à la chimie analytique. Nous sommes conscients du fait que la quantité de N-nitrosamines exogènes volatiles ne constitue qu'une petite fraction de la charge totale des composés N-nitrosés dans l'environnement, mais la détermination de cette quantité semble valable car au stade actuel des connaissances c'est le seul élément que nous puissions détecter.

Nous avons peu d'informations systématiques sur les quantités de composés N-nitrosés existant dans l'environnement. Le service des Cancérogènes de l'Environnement a donc adressé un questionnaire à tous les laboratoires d'analyse qu'il savait s'intéresser à ces recherches. Grâce à ce questionnaire, nous savons maintenant que 27 laboratoires analysent des échantillons d'aliments, 5 d'air, 4 de tabac et 2 d'eau. En outre, 2 laboratoires étudient les sols et les effets des pesticides sur les végétaux. Plusieurs laboratoires n'ont pas indiqué la matière qu'ils étudiaient. Les chimistes du service ont également passé en revue toutes les données disponibles concernant l'analyse des composés N-nitrosés dans les aliments. Ils ont constaté que la plupart des laboratoires se consacrant à l'analyse d'aliments travaillent sur des échantillons de viandes cuites, bien que soient aussi analysés des prélèvements de poisson, des produits laitiers et quelques échantillons de tabac et de boisson. Les principaux composés N-nitrosoés détectés lors de ces études étaient la N-nitrosodiméthylamine, la N-nitrosodiéthylamine, la nitrosopipéridine et la N-nitrosopyrrolidine, et si les quantités décelées variaient de 0 à 1000 µg/kg, elles étaient généralement de l'ordre de 5 à 10 µg/kg.

Malheureusement, on a obtenu ces données en appliquant des méthodes très diverses pour le prélèvement, la conservation, la purification, l'identification et l'estimation, si bien qu'on ignore dans quelle mesure les résultats sont comparables. Aussi la normalisation des méthodes d'identification des composés N-nitroés et la détermination de leur comparabilité constituent-t-elles la première étape à franchir en vue du dosage de ces substances dans l'environnement. D'autre part, il n'est pas possible de donner une définition générale de la sensibilité absolue nécessaire pour la détermination des taux de composés N-nitrosés. Les méthodes devraient être assez sensibles pour permettre de détecter les concentrations qui, malheureusement, sont inconnues à l'heure actuelle. Chez l'animal d'expérience, les composés N-nitrosés agissent à des très faibles doses. Par exemple, une dose de 0,75 à 1,7 mg/kg de diéthylnitrosamine administrée avec les aliments s'avère cancérogène chez le rat, et une dose ayant un effet marginal pourrait être estimée à 0,5 mg/kg [1]. Möhr *et al.* ont établi que la dose de diéthylnitrosamine sans effet chez le hamster correspond à une application unique inférieure à 0,03 mg par animal [19]. Il existe des méthodes permettant de détecter les nitrosamines volatiles jusqu'à la dose du µg/kg [20,21]. Les méthodes de détection des nitrosamines non volatiles en sont encore au stade de la mise au point.

L'analyse de substances pures et de mélanges simples n'a pas suscité de problèmes mais les résultats peuvent être ambigus lorsqu'on a appliqué certaines de ces techniques pour l'analyse d'aliments, de boissons, de fumées et d'échantillons d'air.

Afin de fournir à tous les laboratoires intéressés des informations sur les méthodes d'analyse des nitrosamines volatiles - group de composés le plus accessible et très important biologiquement - le service des Cancérogènes de l'Environnement a organisé une étude collective comportant trois phases:

1. On a fourni aux participants quatre solutions: deux d'eau et deux de dichlorométhane, contenant chacune deux nitrosamines (diéthylnitrosamine et nitrosopyrrolidine) a des concentrations de l'ordre de 10 µg/kg. Les participants

étaient libres de choisir leur propre méthode d'analyse.

Les résultats de cette première phase d'étude ont encouragé le service des Cancérogènes de l'Environnement à la poursuivre.

2. On a envoyé aux participants quatre boîtes de viandes après avoir ajouté à trois d'entre elles des quantités connues (environ 20 µg/kg) de nitrosamines (nitrosodiméthylamine, nitrosodiéthylamine, nitrosodibutylamine et nitrosopyrrolidine). La quatrième boîte, qui ne contenait pas de nitrosamines, a servi de "blanc".

Les résultats de 15 laboratoires ont fait apparaître une reproductibilité entre laboratoires de ± 50% et une reproductibilité dans le même laboratoire de ± 20%.

	NA	NE	NB	NPy
Valeur réelle (µg/boîte)	6,3	5,2	6,6	6,6
Moyenne générale	4,8	4,6	4,9	4,7
Ecart-type de la moyenne à l'intérieur d'un même laboratoire	0,92	0,650	0,898	1,801
Coefficient de variation correspondant	19,5%	14,3%	18,2%	38,1%
Ecart-type des moyennes obtenues par différents laboratoires	2,36	2,14	1,85	2,79
Coefficient de variation correspondant	48,9%	47,0%	37,4%	59,0%

NA - nitrosodiméthylamine; NE - nitrosodiéthylamine;
NB - nitrosodibutylamine; NPy - nitrosopyrrolidine

Tableau II. - Evaluation statistique des résultats de l'étude analytique collective sur les nitrosamines

Aucune méthode ne s'est révélée meilleure que les autres ([20]).

3. Le taux choisi de 20 µg/kg étant considéré comme supérieur à la concentration dans l'environnement en général, on a, pour la troisième phase de l'étude, utilisé de plus petites quantités de nitrosamines, et envoyé aux participants des échantillons de viandes en boîte contenant quatre des mêmes nitrosamines à la dose de 5 µg/kg.

Treize laboratoires ont participé à cette troisième phase et les résultats indiquent que la concordance entre laboratoires demeure de l'ordre de ± 50%, avec une variation assez caractéristique à ce niveau pour les substances à l'état de traces. On peut donc conclure qu'il existe un certain nombre de méthodes adéquates pour l'analyse de petites quantités de nitrosamines volatiles dans les produits existant dans l'environnement, mais les spécialistes de chimie analytique admettent que la spectrométrie de masse à haute résolution est nécessaire pour la confirmation (21). Un détecteur spécifique des nitrosamines (Thermal Energy Analyzer), fondé sur le principe de la rupture sélective de la liaison N-NO, offrait des possibilités exceptionnelles car il élimine la nécessité des opérations de purification. Malheureusement, ce détecteur en est encore au stade de la mise au point et deux laboratoires des Etats-Unis d'Amérique peuvent en disposer (22).

A la récente réunion du Sous-Comité européen pour l'Orientation des Etudes collectives, qui s'est tenue à Tallinn, l'utilité de telles études a été soulignée, et l'on a décidé de les poursuivre sur des matières telles que les viandes épicées et le pain et de les commencer sur les nitrosamines non volatiles. Les participants ont convenus qu'on devrait s'efforcer d'organiser une étude sur la normalisation des procédés de purification qui, pour l'analyse de mélanges complexes, suscitent le plus de difficultés.

Parallèlement aux études concernant l'amélioration des méthodes d'analyse et leur mise au point dans d'autres laboratoires - organisées par le service des Cancérogènes de l'Environnement - le CIRC a entrepris des études visant à mesurer les nitrosamines volatiles dans l'environnement en liaison avec les enquêtes épidémiologiques en cours.

La première étude a porté sur des échantillons d'aliments recueillis en Iran. Les lésions précancéreuses et cancéreuses observées chez les habitants des régions où le cancer oesophagien est endémique ressemblent étroitement, par leur morphologie, à celles des animaux d'expérience ayant reçu des doses de nitrosamines organo-spécifiques. Comme on a démontré que quelques 22 composés nitrosés peuvent provoquer un cancer de l'oesophage chez le rat et le hamster (8), il est logique, semble-t-il, de rechercher ces nitrosamines, diéthylnitrosamine et dibutylnitrosamine, par exemple, dans l'environnement des régions où la morbidité par cancer oesophagien est particulièrement élevée. Cette morbidité étant plus forte sur le littoral oriental de la Caspienne que dans toute autre région du monde, notre objectif est de tenter de trouver à cette incidence élevée une cause environnementale ([26]).

A cette fin, on a recueilli des échantillons d'aliments iranien traditionnels dans deux villages de la région de forte incidence et deux villages de la région de faible incidence (où la morbidité par cancer de l'oesophage est cependant plus forte qu'en Europe et aux Etats-Unis). Cent-quatre vingts échantillons ont été examinés pour la recherche des composés nitrosés volatiles.

Les résultats des analyses des aliments iraniens dans deux laboratoires indiquent que ces substances n'y sont présentes qu'à des doses ne dépassant pas 5 à 10 µg/kg (ppb) (23). Par ailleurs, les différences de dose entre les régions de forte et de faible incidence sont négligeables.

	Forte incidence (88 échantillons)	Faible incidence (91 échantillons)
Résultats négatifs	25 % (22)	35,3 % (32)
Résultats positifs	75 % (66)	64,8 % (59)
<1 µg/kg (ppb)	59 % (52)	51,6 % (47)
>1 <10 µg/kg (ppb)	16 % (14)	13,1 % (12)

Tableau III. - Quantités de diméthylnitrosamine dans 179 échantillons d'aliments iraniens

Les données résultant des enquêtes alimentaires en Iran ne peuvent être évaluées car nous n'avons pas suffisamment d'informations de base sur les nitrosamines volatiles, et nous ne savons pas non plus quelles sont les doses "efficaces" ou "sans effet" pour les nitrosamines. De plus, comme la mesure de nitrosamines non volatiles reste à réaliser, les études jusqu'ici effectuées ne peuvent être considérées que comme une première étape.

Une deuxième étude systématique est en cours dans une autre région de forte morbidité par cancer de l'oesophage, la Bretagne et la Normandie, dans le nord de la France ([27]). A ce jour, on a détecté la présence de nitrosodiméthylamine dans près de 50% des eaux-de-vie de pomme fabriquées à domicile ([23,24]). Ces résultats doivent encore être confirmés par spéctrométrie de masse.

Echantillon	Origine de l'échantillon	NDMA µg/kg ppb
Eau-de-vie de pomme (fabriquée à domicile)	Finistère (Bretagne)	2,5
" " "	" "	2
" " "	" "	3
" " "	" "	N.D.
" " "	" "	1
" " "	" "	N.D.
" " "	" "	1
" " "	" "	N.D.
" " "	" "	10*
" " "	" "	N.D.
" " "	" "	0,5
" " "	" "	2
" " "	" "	2
" " "	" "	N.D.
" " "	" "	N.D.
" " "	" "	N.D.
" " "	" "	N.D.
" " "	" "	N.D.
" " "	" "	N.D.
" " "	" "	3,5
" " "	" "	1
" " "	" "	5
" " "	" "	2
" " "	" "	N.D.
" " "	" "	1
" " "	" "	N.D.
" " "	" "	N.D.
" " "	Quimper-Concarneau (Bretagne)	2
" " "	" "	N.D.

Suite

Echantillon	Origine de l'échantillon	NDMA µg/kg ppb
Eau-de-vie de pomme (commerciale)	Etats-Unis d'Amérique (un seul fabricant)	N.D.
" " "	" "	N.D.
" " "	" "	N.D.
" " "	" "	N.D.
" " "	" "	N.D.
" " "	" "	N.D.
" " "	" "	N.D.
" " "	" "	N.D.
" " "	" "	N.D.
Vinaigre de cidre de pomme	" "	N.D.
Vin de pomme	" "	N.D.
Jus de pomme	" "	N.D.

N.D. - Non détectée
*   ▪ Résultat confirmé par spectrométrie de masse

Tableau IV. - Résultats de l'analyse pour la recherche des nitrosamines dans les eaux-de-vie de pomme et autres produits à base de pomme

    Les données jusqu'ici recueillies sont loin d'être complètes, mais elles sont importantes car elles constituent un premier pas vers l'évaluation du risque que représentent pour la santé les composés N-nitrosés, lesquels sont une fraction de la charge cancérogène totale dans l'environnement humain. De telles informations sont indispensables pour déterminer dans l'avenir des doses maximales admissibles et pour réduire effectivement l'exposition humaine aux cancérogènes [25] (Tableau V).

    Peut-être nos objectifs semblent-ils, pour le moment, irréalistes, mais de tels efforts sont nécessaire. Le statut international du CIRC lui permet de coordonner les études de ce genre, qui ne peuvent réussir que si elles sont menées à l'échelon mondial.

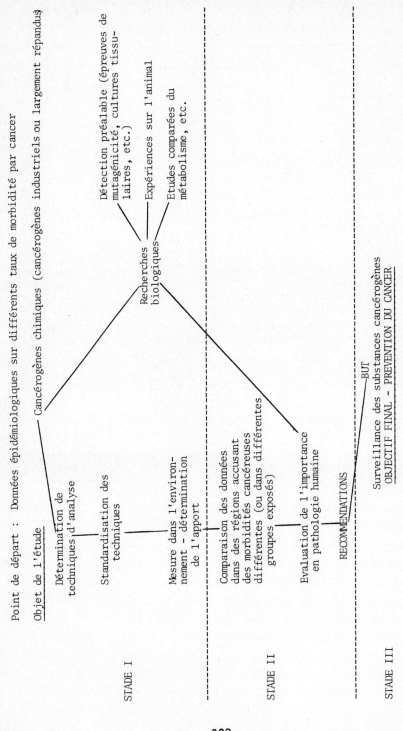

Tableau V. - Evaluation du rôle des cancérogènes chimiques dans les maladies humaines

Summary
N – nitroso compounds – Analysis and possible carcinogenicity in man.

Much work has been carried out on N-nitroso compounds but their role in human pathology has still to be elucidated. We cannot extrapolate experimental data to the human situation but we do have indirect evidence that nitroso compounds can be carcinogens in man. Although some nitrosamines are organ-specific, the nitroso group induces the development of many different types of cancer in animals. It is probable that the same phenomenon occurs in human pathology, and we cannot therefore expect to have special case reports on their carcinogenicity in man. Therefore, an alternative approach to the study of the role of nitrosamines in human pathology would be to establish a correlation between cancer morbidity in some regions and amounts of N-nitroso compounds in the environment.

In view of the complexity of the problem of *in vivo* nitrosamine formation, it is more realistic nowadays to measure exogenous nitrosamines. Many laboratories are currently engaged on studies of N-nitroso compounds, although systematic information on their presence in the environment is scant. Furthermore, the data acquired by different laboratories has been obtained using a variety of methods for sampling, storage, clean-up and identification and estimation, with the result that it is not known to what extent these results are comparable. Therefore, the standardization and determination of comparability of methods for the identification of N-nitroso compounds is the first step towards their quantitation in the environment. Adequate methods are now available for the determination of volatile nitrosamines, down to the µg/kg level, but methods for non-volatile nitrosamines are still in the development stages.

In order to avail all interested laboratories of information on analytical methods for volatile nitrosamines, IARC's analytical chemistry laboratory has organized a three part collaborative study using samples of canned luncheon meat. The results of this study were encouraging, and the European Sub-Committee for the Guidance of Collaborative Studies, at its last meeting, recommended that such studies be continued and extended to include non-volatile nitrosamines.

In parallel with the perfection of analytical techniques, IARC has initiated studies on the measurement of volatile nitrosamines in the environment in conjunction with the epidemiological studies on oesophageal cancer at present being carried out. The data collected up to now is far from being complete but it is important in that it represents the first step towards the evaluation of the risk to health of N-nitroso compounds, which constitute a part of the total carcinogenic load in the human environment.

REFERENCES

1) Preussmann, R.: On the significance of N-nitroso compounds as carcinogens and problems related to their chemical analysis. "N-Nitroso Compounds Analysis and Formation", Bogovski, P., Preussmann, R. and Walker, E.A. eds., 1972, IARC Scientific Publication No. 3, pp. 6-9, Lyon.

2) Clayson, D.B.: Chemicals and environmental carcinogenesis in man. Europ. J. Cancer, 1967, 3, 405-416.

3) Tomatis, L.: The carcinogenic risk for man of environmental chemicals. Proceedings of an International Symposium "Recent Advances in the Assessment of the Health Effects of Environmental Pollution", 1975, ew York (in press).

4) Magee, P.N.: Nitrosamines: ubiquitous carcinogens? New Scientist, 1973, 432-434

5) Montesano, R. and Magee, P.N.: Comparative metabolism *in vitro* of nitrosamines in various animal species including man. Chemical Carcinogenesis Essays, 1974, 39-56.

6) Kolesnichenko, J.: On the sensitivity of human embryonal lung tissue for the action of N-nitrosomethylurea. Proceedings of the 2nd Symposium "Carcinogenic N-Nitroso Compounds - Action, Formation, Detection", Bogovsky, S. and Vinkmann, F. eds., 1975, pp. 73-74, Tallinn, USSR (in Russian).

7) IARC Monographs on the Evaluation of Carcinogenic Risk of Chemicals to Man, 1972-1975, Vols. 1-7, Lyon.

8) Druckrey, H., Preussmann, R., Ivanovic, S. and Schmähl, D.: Organotrope carcinogene Wirkung bei 65 vershiedenen N-nitroso-Verbindungen au BD-ratten. Zeitschrift für Krebsforschung., 1967, 69, 103-201.

9) Linsell, C.A. and Peers, F.G.: The aflatoxins and human liver cancer. "Current Problems in the Epidemiology of Cancer and Lymphomas", Grundmann, E. and Tulinius, H. eds., 1972, pp. 125-129, Springer-Verlag, Berlin.

10) Terracini, B., Magee, P.N. and Barnes, J.M.: Hepatic pathology in rats on low dietary levels of dimethylnitrosamine. Brit. J. Cancer, 1967, 21, 559-565.

11) Mirvish, S.S.: Kinetics of N-nitrosation reactions in relation to tumorigenesis experiments with nitrite plus amines or ureas. "N-Nitroso Compounds Analysis and Formation", Bogovski, P., Preussmann, R. and Walker, E.A. eds., 1972, IARC Scientific Publication No. 3, pp. 104-108, Lyon.

12) Hill, M.J. and Hawksworth, G.: Bacterial production of nitrosamines *in vitro* and *in vivo*. "N-Nitroso Compounds Analysis and Formation", Bogovski, P., Preussmann, R. and Walker, E.A. eds., 1972, IARC Scientific Publication No. 3, pp. 116-121, Lyon.

13) Mirvish, S.S., Wallcave, L., Eagen, M. and Shubik, P.: Ascorbate-nitrite reaction possible means of blocking the formation of carcinogenic N-nitroso compounds, Science, 1972, 177, 65-68.

14) Ziebarth, G. and Scheunig, G.: The effect of some inhibitors on the nitrosation of drugs in human gastric juice. Proceedings of the Fourth Meeting on the Analysis and Formation of N-Nitroso Compounds, Tallinn, Estonian SSR, 1 and 2 October 1975 (in press).

15) Boyland, E.: The effect of some ions of physiological interest on nitrosamine synthesis. "N-Nitroso Compounds Analysis and Formation", Bogovski, P., Preussmann, R. and Walker, E.A. eds., 1972, IARC Scientific Publication No. 3, pp. 124-126, Lyon.

16) Keefer, L.: Promotion of N-nitrosation reactions by metal complexes. Proceedings of the Fourth Meeting on the Analysis and Formation of N-Nitroso Compounds, Tallinn, Estonian SSR, 1 and 2 October 1975 (in press).

17) Pignatelli, B., Castegnaro, M. and Walker, E.A.: The effects of gallic acid and ethanol on formation of nitrosodiethylamine. Proceedings of the Fourth Meeting on the Analysis and Formation of N-Nitroso Compounds, Tallinn, Estonian SSR, 1 and 2 October 1975 (in press).

18) Telling, G., Hoar, D., Caswell, D. and Collings, A.: Studies on the effect of feeding nitrite and secondary amines to Wistar rats. Proceedings of the Fourth Meeting on the Analysis and Formation of N-Nitroso Compounds, Tallinn, Estonian SSR, 1 and 2 October 1975 (in press).

19) Möhr, U., Wieser, O. and Pielsticker, K.: Die minimaldosis für die Wirkung von Diathylnitrosamine auf die Trachea beim Goldhamster. Naturwissenschaften, 1966, 53, 229.

20) Walker, E.A. and Castegnaro, M.: A report on the present status of a collaborative study of methods for the trace analysis of volatile nitrosamines. "N-Nitroso Compounds in the Environment", Bogovski, P. and Walker, E.A. eds., 1974, IARC Scientific Publication No. 9, pp. 57-62, Lyon.

21) Walker, E.A. and Castegnaro, M.: New data on collaborative studies on analysis of volatile nitrosamines. Proceedings of the Fourth Meeting on the Analysis and Formation of N-Nitroso Compounds, Tallinn, Estonian SSR, 1 and 2 October 1975 (in press).

22) Fine, D.H. and Rufeh, F.: Description of the thermal energy analyzer for N-nitroso compounds. "N-Nitroso Compounds in the Environment", Bogovski, P. and Walker, E.A. eds., 1974, IARC Scientific Publication No. 9, pp. 40-44, Lyon.

23) IARC Annual Report 1975 (in press).

24) Bogovski, P., Walker, E.A., Castegnaro, M. and Pignatelli, B.: Some evidence of the presence of traces of nitrosamines in cider distillates. "N-Nitroso Compounds in the Environment", Bogovski, P. and Walker, E.A. eds., 1974, IARC Scientific Publication No. 9, pp. 192-196, Lyon.

25) Shubik, P., Clayson, D.B. and Terracini, B.: The quantification of Environmental Carcinogens", 1970, UICC Technical Report Series, Vol. 4, Geneva.

26) Kmet, J. and Mahboubi, E.: Esophageal cancer in the Caspian Littoral of Iran: Initial studies. Science, 1972, 175, 846-853.

27) Tuyns, A.J. and Massé, G.: Cancer of the esophagus in Brittany: an incidence study in Ille-et-Vilaine. International Journal of Epidemiology, 1975, 4, 1, 55-59.

VIII

INDUSTRIAL AND LEGAL ASPECTS
ASPECTS INDUSTRIELS ET JURIDIQUES

Chairman / *Président :* Dr. R. OWEN

REGULATORY ASPECTS OF OCCUPATIONAL CARCINOGENS : CONTRASTS WITH ENVIRONMENTAL CARCINOGENS

Samuel S. EPSTEIN

Environmental Health Programs
Case Western Reserve Medical School
Cleveland, Ohio 44106, U.S.A.

FAILURE OF PAST REGULATION

New regulatory approaches are critically required for occupational carcinogens. Recent decades have seen a massive and unregulated increase in the numbers and quantities of synthetic organic chemicals manufactured and used, whose human impact in the form of chronic occupational disease, notably cancer, is now becoming manifest. Additional reasons include the following considerations: past regulatory practices, largely based on post hoc epidemiological recognition and regulation, clearly subject workers to massive and involuntary human experiments; there are gross inconsistencies between the protection afforded the general public through environmental standards, and protection afforded workers through occupational standards; and, current regulatory practices for occupational carcinogens appear based, in part, on misconceptions and mythologies, stemming from misapplication of traditional toxicological concepts.

Scientific considerations apart, there are critical deficiencies in legislative and regulatory approaches to occupational carcinogens, including conflicts of interest in the generation and evaluation of data, restrictions on open access to data, and lack of qualified representation of a wide range of concerned viewpoints and interests in decision making processes.

Implications of Recent Trends in the Use of Industrial Chemicals:

The amount of some major classes of synthetic organic chemicals which were produced in the U. S. in 1970, and the increase in U. S. production since 1967 and since 1949, are listed in Table 1.

It is not possible now to estimate the proportion of these novel chemicals which pose carcinogenic, besides other, hazards to man. Except for some special-purpose regulations in the area of pesticides, food additives, and drugs, this massive post-war efflorescence of chemical technology has occurred largely unrestricted by national, much less international, controls. There has been no general requirement for pre-testing of chemicals for carcinogenic or other adverse effects, prior to manufacture or use. As a consequence, it is likely that many carcinogens have been and are in wide use, whose effects may only manifest now, or in the next few years or decades.

TABLE 1

Recent trends in the use of industrial chemicals (1,2)

Class of Chemical	1970 Production Billions of Lbs.	%Increase 1967-1970	%Increase 1949-1970
Cyclic Intermediates	28.3	38	No Data
Plastics, resin material, and plasticizers	20.6	36	1,130
Synthetic rubber and rubber processing	4.7	16	350
Surface-active agents	3.9	12	810
Pesticides	1.0	-2	710
Others	79.7	33	550
Total	138.3	32	No Data

Such increases are likely to have occurred in other industrialized countries, although perhaps later and less dramatically than in the U. S.

The case of vinyl chloride may well be a harbinger of other carcinogens for this generation of materials. Recently recognized as an occupational carcinogen, vinyl chloride was originally introduced into large scale production in the 1950's, and synthesis grew at about 15% per annum until about four billion pounds were manufactured in the U. S. in 1970. Of the vinyl chloride workers identified, as of June, 1974, with confirmed diagnoses of hepatic angiosarcoma, more than half, however, had received their first exposure prior to 1950 (3).

Inconsistency of Environmental and Occupational Standards in the U. S.:

A fundamental dichotomy, both scientific and moral, exists between current approaches to setting of standards for carcinogens, and other toxic chemicals, for the working population and for the population-at-large. In spite of their inadequacies, standards that have been developed for the protection of the population-at-large against adverse effects from exposure to a wide range of carcinogens and other chemical pollutants in consumer products are generally predicated on the availability of an adequate data base, properly developed at the expense of the manufacturer. Necessary information includes chemical composition of such products, labelling and disclosure of ingredient identity, and data on acute and chronic toxic effects in experimental animals prior to their release into commerce. For agents inducing acute and chronic toxicity per se in animals, threshold or no-effect levels are determined and standards are then developed, generally, based on 100-fold safety margins. Agents which induce carcinogenic effects in appropriate animal tests or which are known to be carcinogenic to man are generally banned from commerce, as no level of exposure whatsoever is considered safe. The existence of threshold or safe levels is not recognized for any chemical carcinogen (4-6). This is the explicit basis of the 1958 Delaney Amendment (PL 85-929) to the Food Drug and Cosmetic Act, which imposes a zero tolerance for carcinogenic food additives. This law does not recognize economic or technical feasibility considerations, nor does it allow any regulatory "discretion", except in the defined area of judging whether the animal data on carcinogenicity are based on appropriate tests.

The validity of the Delaney concept has been repeatedly and unanimously endorsed by a wide range of qualified expert national and international committees and bodies. This conclusion has been recently reiterated as the official HEW position, with regard to consumer products, as follows:

> "At present, the Department of Health, Education, and Welfare lacks the scientific information necessary to establish no-effect levels for carcinogenic substances in animals in general and in man in particular. In the absence of such information, we do not believe that detectable residues of carcinogenic animal drugs should be allowed in the food supply" (7).

These general requirements for environmental standards are in striking contrast to those for occupational health. There are no current requirements for "pre-testing" or screening chemicals prior to their manufacture and use in industry, nor are there even general requirements for open disclosure of the identity of chemical agents to which workers are exposed. This is clearly contrary to the intent of the U.S. 1970 Occupational Safety and Health Act, which mandates the provision of a safe and healthy working environment; the Act implicitly provides authority for "pre-testing" chemicals. No such assurances can possibly be made in the absence of information as to the chemical nature and possible biological effects of these exposures. In the absence of "pre-testing", the worker himself or herself, is unwittingly used as an involuntary test subject, to whom test data are not generally made available, if indeed they are ever collected and analyzed. Recognition of adverse effects in epidemiological studies is thus post hoc, and reflects deliberate prospective human experimentation.

Standards exist for only a small number, about 450, of the myriads of chemicals to which workers are exposed. The National Institute for Occupational Safety and Health (NIOSH) Toxic Substances publication, listing approximately 25,000 known chemicals used in industry in 1973, clearly underestimates the number of chemicals to which occupational exposure can occur. Most Occupational Safety and Health Administration (OSHA) standards are based on the approximately 450 Threshold Limit Value (TLV) standards, which have been developed by the industrially-oriented and quasi-consensus American Conference of Government Industrial Hygienists (ACGIH). These TLV standards are often referred to as "proprietary", reflecting their narrowly focused "trade" origin. The concept of adequate safety margins, for chemicals inducing acute and chronic toxicity, is scarcely, if at all, reflected in occupational, in contrast with general environmental standards.

Only few U.S. standards have been promulgated for occupational carcinogens. The asbestos standard, derived from a NIOSH criteria document, still permits continued exposure to a well recognized occupational carcinogen at levels which are likely to result in appreciable cancer mortality. This standard appears to reflect narrowly based economic, rather than scientific, considerations. The underlying assumptions on economic and technical feasibility were not adequately and openly articulated or defined at any stage in the standard setting process. Implicit benefit-risk considerations appear to have been interpreted in terms of maximizing short-term and narrowly defined economic benefits to industry, rather than minimizing substantial risks to the worker.

Mythologies in Carcinogenesis:

Conflicts between crucial social goals, such as reduction in the incidence of human cancer due to environmental and occupational carcinogens, and powerful concentrated economic interests are often joined on supposedly scientific grounds. Thus, a major constraint in the promulgation of scientifically valid and socially responsive standards regulating exposure to occupational carcinogens has been the development of a range of mythologies calculated to minimize the significance or reality of hazards from exposure to particular chemical carcinogens. While these mythologies cannot withstand elementary scientific scrutiny, they nevertheless have been and are being used for effective lobbying in legislative, executive, judicial, and public arenas. Transcripts of the proceedings of the Department of Labor (DOL) 1973 Standards Committee on carcinogens and of the recent U.S. Suspensions Hearings on Dieldrin (EPA and EDF vs. Shell)(8) amply document such mythologies, including the following:

-- "Tumorigens" are less Dangerous than Carcinogens:

The identity of "tumorigens", as opposed to carcinogens, has been vigorously proposed, particularly for chlorinated hydrocarbon pesticides, such as DDT and Dieldrin, which have long been known to induce hepatomas in mice. Presumably on the basis of such alleged distinctions, recent claims have been made, illustratively by a senior HEW spokesman that -- "there is no evidence to my knowledge that DDT is a carcinogen" (9). The invalidity of such alleged distinctions has, however, been repeatedly and unambiguously emphasized by numerous expert national and international committees, which unanimously concluded that the terms tumorigens and carcinogens have synonymous implications (10). This proposition has, however, apparently ceased to be relevant with the recognition of pulmonary metastases resulting from "benign" hepatomas induced in mice, illustratively by "tumorigens" such as Dieldrin, which also induce extra-hepatic neoplasms in both mice and rats (8).

-- "Animal Carcinogens" are less Dangerous than "Human Carcinogens":

This thesis proposes that valid distinctions, from a regulatory standpoint, can be drawn between chemicals shown to be carcinogenic in experimental animals and those known to be carcinogenic in man. It is further proposed that less stringent regulatory standards should be promulgated for "animal" carcinogens such as ethyleneimine, dichlorobenzidine, and 4,4'-methylene(bis)-2-chloroaniline, unless and until their carcinogenic effects can be validated by human experience, based on deliberate continued human exposure.

There is in fact no evidence for the exisitence of "species-specific" carcinogens. All chemicals known to produce cancer in man, with the possible exception of trivalent inorganic arsenic, also produce cancer in experimental animals, generally in rodents. Recent experience with carcinogens such as bis-chloromethyl ether, diethylstilbestrol, and vinyl chloride monomer, moreover, amply confirms the predictive value of animal carcinogenicity data to humans.

-- Human Experience has Demonstrated the Safety of Occupational Exposure to certain Carcinogens:

Such claims have been repeatedly made for a wide range of "animal carcinogens" including Dieldrin, a-naphthylamine, ethyleneimine and dichlorobenzidine, and for "low levels" of exposure to recognized "human carcinogens". These claims are generally made on the basis of lack of positive documentation of excess cancer deaths or on the basis of undisclosed or pratially accessible records on small populations at risk, on undefined turnover rates, and on short periods of follow-up. Clearly, such data do not permit development of valid epidemiological inferences. Such claims fail to recognize inherent limitations and insensitivities of epidemiological techniques and of the sparsity of valid epidemiological studies on occupational carcinogenesis.

-- "Safe Levels" of Exposure to Occupational Carcinogens can be Determined:

It is alleged that no or negligible risks result from exposure to "low levels" of occupational carcinogens; such "low levels" are generally set on the basis of technical and economic expediency, or on other poorly articulated concepts. Illustratively, the ACGIH has assigned "acceptable" levels, TLV's, for asbestos, bis-chloromethyl ether, and nickel carbonyl. However, numerous expert national and international committees and bodies have unanimously attested to the fact that there is no mechanism for determining the existence of biological thresholds for chemical carcinogens, and hence that the TLV concept is totally inapplicable to chemical carcinogens.

RECENT U.S. EXPERIENCES WITH STANDARD SETTING FOR OCCUPATIONAL CARCINOGENS

On February 11, 1974, nearly three years after the effective date, April 28, 1971, of the U. S. Occupational Safety and Health Act, during which time there were no standards at all for occupational carcinogens, other than asbestos, OSHA promulgated standards for 14 chemical carcinogens, based on the report of a 15-member D.O.L. Standards Committee on Carcinogens, and an accompanying recommendation from NIOSH. However, the major key provisions of the Committee, including those endorsed by corporate scientists on the Committee, were disregarded by OSHA, admittedly under strong industrial pressure. These included recommendations of the Committee to ensure effective implementation of the carcinogen standards by instituting sensitive environmental monitoring systems and a permit system. For reasons that appear scientifically invalid, OSHA also excluded from the new standards mixtures containing "low" percentages of carcinogens. The modified standards were issued as 14 separate rules, in further contrast with the recommended single rule proposed by the Committee. These emasculated standards appear cosmetic, affording the illusion, but not the reality, of protection against exposure to chemical carcinogens. Further emasculation has been vigorously pursued through recent industry-initiated suits for stays of promulgated ethyleneimine and MOCA standards.

It is of particular interest to note that at public hearings, September 11, 1973, on the proposed standards for the 14 occupational carcinogens, a lobbyist, representing the major manufacturers of those chemicals, stated that the corporate scientists on the Committee were not qualified in regulatory problems, which were the exclusive responsibility of corporate management.

The debate as to whether to promulgate a vinyl chloride standard based on a "no detectable" level of 1 ppm ($2,600$ ug/m^3), or a level of 1 ppb, as properly demanded by labor and independent carcinogenesis experts, or at some level between 1 ppm and 50 ppm, as demanded by corporate and scientific representatives of the plastic and chemical industry, has been crystallized by recent reports of A.D. Little, under contract to the Society of Plastics Industry, and of Foster D. Snell, Inc., under contract to DOL, which concluded that a "no detectable" level standard was neither economically nor technically feasible. These reports do not appear to have addressed themselves to the costs, economic and otherwise, of carcinogenic and other toxic effects due to human exposure to vinyl chloride monomer. It is of interest to note that the claims of impending financial disaster and of mass unemployment due to promulgation of the current 1 ppm standard have not materialized.

Similar apparent indifference to costs of cancer in workmen has been recently expressed by the American Smelting and Refining Company (ASARCO), which maintains that the present arsenic standard, 0.5 mg/m^3, is adequate to protect from carcinogenic hazards and which objects to consideration of a standard as "restrictive" as that currently recommended by NIOSH, 0.002 mg/m^3.

REGULATORY NEEDS

Benefit-Risk Evaluation:

Since World War II, there has been an exponential and, largely, unregulated increase in the numbers and quantities of synthetic organic chemicals manufactured and used in industrialized countries. The claimed needs to use increasing numbers of new synthetic chemicals makes it essential to recognize and critically evaluate carcinogenic, and other human and environmental, hazards with regard to the real or alleged matching benefits they confer. Such costing must be weighted by factors including the persistence and environmental mobility of the chemical, the size of the population exposed, and the reversibility of the adverse effect. Total national monetary costs in the U.S., both direct and indirect, from cancer are estimated to be approximately $15 billion annually (11), these costs have hitherto been largely externalized or discounted. As the majority of human cancers, both in the general population and in occupational groups, are now considered to be due to chemical carcinogens and hence preventable, there should be clear economic, besides other, incentives to reduce the environmental and occupational burden of chemical carcinogens.

Carcinogenic hazards from a particular synthetic chemical need not necessarily be accepted even when matching benefits appear high, as equally efficacious but nonhazardous alternatives are usually available. The mandatory criterion of efficacy, once extended from therapeutic drugs to other synthetic chemicals, such as deliberate and accidental food additives, feed additives and pesticides, may well simplify such equations, especially for hazards from synthetic chemicals with no demonstrable benefits for the general population. The imposition of a requirement for broad social utility may even further simplify the benefit-hazard equation. Such concepts have been recently emphasized with regard to food additives by a leading industrial representative who recommended that additives be excluded from products unless they either significantly improve the quality or nutritive value of the food or lower its costs as well as being safe (12).

Claims that occupational carcinogens serve industrially unique purposes, must be examined critically by economically disinterested experts with particular recognition of the attendant and generally externalized human costs and the lack of economic incentives to develop similarly efficacious and non-carcinogenic alternatives. In the absence of such alternatives, consideration must be directed to the possible banning of the manufacture and use of the carcinogen or to restrict its use to closed systems which are continuously monitored with instrumentation of maximal sensitivity, and with automated and visible or available read-outs.

Inherent in toxicological and regulatory philosophy and practice is lip service to the concept of balancing benefit, and benefit to the public not to industry, against risk, and risk to public health or environmental integrity and not economic risk to industry. If the chemical in question does not serve a broad socially and economically useful purpose for the general population, why introduce it and force the public-at-large to accept potential hazards without general matching benefits? Such questions should be vigorously directed to carcinogenic, and otherwise hazardous, cosmetic food colouring agents, in particular, and to all food additives, in general. Claims have recently been made (13) that requirements for pre-testing of chemicals prior to their introduction to commerce are acting as disincentives to industrial innovation. These claims have been particularly directed to the manufacture of pesticides (14). Apart from the fact that such claims are predicated on the legitimacy of externalizing public health hazards and costs, they do not bear critical scrutiny even from narrowly defined economic viewpoints. A telling critique of such claims has been expressed by a leading industry spokesman who stated that he

--"emphatically (takes issue with the line of reasoning that) escalating regulatory demands have made the cost of research and development prohibitive, thus drying up any incentive to go develop new agricultural chemicals....

--"...In the first place (argued Dr. Sutherland), new regulations imposed since the creating of EPA affording better protection to fish and wildlife were overdue. More important is the changing aspect of the marketplace, particularly in the pesticide area. Growers now have available to them many first-rate products... many of these are quite inexpensive. What the chemical people are really telling you is that while research costs continue to rise, to come with still better compounds costing no more money than what's already being sold is a tough proposition...; the companies with weak research organization, a shaky financial position, are dropping out. They would rather have FDA and EPA take the rap rather than acknowledge the overall problem"(15).

In any event, information in 1973 suggests that the profitability of pesticides and their development is again rising (16).

It has now become axiomatic that there are major defects in decision-making processes in regulatory practices (17,18). It is clear that the democratic system of checks and balances is largely absent from current regulatory practice. Apart from limited post hoc recourse, the working person, and those who represent his or her interests, scientifically and legally, are virtually excluded from anticipatory involvement in decisions vitally affecting them. The concept of matching benefits against risk has been generally applied to maximize short-term benefits to industry, even though this may entail minimal benefits and maximal risks to workers. Such risks are effectively involuntary, because they are, generally, camouflaged by inadequate unbiased information, and by the illusion of safety created by authoritarian industrial and regulatory practices.

While such approaches are of course detrimental to workers, they are also often detrimental to the long-term interests of industry, which may suffer major economic dislocation when carcinogenic processes and products, to which it has improperly developed premature and major commitments, are belatedly banned from commerce. Such problems are in large measure attributable to crippling constraints which have developed and which still dominate the decision-making process within regulatory agencies. Responsibility for these constraints must be shared with regulatory agencies, by the legislature, by the scientific community, and by workers who have not yet developed adequate mechanisms for protecting their own health and rights.

Promulgation of Standards:

Ideally, the object of standards for occupational carcinogens should be to achieve zero-levels of exposure. From a practical standpoint, interim standards are designed to prevent exposure to the lowest level of carcinogens which can be detected by the most practical and sensitive monitoring methods available. The DOL 1973 Standards Committee on carcinogens recommended that monitoring techniques should be initially and minimally sensitive to 1 ppb, and that standards should be made progressively more rigorous with increasing sensitivity of monitoring techniques. The development of monitoring instrumentation of maximal sensitivity clearly reflects social and industrial priorities and economic considerations. Rohm & Haas and Dow Chemical Company have developed closed systems for analysis and continuous monitoring of bis-chloromethyl ether, with a sensitivity of less than 1 ppb. Such techniques are clearly applicable to all major classes of volatile carcinogens. The high costs of truly "closed systems" and of continuous environmental monitoring should properly act as cogent disincentives to the continued manufacture, processing and use of volatile carcinogens, particularly by small industry, and to the development of alternative products and technologies based on use of non-carcinogenic chemicals.

Implementation of Standards:

Promulgation of standards must be paralleled by the concomitant development of practical and rigorous methods of enforcement for manufacturers, processors and users of chemical carcinogens, based on open registration and mandatory permit systems. Permits should only be granted if the proposed use of the carcinogen serves a critical and unique societal function, if containment and sensitive monitoring technologies can be assured, and if full and informed consent of workers is maintained. The alternative of essentially voluntary compliance is, at best, a random process, which is likely to economically penalize more conscientious industries, and which is likely to be largely ignored by certain industries, particularly the smaller. Rigorous implementation of standards must also be developed to prevent discharge or release of occupational carcinogens, as detectable by the most sensitive practical sampling and analytic methods available, to the external environment -- air, water and solid waste. Pressures on agencies can subvert implementation of standards and of the total regulatory process. This has been well recognized in statements such as the following:

> "It is the daily machine-gun like impact on both agency
> and its staff of industry that makes for industry orientation

on the part of many honest and capable members, as well
as agency staffs" (19).

Nevertheless, appropriate reforms in agency-industry relationships have yet
to be developed. Reforms apart, it is clear that decisions on the use of
toxic agents, such as carcinogenic chemicals in consumer products and in
the workplace, must be made in the open political arena and on the basis
of the evaluation of scientific data that is both expert and unbiased. Industry must be encouraged to avoid preoccupation with short-term economic
interests and the development of premature commitments to products and
processes which have not been adequately tested by competent and independent investigators. Such approaches will minimize or preclude the possibility of economic dislocation which would otherwise ensue when subsequent
challenges necessitate the belated withdrawal of the product or process from
commerce and the workplace. Such approaches also reflect recognition of
the consonance of long-term industrial interests and broadly-based societal
goals and values.

Needs for Carcinogenicity Testing:

In addition to critical needs for promulgating and implementing standards
for agents of known carcinogenicity, there are equally critical needs for
carcinogenicity testing of the myriads of chemicals to which workers are
currently exposed and of new chemicals prior to their introduction into commerce. Such data should be generated and evaluated by economically disinterested experts. A pre-requisite for such testing is complete and open
disclosure of the identity of all known chemicals to which workers are or
will be exposed.

Recent Regulatory Guidelines to Chemical Carcinogenesis:

Stemming from the recent suspension hearings on Dieldrin (8), and the
subsequent decision to ban this pesticide on grounds of imminent carcinogenic
hazard, the underlying principles of chemical carcinogenesis which were the
basis for this decision were crystallized (20) as follows:

1. "A carcinogen is any agent which increases tumor induction in
   man or animals.

2. "Well-established criteria exist for distinguishing between benign
   and malignant tumors; however, even the induction of benign tumors
   is sufficient to characterize a chemical as a carcinogen.

3. "The majority of human cancers are caused by avoidable exposure
   to carcinogens.

4. "While chemicals can be carcinogenic agents, only a small percentage are.

5. "Carcinogenesis is characterized by its irreversibility and long latency period following the initial exposure to the carcinogenic agent.

6. "There is great variation in individual susceptibility to carcinogens.

7. "The concept of a 'threshold' exposure level for a carcinogenic agent has no practical significance because there is no valid method of establishing such a level.

8. "A carcinogenic agent may be identified through analysis of tumor induction results with laboratory animals exposed to the agent, or on a post hoc basis by properly conducted epidemiological studies.

9. "Any substance which produces tumors in animals must be considered a carcinogenic hazard to man if the results were achieved according to the established parameters of a valid carcinogenesis test.

RECOMMENDATIONS

Regulatory standards should be routinely developed to prevent exposure to any level of carcinogen, as detectable by the most sensitive and practical sampling and analytic methods available. The critical premise underlying such standards is that, as overwhelmingly endorsed by the qualified professional community, there is no scientific method for determining safe levels of exposure to carcinogens. These considerations are the basis of the 1958 Delaney Amendment to the U. S. Federal Food, Drug and Cosmetic Act, which imposes a zero tolerance on carcinogenic food additives. It is perhaps no coincidence that the attacks on the Delaney Amendment are mounting at a time when the food chemical industry is poised for a major expansion. The chemical industry predicts that sales of chemical additives are expected to grow from $485 million in 1970 to $750 million by 1980. In providing a framework for evaluating potential hazards of these additives, the Delaney Clause simply ratifies the prevailing expert opinion in the National Cancer Institute and in other professionally qualified groups that there is no practical method to determine safe dietary levels for a carcinogen (5,6). Changing the Delaney Clause to allow regulatory discretion to set tolerances for carcinogens is, therefore, not only scientifically inappropriate, but, administratively foolhardy.

Distinctions between "tumorigens" and carcinogens, and between "animal" and "human" carcinogens, as frequently claimed by representatives of economically concerned interests, are totally invalid and must not be used to justify standards allowing exposure to "low levels" of "tumorigens" and "animal" carcinogens. Similarly, claims that human experience has demonstrated the "safety" of occupational exposure to "tumorigens" and "animal" carcinogens, or to "low levels" of "human carcinogens", have on those occasions when the underlying data exist and are made available, been shown to be spuriously based on inadequate population samples, inadequately followed-up for inadequate periods of time. Standards must be implemented by rigorous registration and "permit" systems for manufacturers, processors, and users of occupational carcinogens; such standards should minimally be as restrictive as those for radioactive chemicals. Small manufacturers or users of carcinogens, unable to fully comply with all requirements of occupational standards should be denied permits. Claims that occupational carcinogens serve industrially unique purposes must be examined critically by economically disinterested experts and with particular recognition of the generally externalized human costs attendant to exposure to carcinogens and the lack of economic incentives for the development of similarly efficacious but non-hazardous alternatives. In the absence of such non-carcinogenic alternatives, the manufacture and use of the carcinogen should, minimally, be restricted to "closed systems", which are continuously monitored with instrumentation of maximal sensitivity and below the ppb range, or preferably banned. No discharge or release of occupational carcinogens, as detectable by the most sensitive practical sampling and analytic methods available, to the external environment -- air, water and solid waste -- should be permitted.

Finally, from an international standpoint, it is critical that in aiding development of industrially lesser developed countries, the industrialized nations do not pass on the occupational health crises they themselves are now experiencing. It is especially imperative that steps be taken to stop the recently-noted trend of industries faced with the imposition of controls relocating in countries where there are no, or grossly inadequate, standards. This clearly is an area in which we may properly expect the WHO to play a much needed role of scientific and moral leadership.

ACKNOWLEDGEMENTS

Based, in part, on a presentation to the XI International Cancer Congress, Florence, October 20-26, 1974 (by kind permission of the editors of Excerpta Medica).

REFERENCES

1. President's Science Advisory Committee, Chemicals and Health. Report of the Panel on Chemicals and Health. Science and Technology Policy Office, National Science Foundation. September 1973, Washington, D. C.: U. S. Government Printing Office.

2. United States Tariff Commission, Synthetic Organic Chemicals - United States Production and Sales, 1970. Tariff Commission Publication 479, 1973. Government Printing Office.

3. World Health Organization, Report of a Working Group on Vinyl Chloride. International Agency for Research on Cancer. June 1974, Internal Technical Report No. 74/005, Lyon, France.

4. Ad Hoc Committee on the Evaluation of Low Levels of Environmental Carcinogens. Evaluation of Environmental Carcinogens. Report to the Surgeon General, 1971, U. S. Government Pringing Office, Washington, D. C.

5. Epstein, S. S. The Delaney Amendment. Prevent. Med., 1973, 2: 140-149.

6. Saffiotti, U. Comments on the Scientific Basis for the "Delaney Clause". Prevent. Med., 1973, 2: 125-132.

7. Weinberger, C. W. (HEW Secretary). Letter to Representative H. O. Staggers, June 18, 1974, Washington, D. C.

8. Epstein, S. S. The Carcinogenicity of Dieldrin. Statement for Testimony at Cancellation Hearings on Aldrin/Dieldrin, 1974, Environmental Protection Agency and Environmental Defense Fund vs., Shell Chemical Co.; The Science of the Total Environment, 1975, 4: 1-52.

9. Tepper, L. Hearings Before a Subcommittee of the House Committee on Appropriations, May 6, 1974, 93rd Congress, p. 63.

10. Epstein, S. S. Environmental Determinants of Human Cancer. Cancer Research, 1974, 34: 2425-2435.

11. National Cancer Program. "The Strategic Plan". D.H.E.W. Publication, January 1973, No. NIH 74-569.

12. Kendall, D. M. "A Summary of Panel Recommendations". Report of a Panel on Food Safety to the White House Conference on Food Nutrition and Health, (19), November 22, 1969.

13. Gehring, P. J., Rowe, V. K., and McCollister, S. B. (Dow Chemical Co.) "Toxicology: Cost/Time", Fd. Cosmet. Toxicol., 1973, 11: 1097.

14. Naegele, J. (Dow Chemical Co.). Testimony to U. S. Congress, House Committee on Agriculture, Hearings on Federal Environmental Pesticide Control Act of 1971, 92nd Congress, 1st Session, Washington, D. C. U. S. Government Printing Office, 1971.

15. Sutherland, G. L. (American Cyanamid). "Agriculture Is Our Best Bargaining Tool", Farm Chemicals, 1972, 135: 44.

16. Bennett I. "Preface", Pesticides Monitoring Journal, 1967, 1: No. 1.

17. Epstein S. S. and Grundy, R. D., eds. The Legislation of Product Safety, Consumer Health and Product Hazards, 1974. Vols. 1 and 2. MIT Press, Cambridge, Mass.

18. Page, J. A. and O'Brien, M. W. 1973. Bitter Wages. Grossman, New York.

19. Landis, J. Report to President-elect Kennedy, 1960.

20. See Recommended Decision on the Suspension of Aldrin and Dieldrin of Herbert L. Perlman, Chief Administrative Law Judge, Environmental Protection Agency. (39 F. R. 37249-65). See also "Opinion of the Administrator, Environmental Protection Agency, on the Suspension of Aldrin/Dieldren", FIFRA Docket Nos. 145 et. al. 39 F. R. 372265-72. See also "In Re: State of Louisiana Request for Emergency Use of DDT on Cotton Order and Determination of the Administrator that Reconsideration of the Agency's Prior Order of Cancellation of DDT for Use on Cotton is Not Warranted". pp. 30-37.

ADMINISTRATIVE ASPECTS ON REGULATION OF CARCINOGENIC HAZARDS IN OCCUPATIONAL ENVIRONMENT

Peter WESTERHOLM

National Board of Occupational Safety and Health,
100 26 Stockholm 34, Sweden

Mr Chairman,

It is a great honour to address an audience of such distinction on a set of problems which we all have in common. My contribution is an attempt to develop the points of view of a medical administrator on the subject at issue. I shall try to make a reasonable compromise between completeness and emphasis on details we have found to be important in my country. It is difficult to satisfy both these objectives in a quick flash of time. It should be remembered that rules and regulations are man-made things. They are a product of the culture, administrative traditions and political climate in any country at a certain period of time. A full understanding of a particular rule requires knowledge about its setting. Rules are not easily imported.

I shall give a brief account of the rule-system in Sweden with due regard to the need to keep the discussion on a level of abstraction where we all are on a common ground. I should think that this best serves the purpose of this conference.

I should like to begin with a quick look backwards in order to remind ourselves that the protection of environment and of man in environment is, from a historical point of view, a relatively new task for the society organs looking after our security. In this field we do not have the traditions and experience we are used to find in other institutions we have erected to take care of other problems we have in common, for instance internal order, a legal system, foreign policy etc. So, we have a late start.

Another point worth remembering is that the very concept of environment is relatively new. Not many really mentioned it some 20 years ago. It is an interesting concept. In it lies a recognition of a totality. Man's environment simply encompasses anything he is exposed to in his home, at his work or elsewhere in the community. So, even if we restrict our scope to the chemical environment we find that the problems have no respect at all for our administrative systems or the traditional borderlines between the areas of competence of our society organs. We find ourselves suddenly involved in close-quarter fighting with a complex set of problems, the nature of which we only partly understand and which has to be tackled comprehensively, taking full use of what we have of common sense, biological and technical knowledge and administrative and other skills.

One of the tools we have at our disposal is our systems of rules or regulations. Let us not be too humble. We shall just have to have a go.

The objective of all rules or rule-systems is to produce safety. What we wish to see as the result of our rules, when enforced, is a disciplined behaviour - let us call it so - in order to avoid a suspected or known carcinogenic risk. This obviously is going to compete or even conflict with other interests in society. An increased safety (= decreased risk) is certainly not going to be produced without some form of economic sacrifice. We must therefore constantly be prepared to prove our case as we go along. As I said earlier we are late starters and what we really do is to bring in a new set of priorities.

When trying to achieve this above mentioned disciplined behaviour, with its resulting technical change of environment, we are wise to keep in mind a few important determinants for the success of our rules. These are in broad terms:

> Technical possibilities
> Economical possibilities
> Attitudes of individuals and groups concerned.

This is what we are up against. We are going to get nowhere with a regulation or rule prescribing a problem solution which is unfeasible or technically not intelligent enough, economically unsound and for various reasons unattractive to the general public. This calls for a pragmatic attitude on our part and appreciation of the comprehensive nature of our problem. It may sometimes lead us to acceptance of a low-grade risk, or uncertainty if that term is preferred.

I know that I am going to be attacked for mentioning economical possibilities. I might as well say already now that I mean economical in a total context. Our problem solution has to be made considering the impact of it on the total life-situation of the individuals concerned. The closing of factories, in this context, should not immediately be accepted as the best solution. In fact, such a decision can be questionable from an individual's health point of view.

Now to our rules which is the term I shall continue to use. we have a rich variety to choose from. We have the whole set of legislation which usually is the product of political organs, the standards and codes of practice commonly produced by administrative and supervising organs and the detailed technical specifications and instructions applying to a particular situation. In occupational environment these latter usually are a product of commercial organs or expert functions in the branch, factory or company concerned.

We have really two different kinds of situation which we wish to control. These are

> entry of carcinogenic risk - known or unknown - into environment
>
> existence of carcinogenic risk - known or unknown - in environment.

I shall try to deal with these situations at the same time.

From a legislative point of view the control of carcinogenic hazards in occupation is based on two law systems:

1. Laws on products hazardous to health and environment
2. Laws on worker's protection.

The first of the above-mentioned systems apply to one particular group of factors in the whole environment. Number two applies to all conceivable kinds of risk factors in one particular sector of the environment.

Next I should like to demonstrate the administrative chain of command regarding control of hazardous and products in general in Sweden.

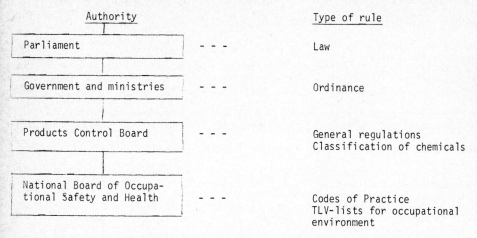

In this list, which is drawn primarily with the interests of the occupational environment in mind, the different levels produce rules of differing magnitude. In the ideal case every rule produced in the bottom of the list is based on a rule produced in the top end. The parliament has adopted an Act on Products Hazardous to Health and Environment. On the basis of this law the government has issued an ordinance on the same products where some more substance is put into the text in order to concretize the objectives of the Act. This legislation is of a recent date. It has been in force since 1st of July 1973. Based on the law. the Products Control Board issues lists classifying chemicals and products as poisons or dangerous substances respectively. Different detailed regulations apply to these two main cathegories. The National Board of Occupational Safety and Health is the government agency supervising and enforcing the rules of the above mentioned authorities within the occupational environment. It has as its main task the supervision of the legislation for the Worker's Protection. The board thus has to model its actions to the two sets of legislation.

It should be noticed here that the above mentioned boards, by tradition common for the Scandinavian countries, have a strong and relatively independent position in the administrative hierarchy. They have administrative powers and responsibilities which normally belong to the ministries and governmental departments in many other countries.

I shall now quote a few important paragraphs of the legislation on Products Hazardous to Health and to the Environment in order to demonstrate the construction of the system.

First we have the paragraph 6 of the Act:

> § 6 If considerations of public health or environment so dictate, the King-in-Council, or such public authority as the King-in-Council may designate, is empowered to order that some particular kind of hazardous product may be handled or imported only by special permission or that the handling or importation of such product shall be subject to other special conditions.
>
> Application for permits according to the first paragraph of this section shall be considered by such public authority as the King-in-Council may designate.

Another quotation is the paragraph 53 of the Ordinance

> § 53 The Products Control Board will issue further regulations according to the first paragraph of §6 of the Act (1973:329) on Products Hazardous to Health and to the Environment insofar as concerns special conditions for the handling or import of hazardous products other than a prohibition to handle or import such products without special permission.

It is to be observed that these two principal regulations open up the possibility to make import and any handling of a product or chemical subject to conditions.

Another interesting pair is the paragraph 5 of the Act

> § 5 Anyone handling or importing a product hazardous to health and to the environment shall take such steps and otherwise observe such precautions as are necessary to prevent or minimise harm to human beings or the environment. It is the particular duty of anyone manufacturing or importing such a product to inquire carefully into its composition and other properties with a bearing on public health or environmental protection. The product shall be clearly marked with information of importance from the standpoint of public health or environmental protection.
>
> The King-in-Council, or such public authority as the King-in-Council may designate, is empowered to promulgate special regulations concerning precautions and the duty of examination and marking as referred to in the first paragraph of this section.

and the corresponding paragraph 50 of the Ordinance

> § 50 The Products Control Board will issue further regulations on precautions and on the duty of examination and marking as laid down in the first paragraph of § 5 of the Act (1973:329) on Products Hazardous to Health and to the Environment.

I wish to draw your attention to a few important principles laid down in these paragraphs. First, it is the principle of allocating the primary responsibility for the production of risk to anyone handling a hazardous product. Thus, the responsibility is not in the first place laid on the public health authorities. Next, it empowers the authorities to enforce a duty of examination.

To go a bit further into details the Ordinance is quoted once again. Paragraph 2.

> § 2 Substances that are classifiable as products hazardous to health and to the environment and are liable to cause harm to man are sub-classified as poisons or dangerous substances.
>
> The poison classification includes substances that are mainly used in the production of pharmaceuticals and that the National Board of Health and Welfare has designated as poisons, and other substances whose handling is associated with very grave hazards to health. Other substances under the main heading are classified as dangerous substances.

and paragraph 5

> § 5 The Products Control Board has a duty to publish reference schedules listing substances covered by § 2.

Once a substance is classified as a poison a whole set of regulations apply to regulate import, sale, manufacture and use. For the so called dangerous substances less rigid rule systems apply. The criteria for being classified as the one or the other are rather ambiguous and I must ask you not to press me on that point. Let us just see it as an administrative decision that the one cathegory is considered to be more hazardous than the other.

This system has not as yet been consistently used to regulate the entry or existence of carcinogenic substances in the occupational setting. I just wish to demonstrate that the system as such offers an administrative preparedness to apply various control measures on practically any chemical.

The control measures which usually are found to be practical are

    a) registration.

This procedure means that the import, sale and use of a product is prohibited unless it is registered and approved of by a competent authority. The decision is taken by the authority after scrutinizing all relevant documentation on toxic and other hazards which may arise from the product. In the present situation in Sweden registration is applied to pesticides with the aid of special provisions in the ordinance. The decisions are based on the properties of the registered products. Requests for registration are dealt with by the Products Control Board.

Registration is a heavy procedure. To do it properly a numerous staff with a suitable toxicologic training is needed and also a good access to qualified scientific expertise.

    b) permission.

This is a somewhat simpler procedure. The term authorization may also be used. It means that the production, importation, use etc. can be made conditional. Without the conditions being satisfied the product can be practically prohibited. There are a number of provisions in the existing ordinance quoted above regulating the conditions attached to the handling of poison and still more can be added. Of course, our view is that in formulating these conditions the authorities must be given large freedom. They can concern anything from the presentation of good scientific documentation and performing of laboratory experiments to more down-to-earth matters such as labelling, packing and marking.

    c) notification

This is the simplest procedure of all. The import, production and use etc. is made notifiable to the proper authority when this is judged to be desirable. It means that the authorities can keep themselves informed about the entry and dissemination of classifiable products or chemicals in their various forms of use in the community. The ordinance contains a large number of provisions to this effect.

Obviously, by a judicious use of these methods, a hazardous agent can be controlled or - if judged necessary - practically eliminated from the occupational environment.

I must admit that in Sweden the public health authorities have not made full use of this control instrument. The explanation for this is partly that it has been adopted relatively resently. But the instrument is there. It is really now a question of making up one's mind as to what to put in the highest order of priority. Where do we start?

Now we go over to the other major domaine of responsibility of the National Board of Occupational Safety and Health. This is to regulate the working environment to be consistent with the objectives of the particular legislation relevant in this regard. When exercising this responsibility the board operates directly under the government (Ministry of Labour). The administrative instrument used here is the code of practice. Since the problems of the occupational environment are pressing hard the board has not found it possible to operate only on the basis of the Products Control legislation. To go directly to the subject-at-issue the board has published a list of Treshold Limit Values to be enforced in the occupational environment from 1st of January, 1975. In this list a number of carcinogenic agents have been included (see appendix). These agents have been divided into groups A, B and C respectively. The TLV-list has the formal status of a Code of Practice.

Group A consists of substances which are not allowed to be used in the working environment. As you can well see there are a few chemicals here which will be difficult to get rid of with a short notice. Our idea here is to put into operation some kind of registration of use. We are prepared to deal with requests for exemptions from this formal prohibition but the user has to prove his case and to conduct a convincing discussion of alternatives. Morover he has to comply with conditions which guarantee a good hygienic standard of working conditions. We intend to be strict here. The requests are starting to drop in in these days and we await further experience. All chemicals in group A have been classified as poisons by the Products Control Board.

In group B are included carcinogenic agents which may be used only after permission by the Regional Factory Inspectorate. These are our regional organs. This means that the use can be made conditional on technical control measures to minimize exposure. The board is now in the process of writing a code of practice in order to ensure uniform procedures, so far as possible, in this regard.

In group C have been included agents for which a formal TLV value for exposure has been defined by an administrative decision. For this group we intend to apply at least a notification. In some cases - for instance asbestos - rather strict rules have been defined in special codes of practice.

I must ask you again not to press me too hard for the criteria for inclusion into one or the other of the groups A, B and C. Originally group A was designed for known carcinogenic risks to man and group B for experimental carcinogens. As you can see, other criteria have entered into the decisionmaking. It must, however, be remembered that the list has a very interimistic quality. It will not look the same as we go along.

The Code of Practice remains our major tool, far superior to any administratively decided working room standard as it may be expressed in a single TLV value.

Implementing a code of practice in reality means that the use of a chemical in the working environment is made conditional. The conditions can be suitably formulated to guarantee technical control measures in order to obtain minimal exposure. This is the practice in many countries and I do not think it is necessary to go into details of this.

The trouble here is that so many products or chemicals merit being regulated in a code of practice that we find it difficult to keep an even pace with development.

Now, I should like to make a few comments about the problems of society interventions with regard to controlling carcinogenic hazards.

So far as rule-making is concerned it is customary to concentrate the efforts to construct a system of principles for society intervention rather than a set of ready-made problem solutions. This is particularly applicable in a field so dominated by uncertainty about further development and appearance of unexpected events as the field of chemical carcinogenesis. These principles ideally should allow a large freedom for detailed action in a particular case.

We can call this type of intervention by means of principles or objectives a "boundary"-type of rule or decision. It may also be called a performance standard. This means the minimum standard or objectives of a technical problem solution is defined leaving it to one or several of the concerned parties to work out the detailed procedures. The definition of a TLV-value for occupational exposure as this is practised in many countries is by nature a performance standard. When efficiently enforced it allows competition, various technical solutions and development within the fixed standard. This is in my opinion of great advantage and I think that we should generally try to keep interventions from public health authorities to be of this type whenever possible. Of course, the impact of these interventions can be increased by attaching a system of sanctions to it. There are various ways to make rule-breaking less attractive than rule-following. I need not comment this further.

Another type of intervention is to prescribe the detailed technical specifications as to how the hazard is to be avoided. This naturally implies assuming of a responsibility that the solution prescribed is good and safe.

How is this applicable to carcinogenic agents? In the present situation I think that most of us agree that our ability to make good risk assessments, particularly in the low-dose levels, for carcinogenic - or potentially carcinogenic agents - is very little developed. Indeed, we find it difficult to fix boundaries of the kind discussed above. If we assume that for these agents there is no real treshold dose under which the risk is absent, we find ourselves without the margin we are used to have when other types of effects are at issue. A treshold for these agents, if it exists, is hard to define. I think that most of us, although unwillingly, accept the theoretical no-treshold concept. This despite the ambiguity of the term carcinogenic and despite the fact that the potency of experimental carcinogens varies so much in the test systems. A fixing of a TLV for a carcinogen means the acceptance of a risk. It may be small and indeed absent from a practical point of view but it has to be accepted as an uncertainty if we choose to operate with this type of standard.

On the other hand a standard of zero exposure gives you very little freedom of action if you mean absolute absence of all exposure. It is in practical life very difficult to reach an absolute zero level for any chemical agent you handle in industry unless you remove it altogether. Our technology is not all that perfect. If we prefer to retain the chemical in our environment we must be prepared to accept and tolerate a minimal and sporadic exposure. Moreover we have the measurement and analytic side of the problem. If you have an analytic tool with a low-grade sensitivity, the measurement result of zero for obvious reasons can not be taken to mean a total absence of the chemical you are after.

The prescribing of a zero exposure as a standard type of rule therefore is of doubtful practical value. It may in fact give rise to confusion about levels of guarantee and such things which can only be solved with a detailed technical penetration of the problem. You may in fact find it difficult to conduct such a discussion simply because it lies mainly beyond your competence. You end up with the question "How safe is safe enough?" which is a tricky one. The point I wish to make is that absolute absence of all exposure is rather unpractical if you see it as a performance standard type of rule. Admittedly, it is an ideal state of affairs worth fighting for. In practical life, however, a minimal exposure is hard to avoid unless the agent is removed from the occupational environment altogether. This means that an uncertainty, or risk, has to be taken.

In Sweden we have found it to be very helpful to involve the employers and the employees in the decision-making process. This is something we have been practising a long time and it is of particular value in situations where a risk acceptance has to be discussed. The codes of practice concerning the occupational environment are usually products of working parties where both employers and employees have been represented. It has many advantages. It adds to our technical and practical knowledge. It results in a joint risk assessment. It also gives us a good insight into the attitudes to the risk-at-issue and what they are based on among the groups directly concerned. This does not change the formal responsibilities for any of the parties involved in the decision-making process. These are legally defined. But certainly, making the rule-production a joint procedure has resulted in a, let us call it, pragmatic touch in all our codes of practice. It must also be appreciated that this kind of consensus has a tremendous power. It guarantees good enforcement.

One point I wish to raise here is that we have found it necessary to increasingly activate ourselves to ask the question "Why?" when coming across a carcinogenic agent in the occupational environment. Why is this product or chemical used here and for what purpose? What alternatives are there? Is it really necessary? If we wish to eliminate carcinogens efficiently we certainly have to enter a discussion of the motives for their use. Admittedly, we have to start doing this in a very amateurish way so long as we do not have enough competence ourselves. This is a new field for us and I think we shall find it advantageous to explore it more. This again calls for a close co-operation with the groups directly interested in the production.

My general comment to this presentation of the rule-system in Sweden is that if we accept a gradual increase in the coverage of our rules as we go along with them and if we are wise in putting priorities we shall most likely have a good deal of success. But we cannot take all carcinogenic risks in one bite, regardless if they can be referred to as new substances or not.

I shall now go on to comment briefly on some administrative aspects of our predictive and supervising systems so far as carcinogens are concerned. For the prevention of new carcinogenic risks to arise in our environment we obviously need to develop good screening systems with a lowest possible rate of false negatives. Today, that seems to me as something of a wish to St.Claus. We are in great need of good scientific guidance to work out suitable test systems and procedures and also principles for the evaluation of our findings. Above all we need validated, to some extent at least, test systems. I think we have to start with agreements on such things as procedures for the testing. This can concern dosages, routes of administration, selection of animal species, length of experiments etc. I must admit that I hesitate to use the powers in our legislation to enforce an extensive system of pre-introductory testing- or post-introductory for that matter before some more effort is put into the definition of such basic criteria for good testing. We need carefully considered performance standards for such testing. I think that this can be produced. I hope that this is not misconstrued to mean that I am against testing. Certainly we have to develop and use predictive tools. Experience has shown that it is very difficult to eradicate a carcinogen once it has entered all the various technological subsystems of our community. But if we are to regulate predictive or confirmatory testing we want to be able to answer questions such as "How?" and "What then?".

One more thing to remember is also that the banning of "new" substances from our environment on the basis of suspected properties alone is something that should not be undertaken lightly. Nor should they be permitted lightly. It may well be so that the chemical may be toxic or carcinogenic under experimental conditions in the laboratory, but in the real life situation it can be controlled to enter our environment in a form which means a low-dose level of exposure. If we react with prohibition only we may find ourselves intervening against product development in a way which may be to our disadvantage. We must remember that product development has an element of safety promotion. We must go on developing, if not for any other reason, in order to produce materials and techniques to replace the hazardous or carcinogenic risks that already exist. I should personally for instance very much welcome many alternatives to the use of asbestos and I should even be prepared to discuss the acceptance of some theoretical uncertainty, call it risk if you wish, in bringing about a change. I have the uneasy feeling that when prohibition of a chemical is made on grounds of suspected properties alone only part of the problem has been assessed. Admittedly, it is perhaps the most important part but still it may not be complete. We must remember that the hazard is a product of the properties of the agent and the exposure to it.

I must here confess that I am not entirely clear in my mind about what I mean when using the entity "New" substance. Do we mean a substance or product containing it which has never existed earlier? Or do we mean something that has existed but not been identified until now or a known or "old" substance in a new setting? Perhaps I should be a bit more cautious in using this term "new" substance. Never mind, we have lots to do, whatever we mean.

I do not wish to expand here on the problems of risk-benefit assessments. This is complicated in many ways and can obviously also be controversial particularly if the groups carrying the risks and drawing the benefits, respectively, are not identical.

But certainly, in any prohibitive decision concerning a chemical, in all fairness, a thought ought to be given the benefit side to that we at least have an idea what we rid ourselves of.

I shall now discuss a few of the problems of supervising the known or suspected carcinogens that we already have around us and which we wish to control in the occupational environment.

In a recent ILO-document including a convention and a recommendation concerning the prevention and control of occupational hazards caused by carcinogenic substances and agents the control is based on technical control measures and the monitoring of environmental and health indices. Regarding environmental monitoring, in the Swedish Worker's Protection Ordinance, paragraph 24, the supervising agencies have been authorized to enforce an extensive monitoring programme whenever this is judged to be necessary.

See Appendix B (Worker's Protection
Ordinance Paragraph 24)

The actual practice of this is regulated in codes of practice issued by the National Board of Occupational Safety and Health.

Here I should like to say that the measurement of environment is not an easy thing. This is true also as regards carcinogenic agents. For many agents we have no measurement methods.

First of all we have to define the objective of the monitoring. Are we monitoring individuals exposed with the purpose of using the information some time in the future for epidemiological studies and à posteriori risk assessments? Or are we to monitor the premises of work in order to control that these keep a hygienic standard as this has been defined in a code of practice?

Whichever of these objectives is chosen there is a need to define a measurement strategy with a selection of measurement points suitably spread in terms of space and time. Ideally we need to satisfy both the above mentioned objectives. We need to enforce our standards and we certainly wish to make risk assessments based on quantitative data in the future. I am certain that the collecting of these data will be advantageous for us. We are, however, here facing big practical problems with the handling of enormous amounts of information which have to be digested, stored and be made ready for convenient retrieval. I must admit that we have to-day no good system for this in Sweden.

It is, however, important that we find a way out of this dilemma. Moreover, we have not much time I am afraid. One of the basic prerequisities for a register, central, regional or local, which is based on a reporting system is good reporting discipline. If there is no confidence for the system we erect and the people supposed to measure and to report do not understand the "Why?" and "How?" the system runs a high risk of failure. As regards medical surveillance schemes I think most of us agree that such things are needed. It is really not responsible to have known carcinogenic hazards in the occupational environment without assessing their effects. We must also keep in mind the lack of reliability in our predictive systems. I shall not go into detailed discussion about the part of the medical schemes which consists of health screening programmes. Only the brief comment can be made that the results of mass screening of well populations for cancer thought to be induced by environment have not been encouraging. Selective screening of high-risk groups has been considered to be better. The screening of dyestuff and rubber workers for bladder cancer can be mentioned as such an example. Further scientific development in this field will be welcomed. Awaiting this, a cautious attitude has been chosen by the responsible authorities in Sweden. In the Swedish Worker's Protection Act the 16th paragraph can be quoted. It runs:

> § 16 The Crown may prescribe special conditions for the employment of persons on work which is considered to involve special danger of accident or injury to health. Is work considered to involve particular danger of injury to health the Crown may order that persons shall not be employd on such work.

This paragraph has not yet been used with regard to carcinogenic agents. But it is there, in case we should find it useful. As you can see, it empowers the government (Crown) to enforce a medical examination scheme, preemployment or periodical, when required. The government may by an ordinance, if required to do so, delegate this responsibility to the National Board of Occupational Safety and Health.

We now turn over to the problem of record-keeping in health monitoring for cancer. In reality this means the establishment of various forms of population registers where information about incident cancer cases is carried in. In Sweden a central cancer registry has been established and it is operated through a code of practice. The reporting discipline of the doctors is very good.

One question which has arisen in this connection is how the individuals in society react to being registered in this way. It has been argued that record-linkage operations where such registers are connected and completed with other registers containing information about the individual or his environment may result in an undue sacrifice of privacy and may indeed have inherent political dangers. It is obvious to me that we here have two needs or objectives in apparent conflict with each other. It is the individual's need of privacy - sometimes called personal integrity - and the needs of those who have a responsibility for public health to use health information for the identification of risks affecting groups of people and for the improvement of environment so that these risk groups are better protected.

Personally I believe that we can find solutions satisfying both these objectives. We have to start with going into a discussion of this concept of privacy or integrity to see what it really consists of in this regard. I have here with me no good definition for the concept of privacy. This is something that can be discussed at length and still not be solved. It is really a difficult one. Let us just put it very simply by saying that the individual fears that information about him in various registers may be in some way used to his disadvantage. This fear seems to increase in proportion with the developing technological possibilities and with the increasing bulk of information that can be built up in registers in various parts of the community.

Still, as you know, in many countries with developed national health or medical services or insurance systems for health care, the registering of this information - large amounts of it really - is already accepted. The principle of registering the actual kind of information is not new. We are accustomed to give up a bit of our privacy in exchange for a better social security for ourselves. I do not, therefore, think that our attitude of mind is altogether logical in this matters. The principle is not new, it is the objective.

We have to investigate if it - for this new objective - will be acceptable for the individual to have his cancer diagnosis carried into a monitoring system where record-linkages are technically possible to other kinds of information sources.

What we, and the general public, need to know is:

> What information is registered
> Why it is registered
> Who is responsible for the register
> Who has access to the registered information, for what purpose and on what conditions
> What guarantees are there that unauthorized persons have no access to the registered information
> What guarantees are there that the information is used only for purposes acceptable for the general public.

If we construct principles for this I am certain we will have a general agreement from everyone concerned. Furthermore, if the supervising organs of society assume a responsibility for the answers and guarantees asked for above I think that we will be able to keep up confidence for the registering operations. It must be remembered though, that this responsibility also implies a duty to see to it that something sensible is made with the information which is collected. This duty, admittedly, is more of a moral character rather than a duty in a legal-technical sense. The public must, however, be confident that the individual's giving up a bit of privacy for the public health objectives is worth while. Registering must not be a passive process. If this condition is not satisfied we run the risk of complete failure.

We have recently had a new legislation introduced in Sweden. On the basis of a Data Act, in force from July 1st, 1973, a set of rules and principles is in the process of being constructed.

Time does not allow me to give a more detailed comment on the problem of information and education. I mentioned earlier that the attitude of individuals and groups concerned are important determinants for the success of preventive programmes. These attitudes can be modified and changed by means of information suitably modelled to reach groups and persons in key positions. A health administrator must keep this in mind when constructing the rules for the public health operations for prevention of risk.

We must remember that the strong association between cigarette smoking and bronchial carcinoma has been known to us for a long time and that we have not really done much about it.

I have in this contribution tried to make the point that carcinoma prevention by means of rules is an interesting field where our knowledge is scarce on many important points. Many of our decisions are made under uncertainty. I should very much like to see some more efforts in the field of health services research to help us evaluate what we really are doing in constructing our regulatory systems.

I would like to end this with the comment that the comprehensive nature of the problems we are dealing with has made it necessary to ask ourselves where we, i.e. the supervising agencies, fit into the system. Are we the guides or leaders of an ignorant public in need of a firm hand, or are we the servants of a well informed public in full possession of good judgement and of other necessary faculties?

The answer is both and everything in between these extremes. We have found it necessary to adapt our decision-making so that we can take full advantage of the technical, economical and practical potentials of the industry and so that we include the concerned groups of individuals in the process of arriving at decisions.

Our rules therefore have got a basic element of consensus. We are part of a team and we have to learn how to excercise our responsibilities as such team members.

As regards carcinogenic agents we have the safety that the team wants to have.

REFERENCES

(1) Act and Ordinance on Products Hazardous to Health and to the Environment.
(Swedish Code of Statutes 1973:329 and 1973:334)
In force from July 1st, 1973.

(2) Act and Ordinance for Worker's Protection
(Swedish Code of Statutes 1973:834 and 1973:841)
In force from January 1st, 1974.

(3) Official Swedish TLV-list for occupational exposure to chemicals
National Board of Occupational Safety and Health
November 1974. Code of Practise No.100/1974.

(4) ILO Convention and Recommendation concerning Prevention and Control of Occupational hazards caused by carcinogenic substances and agents
Record from 59th ILO Conference, Geneva 1974.

(5) Swedish Data Act
(Swedish Code of Statutes 1973:289)
In force from July 1st, 1973.

Appendix A

Carcinogenic chemicals in official list of Treshold Limit Values in Sweden
(To be enforced from January 1st, 1975).

National Board of Occupational Safety and ealth. Code of Practice No.100/1974.

Group A

Chemicals not to be used in occupational environment:

Benzidine
Bis-chloro-methylether
Methyl-chloro-methylether
B-napthylamine
4-nitrodiphenyl
Crocidolite

Group B

Chemicals allowed to be used in occupational environment on conditions
prescribed by Regional Factory Inspectorate. No formal TLV-value for
exposure defined.

2-acetylaminofluorene
Auramine
Dianisidine
Diazomethane
Dimethylminocazobensene
3,3-dichlorobenzidine
Diethylsulphate
Ethylenimine

Ethylenetiocarbamide

Methylene-o-chloro-aniline
a-napthylamine
N-nitrosidimetylamine
1,3 propane-sultone
β-propiolactone
Propylenimine
3,3-diemetenylbenzidine

Group C

Chemicals with a numeric TLV-value

Arsenic and inorganic compounds
Asbestos
Benzene
Beryllium
Dioxane
para-phenylendiamine

Hydrazine
Cobolt
Chromates
Nickel (metal)
Nickel-carbonyl
Vinylchloride

Appendix B

Worker's Protection Ordinance

Paragraph 24

§ 24 In the case of work where dust, smoke, gases or vapours are released in such quantities as to be injurious or offensive to the employees, the working process shall, as far as possible, take place in closed apparatus or the work shall be performed in a separate room or enclosed place. If this is not feasible, satisfactory arrangements for collecting and carrying off the dust, smoke, gases or vapours or otherwise rendering them harmless shall be made as far as possible at the place where such contamination of the air originates and can spread.

Exhaust equipment shall not be installed in such a way that the employees are exposed, through contaminated air from such equipment, to influence of the kind referred to in the first paragraph.

If certain work or certain kind of work is found to cause special risks for bad health on account of the existence of dust, smoke, gases or vapours, the National Board of Occupational Safety and Health may, as condition for employment of persons to such work, prescribe investigation of the conditions of the air. The Board will decide the extent of the investigation and give directives required regarding the accomplishment of the investigation.

An investigation report and any other document necessary for the assessment of the results of the investigation shall be submitted to the Labour Inspectorate if so prescribed by the National Board of Occupational Safety and Health.

All expenses incidental to the investigation shall be paid by the employer, unless by the National Board of Occupational Safety and Health decided, that the expences, when special circumstances render it desirable, shall to a certain extent be paid by Governmental funds.

CONTROL OF CARCINOGENIC HAZARDS IN INDUSTRY

Alexander MUNN

Imperial Chemical Industries Limited, Organics Division,
Blackley, Manchester, M9 3DA, England

My remarks today about the control of carcinogenic hazards in industry are not primarily concerned with the application of industrial hygiene measures to contain known carcinogens, nor with the medical surveillance of workers exposed to carcinogenic hazards, although such measures will certainly be referred to en passant. Nor are they primarily concerned with legislative control, however important that might be. Rather are they oriented towards the duties and responsibilities of industry to be aware of the existence of such hazards, to give consideration to the need to recognize such hazards, to establish mechanisms which will identify such hazards; and towards the considerations which ought to be taken into account in making judgements about such hazards.

The concept of absolute safety in any sphere of human activity is an ideal unlikely ever to be achieved. This is certainly true of industry, where it has long been accepted that in terms of physical safety, in terms of accident prevention, it is industry's duty to take all reasonable and practicable steps to ensure safety. It is not industry's duty to make accidents impossible. Such a counsel of perfection could only be achieved by closing down industry altogether.

Acceptance of this principle implies acceptance of some degree of risk - albeit a small one - and for industry to fulfil its proper function this risk will, in many cases, involve exposure of workmen to carcinogenic agents. There can be few factories where there do not exist some carcinogens.

The word "carcinogen" means different things to different people, and I certainly do not propose in the course of this paper to attempt to define it. The question which demands consideration is not whether such and such an agent is carcinogenic, but whether industrial exposure to that agent constitutes a carcinogenic hazard to workmen; and, if so, whether adequate precautions can be introduced to obviate that hazard. The latter part of the question I do not propose to discuss here. It is essentially a problem of engineering and chemical techniques, associated with environmental and biological monitoring. Suffice it to say that simple hygiene measures - improved cleanliness rather than sophisticated techniques - many years ago led to the virtual disappearance of the chimney sweeps' cancers associated with the name of Sir Percival Pott.

The first part of the question - i.e. whether there is a carcinogenic hazard to workers - demands a judgement. The basis of the judgement can be made under four main headings :

1. Critical evaluation of experimental evidence, both qualitative and quantitative.

2. Critical evaluation of epidemiological evidence - if it exists.

3. Physico-chemical properties of chemical agents, insofar as they

influence systemic absorption.

4. Chemical relationship to other compounds of known hazard or non-hazard.

Experimental Evidence.

As more and more industrial chemicals are tested for carcinogenicity, more and more are found to induce tumours in animals in experimental conditions. In the United States, the National Institute for Occupational Safety and Health recently listed about 1,500 chemical substances which have reportedly produced an observed or suspected carcinogenic response[1]. There is a widespread belief that substances which induce tumours when deliberately introduced into the bodies of experimental animals by a variety of methods of administration, often at very high dose levels, will necessarily induce tumours in workmen exposed in industrial conditions. I hope it is not necessary to point out that this is a non-sequitur. The drug Isoniazid induces lung tumours in the mouse[2]. Cyclamate sweeteners produce bladder tumours in the rat[3], and tannic acid produces liver tumours in the same species[4]. Does anyone seriously believe that these chemicals present a carcinogenic hazard when handled by workmen?

It similarly does not follow that a negative result in a properly conducted animal study necessarily indicates freedom from carcinogenic hazard. For example, there are at least two as yet incomplete experiments in dogs, involving the administration of alpha-naphthylamine, which have been running for more than seven years with no bladder tumours arising so far. Should these experiments be completed without the appearance of bladder tumours, it would, in my view, be totally inappropriate to conclude that alpha-naphthylamine is not carcinogenic. There is plenty of evidence in man of the carcinogenicity of the old commercial quality of alpha-naphthylamine (which contained about 5% of beta-naphthylamine as an impurity), and it seems quite inappropriate to assume that the tumours which have appeared in man have necessarily been caused by the beta impurity.

Both quantitative and qualitative considerations have to be taken into account, i.e. it is not always sufficient to decide whether a chemical which has produced experimental tumours is a potent carcinogen or a weak one, but often one must decide whether the evidence represents true chemical carcinogenesis at all.

Certain types of evidence of carcinogenicity are of limited acceptability. Tumours induced by implantation in the bladder of pellets containing alleged metabolites of suspected bladder carcinogens are certainly of doubtful value in assessing industrial bladder cancer hazards. It has been shown[5] that the vehicle (i.e. the pellet) exerts co-carcinogenic activity in this kind of experiment. If it were a common phenomenon for workmen to have pellets of cholesterol or paraffin wax in the bladder, this kind of experiment might have more relevance to the industrial context.

Subcutaneous sarcomas at the site of repeated subcutaneous injection are also of doubtful significance. A series of elegant experiments by Gangolli, Grasso and Golberg[6] demonstrates that such tumours frequently arise by mechanisms that are not what we commonly mean by chemical carcinogenesis, but result from local tissue damage (caused by physical properties of the formulations), followed by continually reactivated tissue repair. Eventually, control of the mechanisms of repair breaks down, and subcutaneous sarcomas result. In many instances such tumours have little relevance to problems of occupational cancer, although we must realise that true chemical mechanisms may also occur - e.g. with propane sultone. Professor Maltoni the other day suggested that the high incidence of subcutaneous sarcomas induced by lead chromate[7] could reliably have been used to predict

that these pigments presented a carcinogenic hazard to man. I am rather
less certain than Professor Maltoni, but feel that a more appropriate course
has been taken insofar as these findings have led to the epidemiological studies
of lead chromate manufacture that are now being carried out.

Some other experimental techniques are also of doubtful validity.
Experiments on new-born animals should not be used as evidence either of
carcinogenicity or of non-carcinogenicity. The place of tracheal instillation
as a route of administration in the induction of lung tumours is certainly not
adequately evaluated. Certainly the induction of lung tumours by nickel thus
administered has not really helped to resolve the problem of whether the
widespread industrial exposure to nickel (as opposed to nickel carbonyl) truly
represents a carcinogenic hazard. In vitro systems have not yet reached the
stage where they can reliably be used for the prediction of hazards. Caffeine
and formaldehyde are examples of compounds which have given positive results
in such systems, where it is pretty clear that there is no hazard to man.
It would seem more appropriate to use such systems primarily as a basis for
selecting compounds more worthy of mammalian studies.

These qualitative considerations of experimental carcinogenicity have
to be considered along with quantitative considerations, of which the most
important is whether a specific agent possesses high carcinogenic activity, or
weak carcinogenic activity, or none at all. The parameters which have to be
considered are the tumour yield, the length of the latent period, and the dose
required to induce tumours. Misunderstandings with regard to dose are common,
since in many carcinogenicity studies the dose-levels used are the highest
which the animal will tolerate. It does not follow that tumours will not
arise at lower levels, and there is frequent confusion between the dose that
has been used to produce cancer and the dose which is required to produce cancer.
Enormous dose levels such as the 4 g/kilo/day required to induce experimental
bladder tumours in the rat by diethylene glycol[8] seems totally unrelated to
the problems of occupational cancer.

There is a widespread view - not one to which I myself subscribe -
that there is no safe dose for any carcinogenic agent. It would take too long
to advance the arguments against this thesis - which in my view is not merely
theoretical, but speculative, and unsupported by facts - but it is worth
pointing out that a WHO Scientific Group on the Assessment of the Carcinogenicity
of Chemicals has concluded that the possible existence of a threshold to the
effect of chemical carcinogens should be envisaged.[9] Dr.Higginson, whilst
speaking a few weeks ago in Manchester, referred to the carcinogenicity of
alcohol. He stated that among heavy drinkers there was an excess of certain
kinds of cancer, suggesting that three large whiskies or $\frac{1}{2}$ litre of table wine/
day was about the level of intake where the risk became apparent. Below that
level there did not appear to be a risk. However relevant this might be to
drinkers, it suggests to me that industrial workers occupationally exposed to
ethyl alcohol are at no risk whatsoever if they are non-drinkers, i.e.that in
occupational exposure to this carcinogen a threshold effect can be demonstrated.
Furthermore, much was said earlier in the Symposium about the carcinogenic
effects of solar radiation. Even in Manchester, our workers are exposed to
sunlight, but nonetheless we have never felt it necessary to protect them
against this particular carcinogen.

Epidemiological Evidence.

In 1947, the distinguished American pathologist Dr.Fred Stewart
wrote the following wise words[10]: "It is probable that with the development
of the Chemical Industry we will see new chemical cancers, and medicine must

be on the watch for such developments. But it must refrain from ascribing to industry those tumours whose incidence falls within the expectation of the population as a whole, and from making premature conclusions based on lack of appreciation of statistical method." These comments are equally true in respect of establishing both the existence and the absence of occupational cancer hazards. The importance of epidemiological data cannot be over-emphasized. In my generation, the first outstanding example of the use of modern statistical techniques in relation to occupational cancer was the survey of the United Kingdom dyestuffs and rubber industries conducted by Case and his colleagues[11], but many similar examples have emerged since then.

What makes occupational epidemiology particularly difficult is that of necessity it has to be carried out after the event, and is generally dependent on adequate data having been recorded contemporaneously with events. This may make it difficult or impossible to identify the population to be investigated, or to define the chemicals to which a specific population has been exposed. Furthermore, in the investigation of the effects of a specific chemical, bias is liable to be introduced through the effects of exposure to other chemicals, or by dilution of the population under investigation with "non-exposed" individuals.

The need for good epidemiological data is increasingly leading to large chemical companies setting up computerised medical and occupational record systems, which facilitate identification of populations exposed to specific chemicals, as well as ensuring that the chemical exposure throughout the lifetime of individual workers can be instantly recalled when necessary. This represents a great step forward, because there can be little doubt that, despite my earlier reservations, properly-conducted epidemiological studies provide more valid data than any other in recognizing types of cancer caused by industrial chemicals. Such data, furthermore, enables inferences to be drawn in respect of the potential hazards of other chemically-related compounds used in similar industrial conditions. The epidemiological investigation by Veys[12] in respect of more than 4,000 rubber workers exposed for many years to phenyl-beta-naphthylamine (which until a year or two ago contained a beta-naphthylamine impurity generally in the range 20 - 50 p.p.m.) has demonstrated no excess of bladder tumours. This data is invaluable not only in respect of exposure to phenyl-beta-naphthylamine, but also in respect of exposure to other compounds having impurities of carcinogenic aromatic amines. It is well-known that many dyes contain impurities of carcinogenic aromatic amines at levels of a few parts per million. For this reason, the National Union of Dyers, Bleachers and Textile Workers recently commissioned a proportional mortality study of death certificates of their members. The investigation was carried out by Dr.Newhouse of the TUC Centenary Institute of Occupational Health, and no abnormal incidence of bladder cancer - or, indeed, of any cancer - was found. The inferences which may be drawn from these two studies are very important in relation to other products having similar impurities.

Physico-Chemical Properties.

No carcinogenic chemical can exert systemic carcinogenic activity unless and until it has been absorbed into the body. Thus, physico-chemical factors which influence systemic absorption are of profound importance in the assessment of carcinogenic hazards. Workmen do not eat chemicals - they inhale them if they are airborne, or they may absorb them through the skin. A volatile or dusty product is thus more readily absorbed than one which is involatile or non-dusty. A substance which is absorbed through the skin is potentially more hazardous than one which is not absorbed through the skin. The point is, perhaps, best illustrated by an example of acute toxicity. Two highly toxic compounds commonly used in industry are HCN and NaCN. Their acute oral toxicity (both qualitative and quantitative) is very similar, resulting from the cyanide ion circulating round the body. Despite their similar toxicity, however, they present very different toxic hazards. HCN is a volatile liquid, readily absorbed through the skin, and wherever it is used it presents a most serious toxic hazard. Spillage will result in volatilisation, and when a workman inhales the vapour (if it is there at more than a very low atmospheric concentration) he will die. Similarly, quite a small splash on the skin will lead to rapid systemic absorption, and the workman will die. Sodium cyanide, on the other hand, is commonly used in a variety of industries with very much less care than is necessary for HCN. It is an involatile crystalline solid, with very little dust, and is not absorbed through the skin. Occupational poisoning is unknown.

Although the principle I have described has been illustrated by an example of acute toxicity, it remains equally true in respect of chronic toxicity and systemic carcinogenicity. Vinyl chloride, for example, is not really a highly potent carcinogen when compared with beta-naphthylamine or bis(chloromethyl) ether. Its volatility, nevertheless, makes it a significant carcinogenic hazard, insofar as if it is not contained high atmospheric concentrations inevitably arise and ensure that exposed workmen receive high dosage by inhalation. If it were the case that vinyl chloride were an involatile non-dusty solid, not absorbed through the skin, it seems most unlikely that there would be an angiosarcoma problem.

Chemical Relationships.

Carcinogenicity (or non-carcinogenicity) cannot be reliably predicted on the basis of chemical structure, but that is not to say that structure/function relationships should be totally discounted. For example, all of the aromatic amines known to be carcinogenic to man have either fused or conjoined ring structures. No single ring aromatic amine has ever been incriminated in causing industrial cancer. Thus, it seems not unreasonable to give less weight to the experimental carcinogenicity of single ring structures such as ortho-toluidine or paraphenylenediamine, than to the experimental carcinogenicity of, say, dichlorobenzidine, even though the latter also has never been identified as a cause of occupational cancer.

It has been suggested that no new chemical should be introduced into industrial processes without prior testing for carcinogenicity. However desirable this might be, I cannot accept that it is feasible. It is not simply a question of cost - the scientific resources do not exist.

That is not to say that there should be no testing of new products. It is here that chemical relationships must be taken into account. If we wished to introduce the N-acetyl derivative of benzidine, careful carcinogenicity testing would certainly be required. But if we were to introduce a polyoxyethylene derivative of a fatty acid (a great deal is known about the biological properties of compounds coming into this chemical class), it would, in my view, constitute a gross waste of limited resource to devote time and effort to carcinogenicity studies.

Industry has to make judgements of carcinogenic hazards, whether it wishes to do so or not. In order to do so, it has a duty to ascertain what facts are known. Not all industry has the same resources, and it seems to me that the large multinational companies with vast resources at their command have a greater duty than have small companies with few resources. But even the small companies do not have the right to disregard what is widely-known. It was an industrial tragedy that the manufacture of beta-naphthylamine was continued by at least one small company in the United States until only a few years ago.

Having ascertained the relevant facts, industry must consider them carefully and make its judgements about any carcinogenic agent. The judgement may be that the product or process is too dangerous to use in industrial conditions and should be abandoned. This is what we have done with benzidine and its salts. Alternatively, one may take a decision that some danger exists, but that the carcinogen may, nevertheless, be safely used in controlled conditions with suitable industrial hygiene measures to prevent contact, with medical surveillance, and with adequate information, education and training of workers. This is what happens, for example, with ionising radiations. Another conclusion which may be reached is that the evidence of carcinogenicity does not indicate the existence of a hazard. For example, there may or may not be a hazard to the public from the use of cyclamate sweeteners, but it is difficult to believe that their use in soft drinks constitutes a hazard to the workers in the soft drink industry. I must point out, nonetheless, that however real the duty to give careful and sincere consideration to these matters, it is not a legal duty to come to the correct conclusion. Infallibility cannot be prescribed by law.

It will be appreciated from the above that I believe that industry has responsibilities in connection with the control of carcinogenic hazards much greater than merely complying with legislation. Regulatory control of carcinogenic hazards in industry can apply only to clearly recognized and well-defined hazards. It is not possible to banish all carcinogens with an Act of Parliament. We are surrounded by carcinogenic agents, whether it be sunlight or the traces of beta-naphthylamine present as an impurity in products which are themselves safe. We may get rid of the worst of them, but we cannot get rid of them all. We must learn to live with carcinogens rather than die from them.

References.

(1) Federal Register, June 23, 1975, 40, 121, 26390-26496.

(2) Biancifiori C. and Ribacchi R.: Pulmonary tumours in mice induced by oral isoniazid and its metabolites. Nature (Lond.), 1962, 194, 488.

(3) Richardson, H.L., Richardson, M.E., Stewart, H.L., Lethco, E.J. and Wallace, W.C.: Urinary bladder carcinoma and other pathological alterations in rats fed cyclamates. Abstr.Proc.of Amer.Assoc.for Cancer Res., 1972, 2.

(4) Korpassy, B. Cancer Research, 1959, 19, 501.

(5) Bryan, G.T. and Springberg, P.D.: Role of the vehicle in the genesis of bladder carcinomas in mice by the pellet implantation technique. Cancer Research, 1966, 26, 105-109.

(6) Gangolli, S.D., Grasso, P. and Golberg, L.: Physical factors determining the early local tissue reactions produced by food colourings and other compounds injected subcutaneously. Fd.Cosmet.Toxicol., 1967, 5, 601-621.

(7) Maltoni, C., Sinibaldi, C. and Chieco, P.: Subcutaneous sarcomas in rats following local injections of chromium yellow. II International Symposium on Cancer Detection and Prevention, Bologna, Italy, 9-12 April 1973.

(8) Weil, C.S., Carpenter, C.P. and Smyth, H.F.: Urinary bladder calculus and tumor response following either repeated feeding of diethylene glycol or calcium oxalate stone implantation. Industrial Medicine and Surgery, 1967, 36, 55-57.

(9) Evaluation of Certain Food Additives. 18th Report of the Joint FAO/WHO Expert Committee on Food Additives, Technical Report Series No.557, 1974.

(10) Stewart, F.W. Bull.N.Y.Acad.Med., 1947, 23, 145.

(11) (a) Case, R.A.M., Hosker, M.E., McDonald, D.B. and Pearson, J.T.: Tumours of the urinary bladder in workmen engaged in the manufacture and use of certain dyestuff intermediates in the British Chemical Industry. Parts I and II. British Journal of Industrial Medicine, 1954, 11, pages 75-104 and 213-216.

(b) Case, R.A.M. and Hosker, M.E.: Tumour of the urinary bladder as an occupational disease in the Rubber Industry in England and Wales. British Journal of Preventive & Social Medicine, 1954, 8, 2, 39-46.

(12) Veys, C.A.: A study on the incidence of bladder tumours in rubber workers. M.D.Thesis, University of Liverpool, 1973.

SUMMARY

Complete control of carcinogenic hazards in industry is an unrealistic concept. Industry, nevertheless, has a duty to give careful consideration to existing carcinogenic hazards, and to the possibility of new ones. Only well recognized hazards can be subjected to legislative control. Industry must take into account published experimental and epidemiological data, along with physical and chemical considerations related to specific agents, and, where necessary, introduce appropriate preventive measures.

RESUME

Tenter de prévenir totalement les risques cancérogènes dans l'industrie n'est pas réaliste. Néanmoins, l'industrie a le devoir d'examiner avec soin les risques cancérogènes existants, et la possibilité que de nouveaux apparaissent. Seuls les dangers bien reconnus peuvent être soumis à une législation préventive. L'industrie doit tenir compte des données expérimentales et épidémiologiques publiées, ainsi que des considérations physiques et chimiques relatives aux agents particuliers, et, lorsqu'il est nécessaire, prendre les mesures de prévention appropriées.

simple comment upon these factors indicates that our
trial processes are responsible for a great deal of the
n's sickness even without specifying the hazards
ously existing in the processes which include fumes from
icals and gases, fibres and dust. There is little doubt
industry has been well aware of these cancer-producing
ts but more important the medical profession have had
least 200 years' knowledge of the origins of some cancers
n one considers the chimney sweep boys' cancer conditions
ch were noted in 1775.

ilst one sifts through the various arguments which have
ged over the last few years on the question of smoking
d lung cancer and, particularly, some medical practitioners
nking strictly the question of smoking and exposure to
ncer-producing industrial agents, one striking factor
ppears which is that with the combination of one or other,
r both, workers suffer from cancerous conditions and
ther workers involved in the same conditions do not contract
the disease. There have been some suggestions that it is
a known fact that approximately 90 per cent of those
suffering from cancer could be traced back to industrial
origin. We are not in a position to comment upon this.

This brings us to the role of the medical profession and,
indeed, the scientific experts who back up the medical
profession with their knowledge of chemicals and gases, as
to what exactly has been achieved in this field over the
last 200 years. We suppose that we can discount the first
150 years but, clearly, medical knowledge is still at the
stage (apart from one or two exceptions) whereby the only
means of dealing with cancer is either to cut it out or burn
it out. One would have thought that when the system of
screening was developed in the 1950's and 1960's that this
would have been an important fore-runner to developing a
system whereby cancer could be screened and treated as in
the papilloma cases with very good results. At its
infancy stages one would have thought that this system could
have been applied to very many other industries. In the
rubber manufacturing industry there was knowledge of exposure,
there was an effort made to trace those who had been exposed
and many successful results have occurred as a sequel to
screening and treatment. We view with dismay the lack of
further development of this system to many other industries
where it can easily be established that exposure has taken
place.

In 1975 we have, at last, seen some move to set up a
register for those that work in the asbestos industry.
With mesothelioma being one of the disabilities that arise
from this it is probably over-simplification to say that
screening and treatment methods might prevent this
development but the criticism that we have of the medical
profession and of the employers is that this has never
actually been tried to see whether it would be successful
We view with dismay Professor Maltoni's remark that cancer
is irreversable. We have other criticism to make in
respect of the fact that in the chromeate manufacturing
industry it would be easy to identify the groups of workers
who have been exposed. Not being medical experts and,
indeed, being complete laymen in this field we can

426

THE TRADE UNIONIST'S VIEW OF OCCUPATIONAL CANCER

A.C. BLYGHTON

Legal Department Transport and General Workers' Union Transport
Smith Square Westminster, London, SW1P 3JB, England

The Transport and General Workers' Union, since its amalgamation over fifty years ago, has developed programmes to monitor, record and take action upon all matters in relation to industrial health. We are a Union with 1,800,000 members. Our aim is to take vigorous action in the field of accident and disease prevention. A brief summary of part of our experience can be given as follows.

In the late 1920's and 1930's many of our members in the industry of dye-stuff manufacture contracted cancer of the bladder. We still hold files and evidence of inquests confirming that the cause of death was due to contact with beta-naphthylamine.

During that period we had experience of our membership suffering from asbestosis including some cases of cancer of the lung. At that stage whilst there were medical suspicions it was not clearly established that mesothelioma was, in fact, an industrial disease.

In the latter war years an ICI product known as Nonox S was introduced into the rubber manufacturing and cable industries and although this product was banned from use in the years 1949 and 1950 cases of papilloma began to appear in the early 1960's and since that date a regular number of cases have reached the Union for action. As a result of contact with mineral oils we have had cases of cancer of the scrotum. We have also had one case of a woodworker suffering from nasal cancer after 30 to 40 years experience in the industry. We have members working with chromeates suffering from lung cancer. Members exposed to nickel carbonyl processes also suffered from cancers. It would have been thought that this experience alone was sufficient but, recently, a member died who was employed in the processing of vinyl chloride monomer and the cause of his death was diagnosed as angiosarcoma. In addition to this we have had a suspect case of leukaemia where our member had been employed in an atomic energy establishment. Other cancerous conditions arising out of contact with tar, pitch, etc. have been reported.

We recognise that diseases which are diagnosed today result from exposures of 25 years ago or more.

appreciate that there is obviously a difference in treatment of the lungs and of the bladder. We are aware, in some instances, of lungs being partially removed through operative treatment but with an early system of screening we could visualise a great deal of improvement in this field. It is possible that we see the means of dealing with these problems a little more simply than doctors and scientists would accept. Obviously there must be some difficulties but we do not see much effort being made to follow an example that is over 20 years old in order to see whether this could be improved upon.

Bringing the situation up-to-date in regard to cancer of the liver as a result of exposure to vinyl chloride monomer our knowledge of this particular factor indicates that the medical profession has been thrown completely sideways as a result. We did not have any system in this country whereby medical examinations could be carried out to ascertain certain liver diseases; there was no system which could be adopted which would give some early diagnosis of liver malfunctions and even an ultra-sonic system developed by one of our leading hospitals has had doubts cast upon it as to its value.

Vinyl chloride gas was experimented with as an anaesthetic a few years ago. Continued exposure to this process has revealed a risk and, therefore, a further group of substances will have to be very closely investigated as a result. Our Government has been reasonably active enough to produce regulations to counter these hazards to some extent. Many carcinogenic substances mentioned here are subject to these. A good example being the banning of a number of chemicals from use altogether and the lowering of exposure, within the last 12 months, to vinyl chloride monomer from 500 parts per million to 10 parts per million. We were pleased to observe Professor Maltoni's findings in respect of exposures at 50 parts per million. We hope that he will continue with these tests down to 10 parts per million and below. We see a possible improvement being achieved even upon this standard which can be maintained.

We think that scientists and doctors have neglected to provide information to both employers and employees regarding dangerous properties. This field should develop to such an extent that everyone is made aware of hazards in industry. This is now a world-wide problem and there is a clear need for all countries to develop contacts to share knowledge, results of research, medical developments, etc. We must identify sources where hazards are in existence. We must also ensure that once identification of a hazard takes place that everyone who has been exposed should be notified and made aware of medical treatment, screening, etc. which is available. We must develop and perfect techniques by which workers are screened and treated. We also feel that doctors should work strenuously upon the problem as to why some workers contract the disease and others continue without any possibility of the disease developing. We have had no indication that those who are able to reject carcinogenic substances have been epidemiologically tested. Where are the comments upon this phenomena? There are thousands of workers who have been thus exposed some massively so. Clearly, there is a field here for the doctors to develop and work upon rather than being the purveyors of treatment when a cancer is diagnosed.

There is also a duty upon the scientists and research workers to produce substitute materials. The example of the substitute for Nonox S is a clear indication as to what can be done.

We feel that the Tripartite approach to deal with these matters is the most successful.

One further point that we wish to make in regard to cancer in industry is the experience that our Union has encountered where melanoma has occurred at the site of an injury. We have had experience of at least three cases in recent years where this condition has developed as a result of trauma. Medical cases observed over the years seem to support this contention and the point that we have made regarding the make-up of the individual has to be considered. It is accepted, however, in respect of trauma that it may be impossible to carry out any pre-screening tests, but we thought it only right to mention yet another field in which industrial cancer is found.

As laymen we feel somewhat helpless in regard to the present situation because, even now, there must be many suspect processes which could be considered in depth and, indeed, improvements in the working environment could be carried out in order to reduce or remove any slight risk that is in existence.

Industry should be capable of accepting conditions of greater control. The United Kingdom regulations will ensure this and we recognise that we as Trade Unions have our part to play.

We must not lose our perspective. There are industrial diseases outside of carcinogenic conditions which are fatal. For instance, stress causing cardio-vascular conditions. Chest complaints causing pulmonary disorders and many others.

We deplore emotional statements and propaganda such as vinyl chloride experience. Of course, the situation was serious. However, some people making statements in January 1974 were certainly aware of doubts before that time. The same remarks apply to the questions raised on Mesothelioma some 25 years ago.

If you are saying that the risk means nothing until you have proved it after long tests, you are wrong. Equally so, you are wrong if you ban processes indiscriminately. We agree with our Swedish colleague there must be a continuing dialogue tripartite discussions with industry, trade unions and Health and Safety Officials.

There must be a much wider appeal to discourage and ban smoking. We are looking to the experts to give a lead to the world with a view to widening discussion and taking action in this field which causes great misery and despair.

SUMMARY

The Transport and General Workers'Union (UK) has an active programme devoted to monitoring the health of its 1.8 million members.

Files on members who died of bladder cancer after exposure to β-naphthylamine go back to the late 1920's. From the same period come data on members suffering from asbestosis which included some cases of lung cancer before the industrial cause of the disease was recognized.

These and other more recent examples including Nonox S and vinyl chloride amply justify the need for setting up registers of all workers who are at risk from industrial exposure. The scientfic community has a responsibility for communicating data regarding hazards both to employers and employees who can then ensure that all who have been exposed can be notified and screened.

The trade unions have a part to play in ensuring that industry accepts conditions of greater control under the new regulations in the United Kingdom. There must be a continuing tripartite discussion between industrial management, trade unions and the responsible government official.

RESUME

La Transport and General Workers'Union (Royaume-Uni) surveille activement la santé de ses 1 800 000 travailleurs.

Les dossiers de ceux de ses membres morts d'un cancer de la vessie après exposition à la β-naphtylamine remontent à la fin des années 1920. Depuis la même période, on a des données sur les sujets atteints d'asbestose et sur certains cas de cancer du poumon survenus avant que la cause industrielle de la maladie eût été reconnue.

Ces exemples et d'autres plus récents, comme ceux du Nonox S et du chlorure de vinyle, justifient amplement la création de registres de tous les travailleurs industriellement exposés. La communauté scientifique a pour mission de communiquer les informations concernant les risques aux employeurs et aux employés, lesquels peuvent ensuite faire en sorte que toutes les personnes exposées soient averties et dépistées.

Les syndicats ont un rôle à jouer pour veiller à ce que l'industrie accepte les conditions de contrôle plus strict prévues pour la nouvelle réglementation du Royaume-Uni. Une concertation tripartite permanente s'impose entre le patronat, les syndicats et le fonctionnaire responsable.

POINT DE VUE D'UN SYNDICALISTE FRANÇAIS SUR LE CANCER PROFESSIONNEL

J. ESCANEZ

FUC-CFDT
26, rue de Montholon, 75439 Paris, France

Nous voudrions tout d'abord remercier les organisateurs d'avoir bien voulu inviter les syndicalistes que nous sommes, ce qui ne nous semblait pas évident, et nous sommes flattés que la concertation nous vienne des hommes de science.

Devant cette honorable assemblée, nous faisons figure d'ignorants. Cependant, notre intervention se situant sur un autre plan et par rapport à des objectifs précis, avec un peu de chance nous vous apporterons peut-être quelque chose. La courtoisie dont vous faites preuve dans vos débats nous a surpris et fait plaisir. Elle nous a encouragés à nous risquer à notre tour, profitant ainsi de la situation.

Si vous le voulez bien, nous partirons de trois affirmations faites ici :

1)   A propos de l'amiante, on nous dit : "Le fait de fumer multiplie les risques". Doit-on en conclure qu'il suffirait tout simplement d'arrêter de fumer pour que les risques soient minimisés ? N'est-ce pas plutôt une manière comme une autre d'ignorer la véritable cause de l'asbestose ?

2)   A propos du DDT, on nous dit : "L'utilisation de ce produit est indispensable si l'on veut nourrir tout le monde". Nous resterons sceptiques devant une telle affirmation et il doit bien se trouver quelque part dans le monde un savant pour affirmer que les insectes sont friands de DDT.

3)   D'une manière générale, à propos de la fabrication de produits qui s'avèrent dans le temps être cancérigènes, on nous dit : "Les profits sont partagés par la communauté, les risques sont partagés par la communauté".

Nous ne parlerons pas dans cette enceinte des profits, étant persuadés que vous êtes tous au-dessus de ces contingences matérielles. Cette troisième affirmation pourtant donne un sens tout particulier aux deux précédentes et nous fait penser à un phénomène nouveau : la prise d'otages où, pour le bonheur d'un plus grand nombre, on sacrifie quelques sujets - les sujets n'ayant pas eu à choisir. Chacun a le droit de se sacrifier pour sauver l'humanité, encore faut-il que celui qui se sacrifie soit conscient de ce sacrifice et qu'il y consente. Notre avis est que :

1)   Ceux qui prennent des risques, les travailleurs, ne les partagent pas. On ne voit pas bien comment un risque peut se partager. Je regrette mais, tous ici, vous connaissez mieux que moi la valeur des mots. Ensuite, ils n'ont jamais eu à

choisir et bien souvent ils ignorent les risques quotidiens qu'on leur fait courir.

2) Poser le problème en ces termes : produire ou ne pas produire, c'est escamoter une partie importante de ce problème, car il y a :

a) la manière de produire, ce que nous appelons les conditions de travail - qui peut réduire considérablement les risques.

b) l'éducation, l'information de ceux qui produisent, de ceux qui consommen et nous considérons qu'en ce domaine, la responsabilité des chercheurs est engagée.

Si l'on admet que le chercheur est un homme normal, responsable, et non une espèce rare telle qu'on peut le voir au cinéma, planant au-dessus du commun des mortels, on peut penser que, sans être celui qui décide du seuil à ne pas dépasser, il doit être celui qui informe tous les partenaires sociaux qui seraient appelés, eux, à prendre les décisions politiques du type : produire ou ne pas produire, quantité négligeable ou pas. A ce propos, nous pensons qu'on ne peut à la fois refuser de pren dre la décision politique et parler de quantité négligeable. C'est quoi ? C'est qui ? C'est pour qui, la quantité négligeable ? Qui peut affirmer par exemple que 30 ppm de CV dans l'atmosphère ne présentent pas de risques graves pour les travailleurs ?

Le savant, pour nous, ce devrait être aussi celui qui tire la sonnette d'alarme lorsque, par exemple, on vend à des pays du Tiers Monde de grandes quantités de produits que nous avons difficilement réussi à interdire à la vente dans les pays "civilisés". Au fait, un pays civilisé, c'est quoi ? Un pays qui a des savants ou bien un pays qui met tout en oeuvre pour ne pas accélérer la destruction de l'humanité

Nous avons noté, au cours de ce Symposium, certains antagonismes ce que no pourrions appeler des débats d'école. Nous croyons que c'est une très bonne chose en soi, pour la science, et donc pour les travailleurs. Cependant, nous nous risquons à dire : point trop n'en faut, car, pendant ce temps, des millions de travailleurs dans les usines sont exposés tous les jours à toutes sortes de produits et vous êtes leur seul espoir, du moins en matière d'information scientifique.

Vous comprendrez mieux pourquoi nous sommes tant pressés, quand je vous aurai dit que dans une même usine, on fabrique du chlore, du CV, du tri, du per, de l'exachloréthane, du tetra, de l'acide monochloracétique, de l'hexachlorocyclohexane, etc. Tout cela dans une ambiance très dommageable pour les travailleurs et aussi pou l'environnement :

- pour l'eau, que nous polluons régulièrement avec des solvants chlorés, avec du CV et d'autres produits;

- pour la terre, qui ne pourra bientôt plus être travaillée par les paysan des environs;

- pour l'air, dans lequel on envoie tous les produits possibles causant ainsi une gêne à toute la population environnante.

Nous avons accueilli favorablement la communication faite, ce matin, par le Docteur Westerholm, en matière d'élaboration de normes dans son pays, et nous esti mons que ce serait pour nous une amélioration considérable si une telle démarche s'am çait dans notre propre pays.

Les objectifs de la CFDT en matière de produits toxiques sont:

- Avant qu'un produit soit lancé sur le marché, s'assurer par les moyens du moment, qu'il n'est pas dangereux pour la santé des travailleurs et poursuivre la recherche sur ce produit.

- Lorsque l'on découvre qu'un produit est cancérigène, si on estime ne pas pouvoir arrêter sa fabrication, il faut prendre toutes mesures susceptibles de réduire au maximum les risques - encore faut-il savoir qui doit apprécier l'arrêt ou non de cette fabrication.

- Imposer un dossier médical pour chaque travailleur afin que l'on puisse rapidement déterminer les causes d'une maladie éventuelle.

- Ne pas limiter la lutte contre les produits toxiques à l'amélioration de l'atmosphère des ateliers, c'est-à-dire ne pas se contenter d'aspirer l'air pollué pour le rejeter dans l'atmosphère ou dans l'eau.

- Modifier la législation française afin de renforcer le pouvoir des organisations professionnelles qui représentent les travailleurs.

- Renforcer le pouvoir des médecins du travail en même temps que leur autonomie.

LEGISLATIVE FRAMEWORK OF CONTROL OF OCCUPATIONAL CARCINOGENS IN THE UNITED KINGDOM

J.A. CATTON

Health and Safety Executive
Baynards House, London W2 4TF, England

May I firstly take the opportunity of thanking the IARC for inviting me to speak to such a distinguished gathering. May I also as the last formal speaker at the Symposium congratulate the sponsors on the excellent programme they arranged for us. The papers and the discussions have been most stimulating. What I have found particularly useful has been the opportunity to listen not only to the learned exposition of some of the medical/scientific problems by distinguished speakers but also the opportunity to consider the problem of control in a social and economic framework. I think it was Dr Mole earlier this week who suggested a particular role for scientists and doctors and I am sure all would agree that a better understanding of the industrial and legal background can only be of benefit.

I would like to indicate the way we are attempting to deal with the problem in the UK in the field of occupational hazard because in this field we all have much to learn from one another to help us shape controls for the future. These discussions have raised however quite a number of fundamental issues for those like myself who serve in the national authorities, responsible for establishing within the working environment a framework of control over occupational hazards in general not only in the carcinogenic field. Whatever the ideal solution may be we live in a less than ideal world and in exposing our populations to any known, or more likely suspected, hazard we must all, national authorities included, ensure that the full facts are available for consideration accepting that on occasions the balance is not between no possible risk on the one hand and a possible risk on the other but rather the choice of the lesser known risk.

However, let me start by explaining to you briefly the current situation in the UK on the legislation which exists, accepting that the problem of the control of carcinogens or suspect carcinogens is of particular importance to those who may be exposed in the working environment where people have actually died/ on a significant scale.

The general framework of control in Britain of legislation designed to protect employed persons from any occupational hazards they might be confronted with, was fundamentally changed last year. In 1974, the previously fragmented legislation which existed for control of health and safety at work was superseded by a new Health and Safety at Work Act. Its main purpose was to provide for one comprehensive and integrated system of law dealing with the health and safety of all work people, and the health and safety of the public as affected by work activities. The new Act replaced a wide variety of previously existing laws on health and safety in industry, including separate provisions for factories, for mines, for quarries, for offices, for shops as well as for emission standards from plants.

The Act will apply to any work activity being carried out by any person in any situation. No longer are basic legal provisions to be different for those who work in a factory, than for those working in a research establishment, or a hospital, or an educational establishment or any other place of work. If the risks to be found there are similar, then the legal requirements to eliminate or control that risk are to be based on the same provisions.

The Act is designed to secure the health, safety and welfare of persons at work and this is a duty placed on every employer. Without going into detail about its provisions he is required in this context

(a) to provide and maintain plant and systems of work that are so far as is practicable, safe and without risk to health

(b) to ensure the safety and absence of risk to health in connection with the use, handling, storage and transport of substances

(c) to provide such information, training, instruction and supervision as is necessary to ensure the health of his employees

The new Act however, does not seek to cover every eventuality nor does it make any attempt to spell out rules for each and every work situation. It is what we term in the United Kingdom an enabling Act whose foundation is based on the concept of a general duty of care for people associated with work activities. It is a flexible piece of legislation which will allow for Regulations to be made in a wide variety of circumstances.

Carcinogenicity is only one aspect, although an important one, of the general problem of the toxicity of chemicals. The provisions of the Health and Safety at Work Act do not therefore detail provisions aimed at the elimination of occupationally induced cancer but rather provide powers for dealing with problems of toxicity in general. Thus provision is made for Regulations

(a) prohibiting the manufacture, supply, keeping, or using of substances or carrying on of processes

(b) imposing requirements with respect to the carrying out of research

(c) prohibiting the importation of substances

(d) requiring a licence for the carrying on of specific activities

(e) requiring arrangements to be made for atmospheric or biological monitoring

(f) prohibiting or imposing requirements in connection with emissions into atmosphere of any particular substance

One section of the Act is however, so far as the UK is concerned, completely new in its concept with regard to the control of potentially hazardous substances. This is the duty placed not on the user but on the manufacturers, importers and suppliers of any substance for use at work. Their duty, enforceable at law is

(a) to ensure so far as is reasonably practicable that the substance is safe and without risk to health when properly used and

(b) to carry out, or arrange for someone else to carry out, such testing and examination as may be necessary to fulfil the first duty and

(c) to provide adequate information about results of tests which have been carried out in connection with a substance, and about any conditions necessary to ensure that it will be safe and without risk to health when properly used.

These duties do not of course replace the separate duties on users. This concept is quite unique in the UK factory legislation, and is an attempt to place responsibility on those who <u>manufacture</u> substances which may be dangerous including carcinogens or suspected carcinogens, to carry out tests and examinations and to provide information to persons using those substances regarding their possible health hazard so that the user may take adequate precautions. In the past there have been no legal requirements to ensure that anyone manufacturing a known carinogen must notify the people to whom they were supplying the substance that it was in fact carcinogenic, or to indicate the sort of precautions that were essential in its use. This has now been rectified.

The Act has in fact also gone further than this because it also requires and places a duty on people who manufacture substances to carry out any necessary research with a view to the discovery, and so far as is reasonably practicable, the elimination or minimisation of any risk to health or safety to which the substance may give rise. The effect of this again is to place a duty on the manufacturer not only not to supply a material with a known carcinogenic hazard without warning, but also to require him to carry out research which may be necessary for him to identify a substance as being carcinogenic. Although the Act has only been in force for a very few months, it has already led a number of major companies operating in this field, to consider even more carefully the consequences of putting on the market products which have not been subject to detailed research on their possible ill effects

The pressures on inhouse laboratory facilities and independant research laboratories to carry out this type of work, and on toxicologists to evaluate the work, has increased significantly.

The results of this legislation is that already manufacturers, importers and suppliers of material are giving serious consideration to the problem of providing the sort of information to the user required by this section.

The outline I have so far given you is very brief and necessarily was not fully comprehensive. Nevertheless, it is the legal background to the work proceeding in the UK on the control of any toxic substances used in industry including those known or suspected of being carcinogenic. I would now like to be rather more specific firstly in distinguishing the problems in the working environment in contrast to the general environment and secondly, in a more particular approach in the carcinogenic field. The problems of control of carcinogens which gave rise to an environmental one are, I suggest, significantly different and may lead to a different framework of control. This allows occupational carcinogenisis to be dealt with by a system of control geared to the particular requirements in the working environment rather than to be fitted into a system of control basically designed to meet the problems of the general environment.

What is so special about the working environment? Let me give you a few examples. The working environment is to be distinguished from the general environment in a number of ways.

(1) The working environment is separate and largely self-contained whereas the general environment is not.

(2) The concentration of the hazardous substance in the working environment is liable to be greatly in excess of any to be found in the general environment.

(3) The amount of materials and the process used in an industrial undertaking may have the potential to grossly contaminate the working environment eg within a workroom or even within a piece of plant.

(4) The concentrations of toxic material and the closeness of workers to it in the limited space of a workroom make the consequences of loss of control even for a short time, or to a small degree, very serious and indeed in some cases even fatal.

(5) The consequential potential risk to those exposed may be much greater than normally experienced in the general environment.

(6) The possible ill effects may arise in an acute form whereas this is rarely experienced in the general environment.

(7) Not only is the workplace self contained but the source of the contaminant will normally be within that workplace, whereas in the general environment it could arise from almost anywhere.

(8) The pollution of the working environment is normally under the control of a single authority ie those controlling the works, whereas the pollution of the general environment probably arises from a multiplicity of sources and is controlled by a multiplicity of authorities.

(9) The population of the working environment is normally identifiable and somewhat homogeneous in that it excludes the very young, the very old and the obviously sick.

(10) The working population is subject to a greater degree of discipline than the population as a whole.

(11) The period of exposure of the working population can be controlled eg by limiting hours of work, by imposing health standards, by the wearing of protective equipment.

(12) The working population can be monitored and protected by medical examination and biological tests to a much greater extent than the general population. For all these reasons we are entitled and indeed duty bound to distinguish the problems of control in the working environment from the general environment.

It is significant that some of the worst occupational cancer problems of the present and past have arisen with chemicals that had no acute effect on the body and were therefore handled, at the relevant time without any or without adequate control over the degree of exposure of workers to the substances. These included (i) chimney sweeps contact with soot (ii) mule spinners contact with shale oil (iii) rubber workers contact with some anti-oxidants (iv) asbestos workers inhalation of asbestos fibres and (v) plastic control workers inhalation of vinyl chloride monomer.

It is interesting to speculate whether if the strict measures of control now exercised over beta napthylamine, asbestos and VCM for example in the UK had been introduced at an earlier date we should have reaped the bitter harvest which is now being gathered. There is certainly no justification for not enforcing such a policy of control when dealing with any chemical that has been shown to be carcinogenic to man.

I have referred to the differences between the working environment and general environment. If these are as basic as I have indicated what is their effect on possible control measures which might be introduced? The differences make it more necessary, more appropriate, and more practicable to introduce measures to control the pollutant at source.

The setting of control measures is however dependant on the identification and acceptance of levels of exposure whatever these levels may be. If, and it is a very big 'if', an acceptable standard for level of exposure can be set the degree of control can be appropriate to the degree of risk. This may range from total prohibition with or without the possibility of substitution; automation and remote control of the plant; total or partial enclosure of the plant; the provision of effective local exhaust ventilation or in the last resort to personal protection.

In establishing the degree of control appropriate to the risk, the setting of standards is particularly relevant in the working environment even if it may not be possible in the general environment. Similarly the problem of medical examination is one which is capable of being tackled much more readily in the working rather than the general environment. Such medical examinations may include biological tests for an identifiable working population exposed to a known or suspect hazard and the frequency of the medical examination can be predetermined.

I started by a reference to the UK legislation in the general field of health and safety at work and then proceeded to the more particular aspects of that legislation dealing with the control of toxic substances in general and carcinogenic materials in particular and went on to distinguish the problems in the working, as opposed to the general environment. Let me make a few observations on some of the essential control mechanisms which can be legislatively based and which have a particular application in the working environment.

There seems little practical point in adopting special standards, however they may be based and whether they rely on the evidence of animal or human testing, if there are not also stringent methods of ensuring that those standards are met, and where they are not achieved, remedial action is taken. One sometimes gets the impression that more effort goes into the setting of standards than into ensuring that the standards are achieved. It is I think worth pausing to reflect that a standard is only of use for the protection of the population as a whole, or the working population in particular if that standard is and can be achieved.

It is our experience that to ensure that we are able to check standards, a great deal of attention should be paid to the methods by which evaluation is carried out and in particular monitoring is required. The pious expression that a standard should be achieved and monitoring should be undertaken is just not good enough and indeed, in the UK both management and the Trade Unions have recognised this and accepted the need for a much more detailed approach. It is necessary for example to spell out the methods, location and frequency of monitoring and to indicate the role of personal, as opposed to general monitoring. It is only when this is made clear that one can with reasonable assurance check that the standards are being met.

It is similarly necessary to ensure that the methods of recording such information are adequate; that the training of personnel is sufficient; that the problems which the standards are designed to overcome are fully understood by the work force and accepted by them, if there is to be any confidence that any standards being set will achieve the desired result. Added to this is the need for a stringent enforcement policy so that both sides of industry are aware of and respect the fact that the standards have to be achieved.

Last, but unfortunately, in our experience not least important, is the need in the occupational field for control measures to be based on the soundest medical/ scientific knowledge available and to be subject to full and frank discussion with both sides of industry rather than imposed by a statutory authority without consultation, without prior knowledge, without support, without adequate training and education of those who once again on both sides of industry have most both to gain or to lose by such control.

We are perhaps fortunate in the UK in this respect in our new legislation. The new Act, in addition to the powers I indicated to you earlier also set up a Health and Safety Commission composed of representatives of both sides of industry and the Local Authorities. This is the body responsible to Parliament for the efficient observance of the new legislation and it has been made clear that the Health and Safety Executive which is the operational and enforcement arm of the Health and Safety Commission, should conduct much of its business on a similarly tripartite basis.

Thus when the problems of vinyl chloride monomer broke in January 1974 and 3 cases of angio sarcoma related to exposure of VCM were confirmed in the USA immediate action was taken by the Health and Safety Executive in the United Kingdom to identify the hazardous operations and to control exposure at UK plants. The absence of substitutes and the economic importance of the material turned attention to control of the risk rather than to prohibition. Although it was clear that progress was already being made in reducing levels of exposure to VCM, it was also felt desirable that an approach to this problem should be made on a tripartite basis. A Working Group was set up with the employers organisations representing manufacturing industries, with the Trade Unions and with the Factory Inspectorate, the forerunner of the Health and Safety Executive. As a result of their deliberations a Code of Practice for the control of health precautions in the manufacturing and polymerising sections of the industry was prepared. We see this as one method available to the national authority for dealing with this type of problem.

In essence, the Code sets out to:

    (a) define a hygiene standard

    (b) require regular monitoring

(c) outline methods of monitoring

(d) provide for medical supervision

(e) require the keeping of adequate records

(f) provide for medical supervision

(g) provide for joint consultation, education and training

I appreciate that there were criticisms in various quarters including the United Kingdom that the hygiene standards established by this Code of Practice were not sufficiently stringent. I think much of that criticism was ill informed. However, what I want to stress was the method by which the standard was established. I suggest that it was realistic in the light of current medical/scientific knowledge; it took account of industry's practical problems; it was fully discussed and agreed with the representatives of the work people before it was established. The success of this Code has depended primarily on the united efforts of both sides of industry and of Government. Within the terms of the Code it is probably as stringent a standard as can be achieved and enforced at the present time if we are to have regard to the practical problems involved and not merely pay lip service both to measures designed to ensure the continuing good health of people at work and to the provision of a legislative framework of control of occupational carcinogens.

Ladies and Gentlemen may I conclude by reiterating the need for control of chemical substances which may already have proved or in time prove yet to be carcinogenic. If I suggest that this control does not lead in all cases to total prohibition or even to a target of zero concentrations, this is merely an attempt to approach the risk of carcinogenicity from chemical substances in a similar way to the approach to other risks which we all face. The motor car kills thousands or hundreds of thousands of people every year but there is little demand for its more stringent control by an unenforceable 1 mph speed limit let alone its prohibition. Coming closer to the problems of chemical substances there is little demand for the elimination of for example chlorine storage which if released upon an unsuspecting local population could have immediate and catastrophic effects.

Certainly there is a need for a greater understanding of the problem but this is not only of its medical and scientific aspects but also of its social and economic implications. Certainly there is also a need for greater control of the use of chemical substances in general in industry - indeed I have always been amazed at the way in which some science based industries deal so crudely and so unscientifically with some of their raw materials and their processing operations.

We are still at a relatively early stage in legislative control of occupational carcinogens. It is however the formative period of time certainly in much of Europe and America if my discussions with international colleagues are representative and discussions such as we have had can do nothing but good. I suggest that this morning's discussions have indicated some ways in which this control can be exercised and which is both acceptable to the population concerned and enforceable under legislation. It is a way which I suggest is within the spirit of the ILO deliberations on this subject and in which research has a fundamental part to play.

LIST OF PARTICIPANTS
TABLE DES PARTICIPANTS

AUSTRALIE

— Dr F.J. Vett
Medical Director of the Australian Embassy
64, avenue de Iéna
75016 Paris

AUTRICHE

— Professor H. Wrba
Institut für Krebsforschung der Universität Wien
Borschkegasse 8a
1090 Vienne

BELGIQUE

— Mr M. Bonnefoy
Solvay et Cie
33, rue du Prince Albert
1050 Bruxelles

— Mr C. Deckers
Unité de Cancérologie expérimentale
Centre des Tumeurs de l'Université catholique de Louvain
Louvain

— Dr J.H. Kaspersma
European Manager
Air Products and Chemicals, Inc.
92/94 Square Eugène Plasky
1040 Bruxelles

— Dr. M.J. Lefèvre
Solvay et Cie
33, rue du Prince Albert
1050 Bruxelles

— Professeur D. Rondia
Laboratoire de Toxicologie de l'Environnement
Université de Liège
4020 Liège

CANADA

— Professor H.F. Stich
The University of British Columbia
Cancer Research Centre
Vancouver V6T 1W5

— Dr R.S. Thomas
Air Pollution Control Directorate
Environmental Health Centre
Tunney's Pasture
Ottawa K1A OH3

ETATS-UNIS D'AMERIQUE

— Dr D.B. Clayson
Deputy Director
The Eppley Institute for Research in Cancer
The University of Nebraska Medical Center
42nd and Dewey Avenue
Omaha, NE 68105

— Dr S.S. Epstein
Swetland Professor of Environmental Health and Human Ecology
Case Western Reserve University
School of Medicine
Cleveland, OH 44106

— Dr I.T.T. Higgins
Department of Epidemiology
The University of Michigan
School of Public Health
109, Observatory Street
Ann Arbor, MI 48104

— Dr E. Sawicki
Acting Chief, Sampling and Analysis Methods Branch
Atmospheric Chemistry and Physics Division
United States Environmental Research Center
Research Triangle Park, NC 27711

— Dr R.E. Shapiro
Epidemiology Unit
Food and Drug Administration
Bureau of Foods
200 C Street S.W.
Washington DC 20204

— Dr R.G. Tardiff
Director, Organic Contaminants Branch
Health Effects Research Laboratory
US Environmental Protection Agency
Cincinnati, OH 45268

FINLANDE

— Dr O. Elo
The National Board of Health
Department of Epidemiology and Hygiene
Siltasaarenkatu 18 A
00530 Helsinki

— Mr M. Salmenperä
Cabinet Secretary
Department of Labour Protection
Ministry of Social Affairs and Health
Helsinki

— Dr E. Sundqvist
Ministry of Social Affairs and Health
National Board of Labour Protection
Box 536-546
33101 Tampere 10

FRANCE

— Dr Andréani
22, rue Peiresc
83100 Toulon

— Mr G. Astier
FUC-CFDT
26, rue de Montholon
75439 Paris

— Dr J. Aubert
INSERM U.40-CERBOM
Parc de la Côte
1, avenue Jean-Lorrain
06300 Nice

— Mr A. Barbin
CIRC
150, cours Albert Thomas
69008 Lyon

— Dr H. Bartsch
CIRC
150, cours Albert Thomas
69008 Lyon

— Dr P.G. Beau
C.E.A. — C.E.N.
Département de Protection
B.P. No. 6
92260 Fontenay-aux-Roses

— Mr H. Bédouelle
Laboratoire Mutatest
Institut Pasteur
25, rue du Dr Roux
75015 Paris

— Dr J. Berrod
Rhône-Poulenc Industries
25, quai Paul Doumer
92408 Courbevoie

— Professeur H.L. Boiteau
Laboratoire de Toxicologie et d'Hygiène industrielle
UER des Sciences Pharmaceutiques
1, rue Gaston Veil
44000 Nantes

— Mr H. Bordet
Fédération nationale des industries chimiques CGT
33, rue de la Grange-aux-Belles
75010 Paris

— Professeur P. Bourbon
INSERM U.57
B.P. No. 14
31320 Vigoulet-Auzil

— Mlle H. Brésil
CIRC
150, cours Albert Thomas
69008 Lyon

— Mme F. Burnol
Département Colloques et Publications
INSERM — Service Central
101, rue de Tolbiac
75645 Paris Cedex 13

- Mr Challemel du Rozier
  Ministère de la Qualité de la Vie
  Service des Problèmes de l'Atmosphère
  14, Bd du Général Leclerc
  92521 Neuilly-sur-Seine

- Professeur P. Chambon
  Laboratoire de Toxicologie et d'Hygiène industrielle
  UER de Pharmacie
  8, avenue Rockefeller
  69008 Lyon

- Mme R. Chambon
  Maître-Assistante
  Laboratoire de Toxicologie et d'Hygiène industrielle
  UER de Pharmacie
  8, avenue Rockefeller
  69008 Lyon

- Professeur J. Champeix
  Faculté de Médecine
  28, Place Henri-Dunant
  63000 Clermont-Ferrand

- Mlle Y. Chardonnet
  INSERM U.51
  1, Place du Professeur J. Renaut
  69008 Lyon

- Dr Charrier
  Kodak-Pathé
  Zone industrielle
  71100 Chalon-sur-Seine

- Dr J. Chauveau
  Institut de Recherches scientifiques sur le Cancer
  B.P. 8
  94800 Villejuif

- Dr I. Chouroulinkov
  Institut de Recherches scientifiques sur le Cancer
  B.P. 8
  94800 Villejuif

- Professeur P. Chovin
  Directeur du Laboratoire central de la Préfecture de Police de Paris
  39, bis rue de Dantzig
  75015 Paris

- Dr M. Cohen
  Chef du Département d'Hygiène industrielle
  ESSO-S.A.F.
  Cedex No. 2
  92080 Paris-La Défense

- Mr J. Coupé
  FUC-CFDT
  26, rue de Montholon
  75439 Paris

- Mr P.M. Dansette
  C.N.R.S.
  Institut de Biochimie de l'Université de Paris XIe
  75011 Paris

- Dr W Davis
  CIRC
  150, cours Albert Thomas
  69008 Lyon

- Mlle O. Deblock
  CIRC
  150, cours Albert Thomas
  69008 Lyon

- Mr C. Doussain
  Fédération nationale des industries chimiques CGT
  33, rue de la Grange-aux-Belles
  75010 Paris

- Dr. G. Duverneuil
  Rhône-Poulenc Chimie Fine
  Service Médical
  Usine de Saint-Fons
  69190 Saint-Fons

- Mr J. Escañèz
  FUC-CFDT
  26, rue de Montholon
  75439 Paris

- Professeur R. Flamant
  Institut Gustave Roussy
  16, avenue Paul Vaillant Couturier
  94800 Villejuif

  Médecin en Chef R. Fontanges
  Centre de Recherche du Service de Santé des Armées
  Division de Microbiologie
  108, Bd Pinel
  69272 Lyon Cedex 1

— Professeur R. Fournier
INSERM U.26
Hôpital Fernand-Widal
200, rue du Faubourg Saint-Denis
75010 Paris

— Mr M. Gattelet
Caisse Régionale d'Assurance Maladie
Service de Prévention
84, rue du 1er mars
69100 Villeurbanne

— Dr P. Gaucher
Inspection Médicale du Travail et de la main d'œuvre
Ministère du Travail
18, place Fontenoy
75007 Paris

— Mr J. Gaultier
FUC-CFDT
26, rue de Montholon
75439 Paris

— Dr A. Geser
CIRC
150, cours Albert Thomas
69008 Lyon

— Dr L. Griciute
CIRC
150, cours Albert Thomas
69008 Lyon

— Mr D. Guillot
FUC-CFDT
26, rue de Montholon
75439 Paris

— Dr J. Higginson
Directeur
CIRC
150, cours Albert Thomas
69008 Lyon

— Mr M. Huber
Ingénieur de Sécurité
Ciba-Geigy
Usines de Saint-Fons
69190 Saint-Fons

— Dr O. Jensen
CIRC
150, cours Albert Thomas
69008 Lyon

— Mr D. Klein
Ingénieur-chimiste
Laboratoire de Nutrition et des Maladies Métaboliques
Service du Professeur G. Debry
40, rue Lionnois
54000 Nancy

— Mr Lafontaine
Institut National de Recherche et de Sécurité pour la prévention des accidents du travail et des maladies professionnelles
Route de Neufchateau
B.P. 27
54500 Vandœuvre-les-Nancy

— Dr. J. Lafuma
Chef du Service de Pathologie et de Toxicologie expérimentale
Département de Protection du C.E.A.
B.P. 6
92269 Fontenay-aux-Roses

— Professeur R. Latarjet
Institut du Radium
Section de Biologie
26, rue d'Ulm
75231 Paris Cedex 05

— Dr L. Le Bouffant
CERCHAR
B.P. 27
60103 Creil

— Dr Leder
Médecin du Travail
Société des Produits Ugine-Kuhlmann
Division des Colorants
B.P. No 5
38370 Les-Roches-de-Condrieu

— Dr G. Lenoir
CIRC
150, cours Albert Thomas
69008 Lyon

— Dr J.C. Limasset
Chef de la Section des Etudes Générales
Institut National de Recherche et de Sécurité pour la prévention des accidents du travail et des maladies professionnelles
Route de Neufchateau
B.P. No. 27
54500 Vandœuvre-les-Nancy

- Mr C. Malaveille
  CIRC
  150, cours Albert Thomas
  69008 Lyon

- Dr G. Margison
  CIRC
  150, cours Albert Thomas
  69008 Lyon

- Dr R. Masse
  C.E.A.
  Département de Protection
  Service de Pathologie et de Toxicologie expérimentale
  B.P. No. 6
  92260 Fontenay-aux-Roses

- Dr M. Maugras
  Directeur Adjoint du laboratoire d'Hygiène et de Recherche en Santé Publique
  INSERM U.95
  Plateau de Brabois
  54500 Vandœuvre-les-Nancy

- Dr R. McLennan
  CIRC
  150, cours Albert Thomas
  69008 Lyon

- Mr M. Meillon
  FUC-CFDT
  26, rue de Montholon
  75439 Paris

- Professeur R. Monier
  Directeur
  Institut de Recherches scientifiques sur le Cancer
  B.P. 8
  94800 Villejuif

- Professeur M. Mosinger
  Centre d'Explorations et de Recherches médicales
  25, rue des Colonies
  13008 Marseille

- Dr Y. Moulé
  Service de Physiologie
  Institut de Recherches scientifiques sur le Cancer
  B.P. 8
  94800 Villejuif

- Dr C.S. Muir
  CIRC
  150, cours Albert Thomas
  69008 Lyon

- Mr Michel Odet
  Secrétaire de la Fédération nationale des Industries Chimiques CGT
  33, rue de la Grange-aux-Belles
  75010 Paris

- Mme C. Partensky
  CIRC
  150, cours Albert Thomas
  69008 Lyon

- Dr C. Pierre
  Médecin du Travail
  Solvay et Cie
  39500 Tavaux

- Dr Pirot
  Médecin du Travail
  Ato-Chimie
  Balan par Montluel
  B.P. 1
  01120 Montluel

- Mme L. Ploton
  CIRC
  150, cours Albert Thomas
  69008 Lyon

- Dr V. Ponamarkov
  CIRC
  150, cours Albert Thomas
  69008 Lyon

- Dr E. Ravier
  Médecin du travail
  Rhône-Poulenc Pétrochimie
  Usine de Roussillon
  38150 Roussillon

- Dr C. Rosenfeld
  INSERM U.50
  Groupe hospitalier Paul-Brousse
  14, av. P. Vaillant-Couturier
  94800 Villejuif

- Dr B. Simon
  Médecin en Chef de la Direction médicale
  ESSO-S.A.F.
  Cedex 02
  92080 Paris-La-Défense

— Dr G. Smagghe
Chef du Service de Médecine et de Toxicologie
Produits Chimiques Ugine-Kuhlmann
11, bd Pershing
75017 Paris

— Dr M. Stupfel
INSERM U.123
44, chemin de Ronde
78110 Le Vésinet

— Mr J.C. Thomas
Rhône-Poulenc Polymères
25, quai Paul Doumer
93408 Courbevoie

— Dr L. Tomatis
CIRC
150, cours Albert Thomas
69008 Lyon

— Professeur R. Truhaut
Directeur
Université René Descartes
Faculté des Sciences Pharmaceutiques et Biologiques de Paris-Luxembourg
4, avenue de l'Observatoire
75006 Paris

— Mr M. Vialle
FUC-CFDT
26, rue de Montholon
75439 Paris

— Professeur P. Viallet
Directeur du Laboratoire de Chimie-Physique
Centre Universitaire de Perpignan
Avenue de Villeneuve
66025 Perpignan

— Dr F.E. Zajdela
INSERM U.22
Institut du Radium
Bâtiment 110
91405 Orsay

— Dr. L. Zardi
CIRC
150, cours Albert Thomas
69008 Lyon

HONGRIE

— Dr L. Holczinger
Research Institute of Oncopathology
1122 Budapest
Rath Gy. ste. 17

ITALIE

— Dr V.U. Fossato
Istituto di Biologia del Mare
Riva 7 Martiri 1364/A
30122 Venise

— Dr B. Invernizzi
Instituto di Anatomia e Patologia dell' Universita di Torino
Via Santena 7
Turin

— Professor C. Maltoni
Istituto di Oncologia "F. Addarii"
Viale Ercolani 4/2
40138 Bologne

— Dr S. Parodi
Associate Professor
Oncology Department
Istituto di Oncologia dell'Universita di Genova
Gênes

— Dr L. Santi
Professor of Oncology
Istituto di Oncologia dell'Universita di Genova
Gênes

— Dr F. Spreafico
Istituto Mario Negri
Milan

— Professor P. Viola
Istituto Regina Elena per lo Studio e la cura dei Tumori
Viale Regina Elena 9
Rome

PAYS-BAS

— Dr. K.W. Jager
Shell Nederland Raffinaderij B.V.
P.O. Box 7000
Rotterdam

— Dr R. Kummer
Shell Nederland Raffinaderij B.V.
P.O. Box 7000
Rotterdam

POLOGNE

— Dr H. Gadomska
Chef de l'Unité d'épidémiologie
Institut d'Oncologie
rue Wawelska 15
00-973 Varsovie

— Dr. J. Staszewski
Chief, Cancer Epidemiology Unit
Institute of Oncology
ul Stowackiego 50
Gliwice

— Dr Z. Wronkowski
Chief, Cancer Control and Epidemiology Department
Institute of Oncology
Wawelska 15 str.
00-973 Varsovie

REPUBLIQUE DEMOCRATIQUE ALLEMANDE

— Professor R. Engst
Akademie der Wissenschaften der DDR
Forschungszentrum für Molekular Biologie und Medizin
Zentralinstitut für Ernährung
Potsdam-Rehbrücker 12
1505 Bergholz-Rehbrücke
Arthur Scheunert Allee 114-116

— Dr B. Teichmann
Akademie der Wissenschaften der DDR
Krebsforschungszentrum
Lindenberger Weg 80
1115 Berlin-Buch

REPUBLIQUE FEDERALE D'ALLEMAGNE

— Dr H.M. Bolt
Wilhelmstrasse 56
Institut für Toxikologie
D-74 Tübingen

— Dr H. Brune
Arbeitsmedizin
Papenkamp 11
2000 Hambourg 52

— Dr H.P. Friedel
Division of Occupational Diseases
Berufsgenossenschaft der Chemischen Industrie
D-69 Heidelberg
Gaisbergstrasse 11

— Dr M. Habs
Scientific Secretary
Deutsches Krebsforschungszentrum
69 Heidelberg 1
Postfach 101949
Im Neuen heimer Feld 280

— Dr H. Heinrich
Medizinaldirektor
D 5674 Berg.-Neukirchen
Hüscheiderstr. 17

— Professor D. Henschler
Director
Institute of Toxicology
University of Würzburg
Würzburg

— Dr G. Kimmerle
Institut für Toxikologie der Bayer AG
Friedrich-Ebert Str. 217
56 Wuppertal 1

— Dr D. Lorke
Institut für Toxikologie der Bayer AG
Friedrich-Ebert Str. 217
56 Wuppertal 1

— Professor U. Mohr
Director
Medizinische Hochschule Hannover
Abteilung für Experimentelle Pathologie
Karl-Wiechert Allee 9
Postfach 61 0180
300 Hanovre 61

ROYAUME-UNI

— Dr J. Ashby
ICI. Central Toxicology Laboratory
Alderley Park
Nr Macclesfield
Cheshire

— Dr M. Ashley-Miller
Chief Scientist Office
Scottish Home and Health Department
St Andrew's House
Edinburgh, EH1 3DE

— Dr W.H.A. Beverley
Asbestosis Research Council
P.O. Box 40
Rochdale OL 12 7EQ

— Mr A.C. Blyghton
Legal Secretary
Transport and General Workers Union
Transport House
Smith Square
Londres SW1P 3JB

— Dr R.S.J. Buxton
Department of Health and Social Security
Alexander Fleming House
Elephant and Castle
Londres SE1

— Dr J.T. Carter
Senior Medical Officer
The British Petroleum Co., Ltd
Chertsey Road
Sunbury on Thames
Middlesex TW16 7LN

— Mr J.A. Catton
Health and Safety Executive
Baynards House
Westbourne Grove
Londres W2 4TF

— Mr A.A. Cross
Asbestosis Research Council
Environmental Control Committee
114 Park Street
Londres W1Y 4AB

— Dr B. Flaks
Department of Pathology
University of Bristol
University Walk
Bristol 8

— Dr J.C. Gilson
Director, Medical Research Council Pneumoconiosis Unit
Llandough Hospital
Penarth CF6 1XW
Glamorgan

— Dr S. Holmes
Asbestosis Research Council
P.O. Box 40
Rochdale OL12 7EQ

— Dr J.S.P. Jones
Department of Pathology
City Hospital
Nottingham

— Professor P.J. Lawther
Director, Medical Research Council Air Pollution Unit
St. Bartholomew's Hospital Medical College
Charterhouse Square
Londres EC1M 6BQ

— Dr R.C. Lemon
Senior Medical Adviser
Occupational Health
Shell International Petroleum Co., Ltd
Shell Centre, Londres SE1 7NA

— Mr D.W. Meddle
Department of Industry
Laboratory of the Government Chemist
Cornwall House
Stamford Street
Londres SE1 9NQ

— Dr R.H. Mole
Medical Research Council
Radiobiology Unit
Harwell, Didcot
Oxfordshire OW11 ORD

—Dr. A. Munn
ICI Ltd
Hexagon House
P.O. Box 42
Manchester M9 3DA

—Dr R. Owen
Health and Safety Executive
Deputy Director of Medical Services
Baynards House
1 Chepstow Place
Westbourne Grove
Londres W2 4TF

—Mr T.R. Pearson
Management Committee
Association of Scientific, and Managerial Staffs
10-26 A Jamestown Road
Londres NW1 7DT

—Dr I. Purchase
ICI Chemical Industries
Alderley Park
Nr Macclesfield
Cheshire

—Dr C.E. Searle
Senior Lecturer
University of Birmingham
Department of Cancer Studies
The Medical School
Birmingham B15 2TJ

—Dr W.J. Smither
Asbestosis Research Council
P.O. Box 40
Rochdale OL12 7EQ

—Dr S. Venitt
Division of Chemical Carcinogenesis
Institute of Cancer Research
Pollards wood Research Station
Nightingales Lane
Chalfont St Giles
Bucks HP8 4SP

—Professor J.A.H. Waterhouse
Birmingham Cancer Registry
Queen Elizabeth Medical Centre
Birmingham B15 2TH

SUEDE

— Dr P. Westerholm
National Board of Occupational Safety and Health
Medical Department
Fack
S-100 26 Stockholm

SUISSE

— Dr H.L. Küng
Gesundheitsvorsorge
Ciba-Geigy AG
CH-4002 Bâles

— Professor C. Leuchtenberger
Head, Department of Cytochemistry
Swiss Cancer Institute for Experimental Research
Bugnon 21
1011 Lausanne

— Professor R. Leuchtenberger
Chemin du Triolet
1110 Morges

URSS

— Professor L.M. Shabad
Oncological Scientific Center
Academy of Sciences of the USSR
Karshirskoyer shosse 6
115478 Moscou

ORGANISATIONS INTERNATIONALES

— Dr P. Bourdeau
Environmental Research Programme
Directorate General XII for Research, Science and Education
Commission of the European Communities
200, rue de la Loi
1040 Bruxelles

- Dr W.J. Hunter
  Commission of the European Communities
  Health Protection Directorate
  Centre Louvigny
  Avenue Monterey
  Luxembourg
- Dr R. Korneev
  World Health Organization
  1211 Genève 27
- Mr D. Larré
  Administrateur H.C.
  Division Géophysique, Pollution mondiale et Santé
  Programme des Nations Unies pour l'Environnement
  P.O. Box 30552
  Nairobi
- Dr E. Mastromatteo
  Chief, Occupational Safety and Health Branch
  International Labour Office
  CH-1211 Genève 22
- Dr H. Ott
  Environmental Research Programme
  Directorate General XII for Research, Science and Education
  Commission of the European Communities
  200, rue de la Loi
  1040 Bruxelles

I – LIST OF INSERM PUBLICATIONS

II – LIST OF IARC PUBLICATIONS

I – LIST OF INSERM PUBLICATIONS

Les Editions de l'INSERM proposent depuis l'année 1971 une collection de 5 séries de volumes.

The INSERM Publications are divided in 5 main series.

    I   Série SANTE PUBLIQUE
in French

    II Série STATISTIQUES et NOMENCLATURE
in French

    III RAPPORTS ADRESSES au MINISTERE de la SANTE
in French

    IV SERIE COLLOQUES INSERM
INSERM SYMPOSIA SERIES
completely or partly in English

    V INSERM SYMPOSIA Series
published by other Publishers
in English

I. SERIE SANTE PUBLIQUE

Les examens systématiques de santé
Epuisé (*)

Les assistantes de service social
Contribution à la sociologie d'une profession. Année 1970
203 pages / 15 francs

Les lycéens devant la drogue et les autres produits psychotropes. Année 1973
207 pages / 40 francs

Morbidité et mortalité par suicide (1975)
90 pages / 23 francs

Problèmes de santé dans une ville en mutation (1975)
164 pages / 40 francs

Dépistage du cancer du col de l'utérus (1975)
99 pages / 40 francs

Epidémiologie, examens de santé et prévention (1975)
145 pages / 30 francs

Diffusion des polluants en mer, 4 tomes (1975)
Prix : 10 francs par tome

II. SERIE STATISTIQUES

Statistique des causes médicales de décès. Tome I : Résultats France entière.
Années : 1968 - 1969 - 1970 - 1971 - 1972 – *Epuisé* (*)
Année : 1973 – 60 francs

Statistique des causes médicales de décès. Tome II : Résultats par région
Années : 1968 - 1969 - 1970 - 1971 - 1972
Prix : 60 francs par volume

Rapport sur l'état de santé de la population française
Années : 1971 - 1972
Epuisé (*)

Statistiques médicales des établissements psychiatriques
Années : 1969 - 1970 - 1971 - 1972 - 1973
Prix : 30 francs par volume

(*) Disponible auprès de la Division de la Recherche médico-sociale, INSERM, Le Vésinet.

Mortalité par Cancer
Années 1968-1970

Prix : 60 francs par volume

Etude de morbidité hospitalière

108 pages/40 francs

Enquête permanente Cancer. Survie à long terme

288 pages / 60 francs

Nomenclature

Nomenclature clinique et anatomique des cancers (1968)

61 pages / 9 francs

Code histopathologique des tumeurs humaines (1971)

138 pages / 50 francs

Langage chirurgical en carcinologie (1974)

140 pages / 15 francs

III. RAPPORTS ADRESSES AU MINISTERE DE LA SANTE

Les contraceptifs oraux (1971)

330 pages / 40 francs

Rapport d'activité 1971-1974 ; Caisse Nationale de l'Assurance Maladie et INSERM
Tome I : 440 pages ; Tome II : 600 pages.

IV. COLLOQUES ET SEMINAIRES TECHNOLOGIQUES

ANNEE 1971

Vol 1 : La Cinétique de Prolifération cellulaire / *The kinetics of cellular proliferation*
Jan. 1971 – M. Tubiana Ed.

419 pages / 50 francs

Vol. 2 : Développements récents dans l'étude chimique de la structure des protéines /
Recent developments in the chemical study of protein structures
Sept. 1971 – A. Previero and J.F. Pechère Eds.
Epuisé (**)

Vol. 3 : Pré-adaptation et adaptation génétique
Oct. 1971 – J. Ruffié Ed.

316 pages / 50 francs

(**) Disponible auprès des sections de médecine et de pharmacie des bibliothèques interuniversitaires, ou auprès de l'organisateur du colloque.

Available from the INSERM Microfiche Service, 44, chemin de Ronde – 78110 – Le Vésinet – France

Vol. 4 : Méthodologie expérimentale en physiologie et en physio-pathologie thyroïdiennes
Oct. 1971 – R. Mornex et J. Nunez Eds.
432 pages / 50 francs

Vol. 5 : Immunodépression, bases expérimentales, résultats thérapeutiques
Nov. 1971 – F. Delbarre et B. Amor Eds.
351 pages / 50 francs

COLLOQUES ET SEMINAIRES TECHNOLOGIQUES

ANNEE 1972

Vol. 6 : Les surcharges cardiaques / *Heart overloading*
March 1972 – P.Y. Hatt Ed.
433 pages / 50 francs

Vol. 7 : Techniques Radioimmunologiques / *Radioimmunoassay*
May 1972 – G. Rosselin Ed.
Epuisé (**)

Vol 8 : Hormones pancréatiques ; Hormones de l'eau et des électrolytes / *Pancreatic hormones ; Hormones of the electrolytic homeostasis*
May 1972 – P. Freychet Ed.
Epuisé (**)

Vol. 9 : Hormones glycoprotéiques hypophysaires / *Pituitary glycoprotein hormones*
May 1972 – M. Jutisz Ed.
Epuisé (**)

Vol. 10 : La transmission cholinergique de l'excitation / *Cholinergic transmission of the nerve impulse*
June 1972 – M. Fardeau, M. Israël and R. Manaranche Eds.
Epuisé (**)

Vol. 11 : Recherches fondamentales sur les tumeurs mammaires / *Fundamental research on mammary tumours*
June 1972 – O. Mühlbock and J. Mouriquand Eds.
485 pages / 50 francs

Vol. 12 : Activités évoquées et leur conditionnement chez l'homme normal et en pathologie mentale / *Average evoked responses and their conditioning in normal subjects and psychiatric patients*
Sept. 1972 – G. Lelord Ed.

Epuisé (**)

Vol. 13 : L'Etude phylogénique et ontogénique de la réponse immunitaire et son apport à la théorie immunologique / *Phylogenic and ontogenic study of the immune response and its contribution to the immunological theory*
Oct. 1972 – P. Liacopoulous and J. Panijel Eds.
364 pages / 50 francs

Vol. 14 : L'impédance bioélectrique : recherche et applications cliniques
Dec. 1972 — C. Fourcade Ed.
261 pages / 30 francs

Vol. 15 : Etude des fonctions des plaquettes sanguines
Dec. 1972 — J. Caen Ed.
Epuisé (**)

COLLOQUES ET SEMINAIRES TECHNOLOGIQUES

ANNEE 1973

Vol. 16 : Circulation osseuse / *Bone circulation*
April 1973 — P. Ficat and J. Arlet Eds.
331 pages / 50 francs

Vol. 17 : Synthèse normale et pathologique des protéines chez les animaux supérieurs / *Normal and pathological protein synthesis in higher organisms*
May 1973 — G. Schapira Ed.
Epuisé (**)

Vol. 18 : Graisses du régime et thrombose
June 1973 — S. Renaud et P. Gautheron Eds.
246 pages / 50 francs
La version anglaise de ce volume "Dietary fats and thrombosis" est éditée par S. Karger, Bâle, Suisse

Vol. 19 : Différenciation des cellules eucaryotes en culture / *Différentiation of eukaryotic cells in culture*
June 1973 — M. Prunieras, L. Robert and C. Rosenfeld Eds.
309 pages / 50 francs

Vol. 20 : Cryoconservation des cellules normales et néoplasiques / *Cryopreservation of normal and neoplastic cells*
June 1973 — R. Weiner, R. Oldham and L. Schwarzenberg Eds.
217 pages / 50 francs

Vol. 21 : Epidémiologie et prévention des maladies cardio-vasculaires / *Epidemiology and prevention of cardiovascular diseases*
Sept. 1973 — P. Ducimetière Ed.
280 pages / 50 francs

Vol. 22 : Neuro-endocrinologie de l'axe corticotrope / *Brain adrenal interactions*
Sept. 1973 — P. Dell Ed.
354 pages / 50 francs

Vol. 23 : Les accidents chromosomiques de la reproduction / *Chromosomal errors in relation to reproductive failure*
Sept. 1973 — A. Boué and C. Thibault Eds.
425 pages / 50 francs

Vol. 24 : Physiologie appliquée de la ventilation assistée chez le Nouveau-Né / *The applied physiology of ventilatory assistance in the newborn*

Sept. 1973 – J.P. Crance and P. Vert Eds.

290 pages / 50 francs

Vol. 25 : Prostaglandines 1973 / *Prostaglandins 1973*

Oct. 1973 – E.E. Baulieu and F. Bayard Eds.

Epuisé (**)

Vol. 26 : Transport, survie et pouvoir fécondant des spermatozoïdes chez les vertébrés.

Nov. 1973 – E. Hafez and C. Thibault Eds.

586 pages / 50 francs

La version anglaise de ce volume "Sperm transport, survival and fertilizing ability" est éditée par S. Karger, Bâle, Suisse.

Vol. 27 : Mécanismes du vieillissement moléculaire et cellulaire / *Molecular and cellular mechanisms of aging*

Dec. 1973 – F. Bourlière and Y. Courtois Eds.

276 pages / 50 francs

COLLOQUES ET SEMINAIRES TECHNOLOGIQUES

ANNEE 1974

Vol. 28 : L'alpha-fœto-protéine / *Alpha-feto-protein*

March 1974 – R. Masseyeff Ed.

607 pages / 80 francs

Vol. 29 : Réactions bronchopulmonaires aux polluants atmosphériques

Jan. 1974 – J. Chrétien, P. Sadoul et C. Voisin Eds.

376 pages / 30 francs

Vol. 30 : Physiologie du néphron : mécanismes et régulation / *Nephron physiology : mechanism and regulation*

May 1974 – J.P. Bonvalet Ed.

Epuisé (**)

Vol. 31 : Cytotoxicité des cellules immunocompétentes et des antisérums

May 1974 – D. Oth Ed.

Epuisé (**)

Vol. 32 : Endocrinologie sexuelle de la période périnatale / *Sexual endocrinology of the perinatal period*

May 1974 – M.G. Forest and J. Bertrand Eds.

439 pages / 50 francs

Vol. 33 : Les mycoplasmes / *Mycoplasmas*

Sept. 1974 – J.M. Bové and J.F. Duplan Eds.

449 pages / 70 francs.

Vol. 34 : Application de la vélocimétrie ultrasonore Doppler à l'étude de l'écoulement sanguin dans les gros vaisseaux / *Ultrasonic Doppler Velocimetry*
Oct. 1974 — P. Peronneau Ed.
269 pages / 30 francs.

Vol. 35 : Stimulation blastique des lymphocytes par les mitogènes
Nov. 1974 — B. Serrou Ed.
Epuisé (**)

Vol. 45 : New concepts in human placental biology
Oct. 1974 — L. de Ikonicoff and L. Cedard Eds.
123 pages / 30 francs

ACTIONS THEMATIQUES — RAPPORTS DE SYNTHESE

Vol. 36 — *Rapport N° 1 :* Interaction cellulaires (Années 1971-1972-1973)
Colloque de synthèse. Sept. 1974 — M. Boiron Ed.
328 pages / 20 francs

Vol. 37 — *Rapport N° 2 :* Actions physiopathologiques des huiles de crucifères (Années 1971-1972-1973)
Colloque de synthèse. Avril 1974 — A. François Ed.
389 pages / 30 francs

Vol. 38 — *Rapport N° 3 :* Biologie du comportement (Années 1971-1972-1973)
J. Scherrer Ed.
155 pages / 20 francs

Vol. 39 — *Rapport N° 4 :* Immunopathologie du système nerveux (Années 1971-1972-1973)
Colloque de synthèse. Nov. 1974 — F. Lhermitte et E. Schuller Eds.
317 pages / 30 francs

Vol. 41 — *Rapport N° 5 :* Prothèse totale d'épaule — Bases expérimentales et premiers résultats cliniques.
J.Y de la Caffinière, F. Mazas, Y. Mazas, F. Pelisse et D. Présent Eds.
113 pages / 40 francs

Vol. 48 — *Rapport N° 6 :* Physiopathologie de l'articulation (Années 1972-1973-1974)
Colloque de synthèse. Juin 1975 — F. Delbarre Ed.
233 pages / 30 francs

COLLOQUES ET SEMINAIRES TECHNOLOGIQUES

ANNEE 1975

Vol. 40 : Alpha-l-Antitrypsine et le système Pi
Jan. 1975 — J.P. Martin Ed.
200 pages / 30 francs

Vol. 41 : *Se reporter à la série "Action thématique"*

Vol. 42 : Cultures de cellules hématopoïtiques in vitro
Mars 1975 — D. Hollard et R. Berthier Eds.
189 pages / 30 francs

Vol. 43 : Aspects of neural plasticity
April 1975 — F. Vital-Durand and M. Jeannerod Eds.
275 pages / 50 francs

Vol. 44 : La Drépanocytose / *Sickle-cell anaemia*
Jan. 1975 — R. Cabannes Ed.
440 pages / 50 francs

Vol. 45 : *se reporter à la série "Colloque Année 1974"*

Vol. 46 : Les Mycotoxines
Mai 1975 — J. Biguet Ed.
155 pages / 25 francs

Vol. 47 : In vitro transcription and translation of viral genomes
July 1975 — A.L. Haenni and G. Beaud Eds.
450 pages / 60 francs.

Vol. 48 : *se reporter à la série "Action thématique"*

Vol. 49 : Liver diseases in Children
June 1975 — D. Alagille Ed.
103 pages / 30 francs

Vol. 50 : Smooth muscle pharmacology and physiology
July 1975 — M. Worcel and G. Vassort Eds.
477 pages / 50 francs

Vol. 51 : Distribution of pulmonary gas exchange
C. Hatzfeld Ed.
475 pages / 50 francs

Vol. 52 : Environmental pollution and carcinogenic risks
W. Davis and C. Rosenfeld Eds.
478 pages / 50 francs

Vol. 53 : Réanimation entérale à faible débit continu
E. Lévy Ed.
A paraître

Vol. 54 : Symposium franco-britannique sur l'alcoolisme / *Anglo-French Symposium on alcoholism*
Nov. 1975 — R.J. Royer and J. Levi Eds.
183 pages / 30 francs

Vol. 55 : Hormones and breast cancer
May 1975 — M. Namer and C. Lalanne Eds.
267 pages / 50 francs

Prenatal-diagnosis of genetic disorders of the fœtus
European Medical Research Councils Conference Stockholm, June 12-13, 1975
Edited by Jan Lindsten, Rolf Zetterström and Malcolm Ferguson-Smith
99 pages / 30 francs

COLLOQUES ET SEMINAIRES TECHNOLOGIQUES

ANNEE 1976

Vol. 56 : Etude des plaquettes sanguines
April 1976, S. Levy-Toledano Ed.
39 pages / 10 francs

Vol. 57 : Techniques de séparation et d'identification des lymphocytes humains / *Techniques of separation and characterization of human lymphocytes*
May 1976, D. Sabolovic and B. Serrou Eds.
329 pages / 30 francs

Vol. 58 : HLA and Disease
Abstracts, June 1976, J. Dausset and L. Degos Eds
333 pages / 30 francs

Vol. 59 : Respiratory centers and afferent systems
March 1976, B. Duron Ed.
350 pages / 50 francs

Vol. 60 : Biology of the epithelial lens cells
March 1976, Y. Courtois and F. Regnault Editors
On the press

Vol. 61 : Prenatal Diagnosis 1976
June 1976, A. Boué Ed.
330 pages / 50 francs

Quelques livres à paraître / *Some titles to be published*

Vol. 62 : Cryoimmunology, June 1976, J.M. Turc Ed.

Vol. 63 : Anthropologie et biologie des populations andines, 30-31 août-1er septembre 1976, J. Ruffié et J.C. Quilici Eds.

Vol. 64 : Sphéroplastes, protoplastes et Formes L des Bactéries, Septembre 1976, J. Roux et et R. Caravano Eds.

Vol. 65 : *Rapport d'actions thématiques n° 7 :* Pharmacologie clinique des hormones hypothalamohypophysaires — Neuromédiateurs et polypeptides hypothalamiques à action relachante ou inhibitrice
Colloque de synthèse. Octobre 1976 — R. Mornex et J. Barry Eds.

V. AUTRES PUBLICATIONS DE L'INSERM
V. OTHER INSERM PUBLICATIONS

PUBLISHED BY S. KARGER, BASEL :

Dietary fats and thrombosis
INSERM Symposium, June 1973
Edited by S. Renaud and P. Gautheron

Sperm transport, survival and fertilizing ability
INSERM symposium, Nov. 1973
Edited by E. Hafez and C. Thibault

PUBLISHED BY SPRINGER-VERLAG, HEIDELBERG

Recent advances in cerebral angiography
INSERM symposium, May 1975
Edited by G. Salamon

Preleukemic states
INSERM symposium, September 1975
in Blood Cells, vol. 2, n° 1, 1976
Edited by Marcel Bessis

PUBLISHED BY NORTH-HOLLAND/AMERICAN ELSEVIER, AMSTERDAM

Membrane receptors of lymphocytes
INSERM symposium 1, May 1975
Edited by M. Seligmann, J.L. Preud'Homme and F.M. Kourilsky

Immunoenzymatic Techniques
INSERM symposium 2, April 1975
Edited by G. Feldmann, S. Avrameas, J. Bignon and P. Druet

On the press :

Hormonal receptors in digestive tract physiology
INSERM Symposium 3, September 1976
Edited by S. Bonfils, P. Fromageot, G. Rosselin

Radiation induced leukemogenesis and related viruses
INSERM symposium 4, December 1976
Edited by J.F. Duplan

First European Symposium on Hormones and Cell regulation
INSERM Hormones and Cell regulation Series — N° 1, September 1976
Edited by J. Nunez and J. Dumont

POINTS DE VENTE / *SALES AGENTS*
INSERM
Series I to IV

INSERM

Les Commandes peuvent être adressées directement, accompagnées du règlement, à l'adresse suivante :.
INSERM Publications are available
— if requests are accompanied by a money order — from :

INSERM — Section Vente des Publications
101, rue de Tolbiac — 75645 PARIS Cedex 13, Tél. : 584.01.41.

Règlement par chèque postal (Paris 9062-38) ou par chèque bancaire à établir à l'ordre de M. l'Agent Comptable de l'INSERM / *Payment by check in the order of INSERM.*

LIBRAIRIES / *BOOKSELLERS*

EN FRANCE	Lille	— Librairie Le Furet du Nord — Place du Général de Gaulle 59000 Lille — Tél. 54.12.34
	Lyon	— Librairie Flammarion — 19, place Bellecour 69002 Lyon — Tél. 37.40.31
	Marseille	— Librairie Maupetit — 142, La Canebière 13232 Marseille Cedex 1 — Tél. 47.63.80
	Nancy	— Librairie Didier — 6, rue Gambetta 54000 Nancy — Tél. 52.25.80
	Paris	— Librairie Le François — 91, boulevard Saint-Germain 75006 Paris — Tél. 326.55.45
		— Librairie Offilib — 48, rue Gay Lussac 75005 Paris — Tél. 325.31.92
		— Librairie Arnette — 2, rue Casimir Delavigne 75006 Paris — Tél. 326.09.60
	Toulouse	— Librairie Marqueste — Place Rouaix 31000 Toulouse — Tél. 52.08.15
EN BELGIQUE		— Grande Librairie des Facultés — 148, rue Berckmans 1060 BRUXELLES — Tél. 37.18.70
EN GRANDE-BRETAGNE		— Medical Book store Lewis — 136 Gowen Street LONDON WC 1
EN HOLLANDE		— Swetz Zeitlinger — Zeisersgracht 471 487 AMSTERDAM — Tél. 22.36.26
EN REPUBLIQUE FEDERALE D'ALLEMAGNE		— Dokumente Verlag-Postfach 1340 D76 OFFENBURG — Tél. 781.27.42
EN SUISSE		— Karger Libri — Petersgraben 31 CH-401 1 BASEL — Tél. (061) 39.08.80
		— Librairie Payot — 1, rue du Bourg 1003 LAUSANNE — Tél. 20.33.31
AUX ETATS-UNIS		— Institute for Scientific Information Current Book Contents 325 Chestnut Street, PHILADELPHIA Pennsylvania 191-6, U.S.A.
		— Scientific and Medical Publications of France 14 East 60th street NEW YORK N.Y. 10022, U.S.A.
AU CANADA		— Librairie Dussault — 8955 boulevard Saint-Laurent MONTREAL 354 (Québec)

II – PUBLICATIONS OF THE INTERNATIONAL AGENCY FOR RESEARCH ON CANCER

Annual Report, 1970 *(1971)*
English and French editions .... Sw.fr. 3.-   US$1.00

Annual Report, 1971 *(1972)*
English and French editions .... Sw.fr. 3.-   US$0.75

Annual Report, 1972-73 *(1973)*
English and French editions .... Sw.fr. 6.-

Annual Report, 1974 *(1974)*
English and French editions Sw.fr.10.-

Annual Report, 1975 *(1975)*
English and French editions Sw.fr.12.-

IARC Monographs on the Evaluation of Carcinogenic Risk of Chemicals to Man:

Volume 1 *(1972)*
English edition only .... Sw.fr.12.-   US$4.20

Volume 2 *(1973)*
Some inorganic and organometallic compounds
English edition only .... Sw.fr.12.-   US$3.60

Volume 3 *(1973)*
Certain polycyclic aromatic hydrocarbons and heterocyclic compounds
English edition only .... Sw.fr.18.-   US$5.40

Volume 4 *(1974)*
Some aromatic amines, hydrazine and related substances, N-nitroso compounds and miscellaneous alkylating agents
English edition only Sw.fr.18.-

Volume 5 *(1974)*
Some organochlorine pesticides
English edition only Sw.fr.18.-

Volume 6 *(1974)*
Sex hormones
English edition only Sw.fr.18.-

Volume 7 *(1974)*

Some anti-thyroid and related substances, nitrofurans and industrial chemicals
English edition only
Sw.fr.32.-

Volume 8 *(1975)*

Some aromatic azo compounds
English edition only
Sw.fr.36.-

Volume 9 *(1975)*

Some aziridines, N-, S- and O-mustards and selenium
English edition only
Sw.fr.27.-

TITLES IN PRESS

Volume 10   *Some naturally occurring substances*

Volume 11   *Some epoxides and miscellaneous chemicals of industrial importance*

Volume 12   *Some carbamates and thiocarbamates*

IARC SCIENTIFIC PUBLICATIONS :

No. 1   Liver Cancer. Proceedings of a Working Conference held at the Chester Beatty Research Institute, London, England, 30 June to 3 July 1969 (1971)
English edition only
Sw.fr. 30.-   US$10.00

No. 2   Oncogenesis and Herpesviruses. Edited by P. N. Biggs, G. de Thé & L. N. Payne. Proceedings of a Symposium held at Christ's College, Cambridge, England 20 to 25 June 1971 (1972)
English edition only
Sw.fr.100.-   US$25.00

No. 3   N-Nitroso Compounds. Analysis and Formation. Edited by P. Bogovski, R. Preussmann & E.A. Walker. Proceedings of a Working Conference held at the Deutsches Krebsforschungszentrum, Heidelberg, Federal Republic of Germany 13-15 October 1971 (1973)
English edition only
Sw.fr. 25.-   US$6.25

No. 4   Transplacental Carcinogenesis. Edited by L. Tomatis & U. Mohr. Proceedings of a Meeting held at the Medizinische Hochschule, Hannover, Federal Republic of Germany 6-7 October 1971 (1973)
English edition only
Sw.fr. 40.-   US$12.00

No. 5   Pathology of Tumours in Laboratory Animals. Volume I. Tumours of the Rat, part 1. Editor-in-Chief V.S. Turusov. (1973)
English edition only
Sw.fr. 50.-   US$15.00

No. 6   Pathology of Tumours in Laboratory Animals. Volume I. Tumours of the Rat, part 2. Editor-in-Chief V.S. Turusov. (1976) (in press)

No. 7   Host Environment Interactions in the Etiology of Cancer in Man. Edited by R. Doll & I. Vodopija. *Proceedings of a meeting held at Primosten, Yugoslavia, 27 August to 2 September 1972 (1973)*

English edition only   Sw.fr.100.-

No. 8   Biological Effects of Asbestos. Edited by P. Bogovski, J. C. Gilson, V. Timbrell & J. C. Wagner. *Proceedings of a Working Conference held at the International Agency for Research on Cancer, Lyon, France 2-6 October 1972 (1973)*

English edition only   Sw.fr. 80.-

No. 9   N-Nitroso Compounds in the Environment. Edited by P. Bogovski & E. A. Walker. *Proceedings of a Working Conference held at the International Agency for Research on Cancer, Lyon, France, 17-20 October 1973 (1974)*

English edition only   Sw.fr. 50.-

No. 10   Chemical Carcinogenesis Essays. Edited by R. Montesano & L. Tomatis. *Proceedings of a Workshop on Approaches to Assess the Significance of Experimental Carcinogenesis Data for Man organized by IARC and the Catholic University of Louvain, Brussels, Belgium, 10-12 December 1973 (1974)*

English edition only   Sw.fr. 50.-

No. 11   Oncogenesis and Herpesviruses II. Edited by G. de Thé, M. A. Epstein & H. zur Hausen. *Proceedings of the Second International Symposium on Oncogenesis and Herpesviruses held in Nuremberg, Federal Republic of Germany, 14-16 October 1974 (1975)*

    Part 1: *Biochemistry of viral replication and in vitro transplantation*

English edition only   Sw.fr.100.-   US$38.00

    Part 2: *Epidemiology, host response and control*

English edition only   Sw.fr. 80.-   US$30.00

No. 12   Screening Tests in Chemical Carcinogenesis. Edited by R. Montesano, H. Bartsch & L. Tomatis. *Proceedings of a Workshop on Rapid Screening Tests to Predict Late Toxic Effects of Environmental Chemicals organized by IARC and the Commission of the European Communities held in Brussels, Belgium, 9-12 June 1975 (1976) (in press)*

No. 13   Environmental Pollution and Carcinogenic Risks. Edited by C. Rosenfeld & W. Davis. *Proceedings of a Symposium organized by IARC and the French National Institute of Health and Medical Research held at the International Agency for Research on Cancer, Lyon, France, 3-5 November 1975 (1976)*   Sw.fr. 50.-   US$20.00

No. 14   Environmental N-Nitroso Compounds. Analysis and Formation. Edited by E. A. Walker, P. Bogovski & L. Griciute. *Proceedings of the Fourth Meeting on the Analysis and Formation of N-Nitroso Compounds held in Tallinn, Estonian SSR, 1-2 October 1975 (1976) (in press)*

No. 15   Cancer Incidence in Five Continents, Vol. III. Edited by C. S. Muir, J. A. H. Waterhouse & P. Correa. *(1976) (in press)*

Prices are subject to change without notice

WHO publications may be obtained, direct or through booksellers, from:

ALGERIA	Société Nationale d'Edition et de Diffusion, 3 Bd Zirout Youcef, Algiers.
ARGENTINA	Librería de las Naciones, Cooperativa Ltda., Alsina 500, Buenos Aires — Editorial Sudamericana S.A., Humberto 1 - 545, Buenos Aires.
AUSTRALIA	Mail Order Sales, Australian Government Publishing Service, P.O. Box 84, Canberra A.C.T. 2600; *or over the counter from* Australian Government Publications and Inquiry Centres at: 113 London Circuit, Canberra City; 347 Swanston Street, Melbourne; 309 Pitt Street, Sydney; Mr. Newman House, 200 St. George's Terrace, Perth; Industry House, 12 Pirie Street, Adelaide; 156-162 Macquarie Street, Hobart — Hunter Publications, 58A Gipps Street, Collingwood, Vic. 3066.
AUSTRIA	Gerold & Co., I. Graben 31, Vienna 1.
BANGLADESH	WHO Representative, G.P.O. Box 250, Dacca 5.
BELGIUM	Office international de Librairie, 30 avenue Marnix, Brussels.
BRAZIL	Biblioteca Regional de Medicina OMS/OPS, Unitad de Venta de Publicaciones, Caixa Postal 20.381, Vila Clementino, 01000 São Paulo — S.P.
BURMA	*see* India, WHO Regional Office.
CANADA	Information Canada Bookstore, 171 Slater Street, Ottawa, Ontario K1A OS9; Main Library, University of Calgary, Calgary, Alberta; 1683 Barrington Street, Halifax, N.S. B3J 1Z9; 640 Ste Catherine West, Montreal, Quebec H3B 1 B8; 221 Yonge Street, Toronto, Ontario M5B 1 N4; 800 Granville Street, Vancouver, B.C. V6Z 1 K4; 393 Portage Avenue, Winnipeg, Manitoba R3B 2C6. *Mail orders to* 171 Slater Street, Ottawa, Ontario K1A OS9.
CHINA	China National Publications Import Corporation, P.O. Box 88, Peking.
COLOMBIA	Distrilibros Ltd, Pío Alfonso García, Carrera 4a, Nos 36-119, Cartagena.
COSTA RICA	Imprenta y Librería Trejos S.A., Apartado 1313, San José.
CYPRUS	MAM, P.O. Box 1674, Nicosia.
CZECHOSLOVAKIA	Artia, Smecky 30, 111 27 Prague 1.
DENMARK	Ejnar Munksgaard, Ltd, Nørregarde 6, Copenhagen.
ECUADOR	Librería Científica S.A., P.O. Box 362, Luque 223, Guayaquil.
EGYPT	Nabaa El Fikr Bookshop, 55 Saad Zaghloul Street, Alexandria — Anglo-Egyptian Bookshop, 165 Mohamed Farid Street, Cairo.
EL SALVADOR	Librería Estudiantil Edificio Comercial B No 3, Avenida Libertad, San Salvador.
FIJI	The WHO Representative, P.O. Box 113, Suva.
FINLAND	Akateeminen Kirjakauppa, Keskuskatu 2, Helsinki 10.
FRANCE	Librairie Arnette, 2, rue Casimir-Delavigne, Paris 6e.
GERMAN DEMOCRATIC REPUBLIC	Buchhaus Leipzig, Postfach 140, 701 Leipzig.
GERMANY, FEDERAL REPUBLIC OF	Govi-Verlag GmbH, Ginnheimerstrasse 20, Postfach 5360, 6236 Eschborn — W.E. Saarbach, Postfach 1510, Follerstrasse 2, 5 Cologne 1 — Alex. Horn, Spiegelgasse 9, 62 Wiesbaden.
GREECE	G. C. Eleftheroudakis S.A., Librairie internationale, rue Nikis 4, Athens (T. 126).
HAITI	Max Bouchereau, Librairie « A la Caravelle », Boîte postale 111-B, Port-au-Prince.
HUNGARY	Kultura, P.O.B. 149, Budapest 62 — Akadémiai Könyvesbolt, Väci utca 22, Budapest V.
ICELAND	Snaebjörn Jonsson & Co., P.O. Box 1131, Hafnarstraeti 9, Reykjavik.
INDIA	WHO Regional Office for South-East Asia, World Health House, Indraprastha Estate, Ring Road, New Delhi 1 — Oxford Book & Stationery Co., Scindia House, New Delhi; 17 Park Street, Calcutta 16 (Sub-agent).
INDONESIA	*see* India, WHO Regional Office.
IRAN	Iranian Amalgamated Distribution Agency, 151 Khiaban Soraya, Teheran.
IRELAND	The Stationery Office, Dublin.
ISRAEL	Heiliger & Co., 3 Nathan Strauss Street, Jerusalem.
ITALY	Edizioni Minerva Medica, Corso Bramante 83-85, Turin; Via Lamarmora 3, Milan.
JAPAN	Maruzen Co., Ltd, P.O. Box 5050, Tokyo International, 100-31.
KENYA	The Caxton Press Ltd, Head Office: Gathani House, Huddersfield Road P.O. 1742, Nairobi.
KUWAIT	The Kuwait Bookshops Co. Ltd, Thunayan Al-Ghanem Bidg, P.O. Box 2942, Kuwait.
LAO PEOPLE'S DEMOCRATIC REPUBLIC	The WHO Representative, P.O. Box 343, Vientiane.
LEBANON	Documenta Scientifica/Redico, P.O. Box 5641, Beirut.
LUXEMBOURG	Librairie du Centre, 49 bd Royal, Luxembourg.
MALAYSIA	The WHO Representative, Room 1004, Fitzpatrick Building, Jalan Raja Chulan, Kuala Lumpur 05-02 — Jubilee (Book) Store Ltd, 97 Jalan Tuanku Abdul Rahman, P.O. Box 629, Kuala Lumpur — Parry's Book Center, K.L. Hilton Hotel, Kuala Lumpur.

IMPRIMERIE LOUIS-JEAN
Publications scientifiques et littéraires
TYPO - OFFSET
05002 GAP - Téléphone 51 35 23 -
Dépôt légal 444-1976

WHO publications may be obtained, direct or through booksellers, from:

MEXICO	La Prensa Médica Mexicana, Ediciones Científicas, Paseo de las Facultades, 26, Mexico City 20, D.F.
MONGOLIA	see India, WHO Regional Office.
MOROCCO	Editions La Porte, 281 avenue Mohammed V, Rabat.
NEPAL	see India, WHO Regional Office.
NETHERLANDS	N.V. Martinus Nijhoff's Boekhandel en Uitgevers Maatschappij, Lange Voorhout 9, The Hague.
NEW ZEALAND	Government Printing Office, Government Bookshops at: Rutland Street, P.O. Box 5344, Auckland; 130 Oxford Terrace, P.O. Box 1721, Christchurch; Alma Street, P.O. Box 857, Hamilton; Princes Street, P.O. Box 1104, Dunedin; Mulgrave Street, Private Bag, Wellington — R. Hill & Son Ltd, Ideal House, Cnr, Gilles Avenue & Eden St., Newmarket, Auckland S.E. 1.
NIGERIA	University Bookshop Nigeria, Ltd, University of Ibadan, Ibadan.
NORWAY	Johan Grundt Tanum Bokhandel, Karl Johansgt. 43, Oslo 1.
PAKISTAN	Mirza Book Agency, 65 Shahrah Quaid-E. Azam, P.O. Box Box 729, Lahore 3.
PARAGUAY	Agencia de Librerías Nizza S.A., Estrella No. 721, Asunción.
PERU	Distribuidora Inca S.A., Apartado 3115, Emilio Althaus 470, Lima.
PHILIPPINES	World Health Organization, Regional Office for the Western Pacific, P.O. Box 2932, Manila — The Modern Book Company Inc., P.O. Box 632, 926 Rizal Avenue, Manila.
POLAND	Składnica Księgarska, ul. Mazowiecka 9, Warsaw *(except periodicals)* — BKWZ Ruch, ul. Wronia 23, Warsaw *(periodicals only)*.
PORTUGAL	Livraria Rodrigues, 186 Rua Aurea, Lisbon.
REPUBLIC OF KOREA	The WHO Representative, Central P.O. Box 540, Seoul.
SINGAPORE	The WHO Representative, 144 Moulmein Road, G.P.O. Box 3457, Singapore 1.
SOUTH AFRICA	Van Schaik's Bookstore (Pty) Ltd, P.O. Box 724, Pretoria.
SPAIN	Comercial Atheneum S.A., Consejo de Ciento 130-136, Barcelona 15; General Moscardó 29, Madrid 20 — Librería Díaz de Santos, Lagasca 95, Madrid 6.
SRI LANKA	see India, WHO Regional Office.
SWEDEN	Aktiebolaget C.E. Fritzes Kungl. Hovbokhandel, Fredsgatan 2, Stockholm 16.
SWITZERLAND	Medizinischer Verlag Hans Huber, Länggass Strasse 76, 3012 Berne 9.
THAILAND	see India, WHO Regional Office.
TUNISIA	Société Tunisienne de Diffusion, 5 avenue de Carthage, Tunis.
TURKEY	Librairie Hachette, 469 avenue de l'Indépendance, Istanbul.
UGANDA	see address under Kenya.
UNITED KINGDOM	H. M. Stationery Office: 49 High Holborn, London WC1V 6HB; 13a Castle Street, Edinburgh EH2 3AR; 109 St Mary Street, Cardiff CF1 1JW; 80 Chichester Street, Belfast BT1 4JY; Brazennose Street, Manchester M60 8AS; 258 Broad Street, Birmingham B1 2HE; 50 Fairfax Street, Bristol BS1 3DE. *All mail orders should be sent to* P.O. Box 569, London SE1 9NH.
UNITED REP. OF TANZANIA	see address under Kenya.
UNITED STATES OF AMERICA	*Single and bulk copies of individual publications (not subscriptions):* Q Corporation, 49 Sheridan Avenue, Albany, NY 12210. *Subscriptions:* Subscription orders, accompanied by check made out to the Chemical Bank, New York, Account World Health Organization, should be sent to the World Health Organization, P.O. Box 5284, Church Street Station, New York, NY 10249. Correspondence concerning subscriptions should be forwarded to the World Health Organization, Distribution and Sales Service, 1211 Geneva 27, Switzerland. *Publications are also available from the* United Nations Bookshop, New York, NY 10017 *(retail only)*.
USSR	*For readers in the USSR requiring Russian editions:* Komsomolskij prospekt 18, Medicinskaja Kniga, Moscow—*For readers outside the USSR requiring Russian editions:* Kuzneckij most 18, Meždunarodnaja Kniga, Moscow G-200.
VENEZUELA	Editorial Interamericana de Venezuela C.A., Apartado 50785, Caracas—Librería del Este, Av. Francisco de Miranda 52, Edificio Galipán, Caracas.
YUGOSLAVIA	Jugoslovenska Knjiga Terazije 27/II, Belgrade.

Orders from countries where sales agents have not yet been appointed may be addressed to: World Health Organization, Distribution and Sales Service, 1211 Geneva 27, Switzerland, but must be paid for in pounds sterling, US dollars, or Swiss francs.